INTRODUCTION TO DISTANCE SAMPLING

Introduction to Distance Sampling

Estimating Abundance of Biological Populations

S. T. BUCKLAND
University of St. Andrews

D. R. ANDERSON
Colorado Cooperative Fish and Wildlife Research Unit

K. P. BURNHAM
Colorado Cooperative Fish and Wildlife Research Unit

J. L. LAAKE
National Marine Mammal Laboratory, Seattle

D. L. BORCHERS
University of St. Andrews

L. THOMAS
University of St. Andrews

OXFORD
UNIVERSITY PRESS

Great Clarendon Street, Oxford OX2 6DP

Oxford University Press is a department of the University of Oxford.
It furthers the University's objective of excellence in research, scholarship,
and education by publishing worldwide in

Oxford New York

Auckland Cape Town Dar es Salaam Hong Kong Karachi
Kuala Lumpur Madrid Melbourne Mexico City Nairobi
New Delhi Shanghai Taipei Toronto

With offices in

Argentina Austria Brazil Chile Czech Republic France Greece
Guatemala Hungary Italy Japan South Korea Poland Portugal
Singapore Switzerland Thailand Turkey Ukraine Vietnam

Oxford is a registered trade mark of Oxford University Press
in the UK and in certain other countries

Published in the United States
by Oxford University Press Inc., New York

First published 2001

A catalogue record for this title is available from the British Library

Library of Congress Cataloging in Publication Data

Introduction to distance sampling : estimating abundance of biological populations /
S. T. Buckland . . . [et al.].
Includes bibliographical references
1. Animal populations–Statistical methods. 2. Sampling (Statistics)
I. Buckland, S. T. (Stephen T.)
QL752 .I59 2001 591.7′88′0727–dc21 2001033232

ISBN 978-0-19-850927-1 (Pbk)

10 9 8 7 6 5

Typeset by Newgen Imaging Systems (P) Ltd., Chennai, India
Printed in Great Britain
on acid-free paper by
Biddles Ltd., King's Lynn, Norfolk

Preface

This book is about the use of distance sampling to estimate the density or abundance of biological populations. Line and point transect sampling are the primary methods that sample distances: lines or points are surveyed in the field and the observer records a distance to those objects of interest that are detected. The key sample data are the set of distances of detected objects from the line or point. Effort (total line length or number of visits to points) is also recorded, as are any relevant covariates or stratification factors. Many objects may remain undetected even within the surveyed strips or circles during the course of the survey. Distance sampling provides a way to obtain reliable estimates of density of objects under fairly mild assumptions. Distance sampling is an extension of plot or quadrat sampling methods for which it is assumed that all objects within sample plots are counted.

The objects of interest are usually vertebrate animals, including those that typically occur in 'clusters' (flocks, schools, etc.). The surveyed objects are sometimes inanimate such as bird nests, mammal burrows, deer dung or dead animals. The range of application is broad; distance sampling has been used for a wide range of songbirds, seabirds, gamebirds, terrestrial mammals, marine mammals, fish and reptiles. Distance sampling often provides a practical, cost-effective class of methods for estimating population density. For objects distributed sparsely across large geographic areas, there are often no competing methods.

The objective of this book is to provide a comprehensive introduction to distance sampling theory and application. Most of the development of this subject occurred in the last quarter of the 20th century. A 1980 monograph, written by KPB, DRA and JLL, was the first synthesis of the topic, followed in 1993 by the first 'distance sampling' book, written by STB, DRA, KPB and JLL. This is essentially an update to that book. We are also working on 'Advanced Distance Sampling', a book that will complement this one, and cover issues such as spatial distance sampling; double-platform methods for when detection at the line or point is not certain; general covariate methods; automated design algorithms; adaptive distance sampling; and use of Geographic Information Systems in distance sampling surveys. Some of the more theoretical material from the 1993 book will also appear in the advanced book.

This book covers the theory and application of distance sampling with emphasis on line and point transects. Specialized applications are noted briefly, such as trapping webs and cue counts. General considerations are given to the design of distance sampling surveys. Many examples are provided to illustrate the application of the theory. The book is written for both statisticians and biologists.

The book contains eight chapters. Chapters 1 and 2 are introductory. Chapter 3 presents the general theory for both line and point transect sampling, including modelling, estimation, testing and inference. Chapters 4 and 5 describe the application of the theory for line and point transects, respectively. These chapters are meant to stand alone, thus there is some duplication of the material. Several related methods and slight modifications to the standard methods are given in Chapter 6. Chapter 7 discusses the design and field protocol for distance sampling surveys. The emphasis is on ways to ensure that the key assumptions are met. Field methods are discussed for a range of survey types. Chapter 8 provides several comprehensive examples. Finally, around 600 references to the published literature are listed.

The main concepts in this book are not complex; however, some of the statistical theory may be difficult for non-statisticians. We hope biologists will not be deterred by the quantitative theory chapter and hope that statisticians will understand that we are presenting methods intended to be useful and usable given all the practicalities a biologist faces in field sampling. We assume that the reader has some familiarity with basic statistical methods, including point and variance estimation. Knowledge of sampling theory would be useful, as would some acquaintance with numerical methods. Some experience with likelihood inference would be useful. The following guidelines are provided for a first reading of the book.

Everyone should read Chapters 1 and 2. While statisticians will want to study Chapters 3 and 6, Chapters 4 (line transects) and 5 (point transects) will be of more interest to biologists. Biologists should study Chapter 7 (design) in detail. Everyone might benefit from the illustrative examples and case studies in Chapter 8, where readers will find guidance on advanced applications involving several data sets.

Our interest in these subjects dates back to 1966 (DRA), 1974 (KPB), 1977 (JLL), 1980 (STB), 1987 (DLB) and 1997 (LT). We have all contributed to the theory, been involved with field sampling, and have wide experience of the analysis of real sample data. All of us contribute regularly to training workshops on the topic. Computer software packages TRANSECT and early versions of DISTANCE were the domain of JLL, while LT is largely responsible for the current versions of Distance (3.5 and 4.0).

Steve Buckland acknowledges the support from the University of St Andrews and, formerly, of the Scottish Agricultural Statistics Service (now BioSS). David Anderson and Ken Burnham are grateful to the US Fish and Wildlife Service, and more recently to the U.S. Geological Survey for support and freedom in their research. Jeff Laake and the early development of DISTANCE were funded by the Colorado Division of Wildlife and the U.S. National Marine Fisheries Service. David Borchers' contribution was funded by the University of St Andrews, and Len Thomas developed the windows versions of Distance with sponsorship from the U.K. Engineering and Physical Sciences Research Council, the U.K. Biotechnology and Biological Sciences Research Council, the U.S. National Marine Fisheries Service and the U.S. National Parks Service. Thomas Drummer, Eric Rexstad, Tore Schweder, David Bowden, Robert Parmenter and George Seber provided review comments on a draft of the 1993 book. Several biologists generously allowed us to use their research data as examples: Roger Bergstedt, Colin Bibby, Eric Bollinger, Graeme Coulson, Fritz Knopf and Robert Parmenter. We also gratefully acknowledge the following organizations for funding research to address the practical problems of distance sampling and for allowing us to use their data: the Inter-American Tropical Tuna Commission; the International Whaling Commission; the Marine Research Institute of Iceland; and the U.S. National Marine Fisheries Service. We have all benefited from the use of Les Robinette's data sets. David Carlile provided the photo of the DELTA II submersible, Fred Lindzey provided photos of aircraft used to survey pronghorn in Wyoming, and John Reinhardt allowed the use of a photo of the survey aircraft shown in Chapter 7. Tom Drummer and Charles Gates helped us with their software, SIZETRAN and LINETRAN, respectively. David Gilbert provided help with the Monte Vista duck nest data. Karen Cattanach carried out some of the analyses in the marine mammal examples and generated the corresponding figures. Finally, we acknowledge Barb Knopf's assistance in manuscript preparation for the 1993 book and Eric Rexstad's help with many of the figures.

We plan to continue our work and interest in distance sampling issues. We welcome comments and suggestions from those readers who share our interests.

December 2000

S. T. B.
D. R. A.
K. P. B.
J. L. L.
D. L. B.
L. T.

Contents

1

Introductory concepts

1.1 Introduction

Ecology is the study of the distribution and abundance of plants and animals and their interactions with their environment. Many studies of biological populations require estimates of population density (D) or size (N), or rate of population change $\lambda_t = D_{t+1}/D_t = N_{t+1}/N_t$. These parameters vary in time and over space as well as by species, sex and age. Population dynamics, and hence these parameters, often depend on environmental factors. Distance sampling can be an effective approach for estimating D and N.

This book covers the design and analysis of standard distance sampling methods, for which the fundamental parameter of interest is density (D = number per unit area). Density and population size are related as $N = D \cdot A$ where A is the size of the study area. Thus, attention can be focused on D. Buckland *et al.* (in preparation) deals with more advanced topics and areas of current research, although the focus is still on methodologies of direct relevance to wildlife managers and conservation biologists. These two publications are intended to replace Buckland *et al.* (1993a). The current book is largely a rewrite of that book, with the addition of some new sections, and updating of other sections. Some of the more advanced material of that book will appear in Buckland *et al.* (in preparation), together with a substantial body of work reflecting many of the advances in methodology since the appearance of the 1993 book.

1.2 Distance sampling methods

1.2.1 *Quadrat sampling*

Although not itself a distance sampling method, the various types of distance sampling are extensions of quadrat sampling. It is useful therefore to formulate quadrat sampling at this stage.

Figure 1.1 shows five $1\,\mathrm{m}^2$ quadrats, selected randomly, in which $n = 10$ objects are counted. The total area sampled is $a = 5\,\mathrm{m}^2$, so the estimate

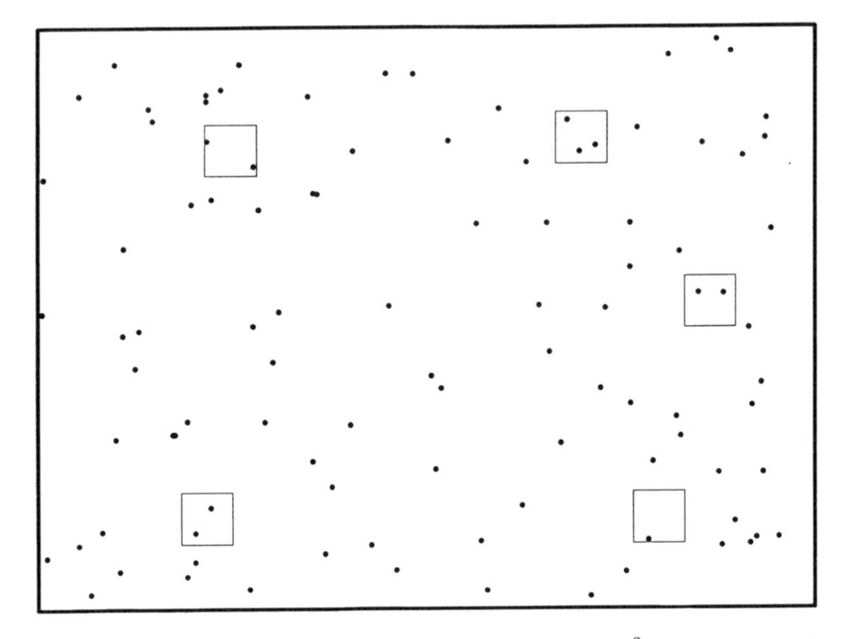

Fig. 1.1. Finite population sampling approach with five $1\,\mathrm{m}^2$ quadrats placed at random in a population containing 100 objects of interest. $\sum a_i = 5$, $\sum n_i = 10$ and $\hat{D} = 2$ objects/m^2. In this illustration, the population is confined within a well-defined area.

$\hat{D}$ of animal density D is

$$\hat{D} = \frac{n}{a} = 2 \text{ objects/m}^2 \tag{1.1}$$

If the total study area covered is $A = 125\,\mathrm{m}^2$, then we estimate population size as $\hat{D} \cdot A = 2 \times 125 = 250$ objects. Finite population sampling theory (Cochran 1977) is used to estimate the precision of the estimate. This method is often useful for estimating density or abundance of plants, but can be problematic for mobile animals. For efficient sampling of animal populations, quadrats typically need to be large, in which case it may be difficult to ensure that all animals within a quadrat are counted. This difficulty is compounded if animals are moving in and out of the quadrat during the count.

1.2.2 *Strip transect sampling*

This is a natural modification of quadrat sampling, to ease the task of counting all objects in the quadrat. The quadrats are long, narrow strips, so that the observer can travel down the centreline of each strip, counting all objects within the strip. Suppose there are k strips, each of width $2w$ (corresponding to a distance w either side of the centreline), with strip i of length l_i. If we denote the total length $\sum_{i=1}^{k} l_i = L$, then the sampled area is $a = 2wL$. If the total number of objects counted is $\sum_{i=1}^{k} n_i = n$, then density D is estimated by

$$\hat{D} = \frac{n}{2wL} \tag{1.2}$$

Strip transect sampling is based on the critical assumption that all objects within the strip are detected. Generally, no data are collected to test this assumption. To ensure that it holds to a good approximation in all habitats and conditions, it may be necessary to use narrow strips. Thus the method can be very inefficient (Burnham and Anderson 1984), as many objects beyond the strip will be detected but ignored.

Sometimes in woodland habitats, the width of the strip is estimated by selecting a sample of points along the line and measuring the distance out to which an object is visible. The average of these distances is taken as the width of the strip. This method, though conceptually appealing, can give large bias (e.g. Pierce 2000). For example, an object well beyond the visible distance, as judged by an observer from the point of closest approach along the line, might be easily visible to the observer from other points along the line. Additionally, assessment of the visible distance is often subjective and prone to bias.

1.2.3 *Line transect sampling*

In line transect sampling, as for strip transect sampling, the observer travels along a line, recording detected objects. However, instead of counting all objects within a strip of known width, the observer records the distance from the line to each object detected (Fig. 1.2). In the standard method, all objects on or near the line should be detected, but the method allows a proportion of objects within a distance w of the line to be missed. Thus a wider strip can be used than for strip transect sampling; indeed, all detected objects may be recorded, regardless of how far they are from the line. For sparsely distributed objects, the method is typically more efficient than strip transect sampling, as sample size is larger for the same amount of effort.

As for strip transects, the design comprises a number of randomly positioned lines, or a grid of systematically spaced lines randomly superimposed on the study area (Fig. 1.3). The method is a distance sampling method because distances of objects from the line are sampled. We denote these n sampled distances by $x_1, x_2, \ldots, x_n$. Unbiased estimates of density, and hence abundance, can be obtained from these distance data if certain assumptions are met. Conceptually, we can think in terms of an effective strip half-width μ, which is the distance from the line for which as many objects are detected beyond μ as are missed within μ of the line. With this

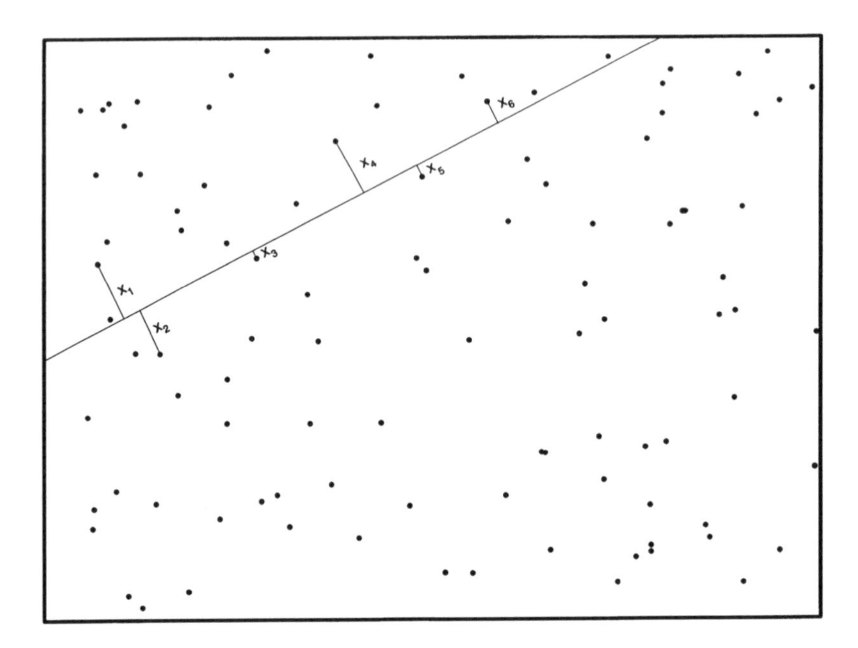

Fig. 1.2. Line transect sampling approach with a single, randomly placed, line of length L. Six objects ($n = 6$) were detected at distances $x_1, x_2, \ldots, x_6$. Those objects detected are denoted by a line showing the perpendicular distance measured. In practical applications, several lines would be used to sample the population.

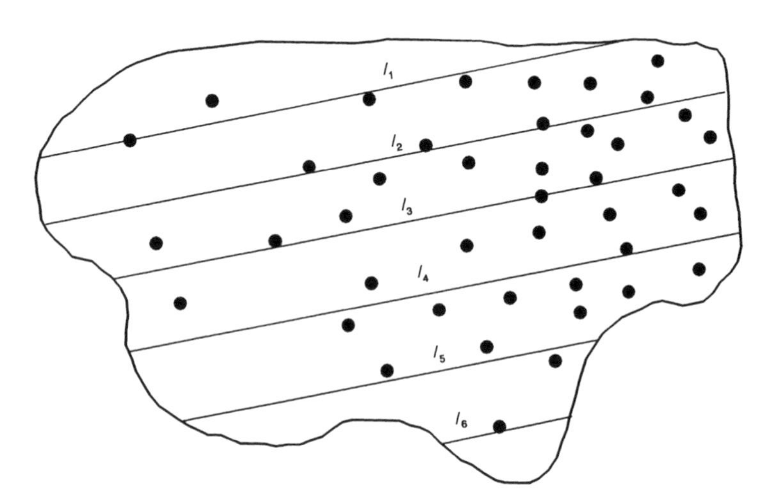

Fig. 1.3. A population of objects with a gradient in density is sampled with lines parallel to the direction of the gradient. In this case, there are $k = 6$ lines of length $l_1, l_2, \ldots, l_6$, and $\sum l_i = L$.

definition, we can estimate density D by

$$\hat{D} = \frac{n}{2\hat{\mu}L} \tag{1.3}$$

Note the similarity with eqn (1.2); the known strip half-width w is replaced by an estimate of the effective half-width μ.

Perhaps a better way to view this is to note that we expect to detect a proportion P_a of the objects in the strip of length L and width $2w$, so that density D is estimated by

$$\hat{D} = \frac{n}{2wL\hat{P}_a} \tag{1.4}$$

We show later how estimation of P_a, or equivalently μ, may be reformulated as a more familiar statistical estimation problem.

It is often convenient to measure the sighting or 'radial' distance r_i and sighting angle θ_i, rather than the perpendicular distance x_i, for each of the n objects detected (Fig. 1.4). The x_i are then found by simple trigonometry: $x_i = r_i \cdot \sin(\theta_i)$.

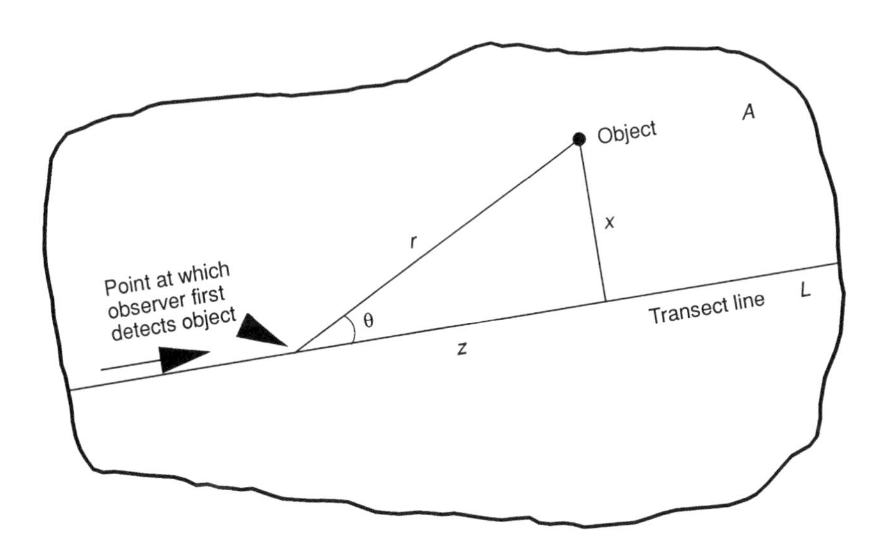

Fig. 1.4. Basic measurements that can be taken in line transect surveys. Here an area of size A is sampled by a single line of length L. If sighting distances r are to be taken in the field, one should also measure the sighting angles θ, to allow analysis of perpendicular distances x, calculated as $x = r\sin(\theta)$. The distance of the object from the observer parallel to the transect at the moment of detection is $z = r\cos(\theta)$.

1.2.4 *Point counts*

Point counts are widely used in avian studies. Frequently, all detections are recorded, regardless of distance from the point. Increasingly, sighting or radial distances to detected objects are estimated, but in this section, we consider the case that distances are not estimated. Point counts to infinity, though commonly used, are of limited value. They provide distributional information, but do not allow estimation of abundance. Frequently it is claimed that the counts are a relative abundance index, which may be used to estimate spatial and temporal trends in abundance. In fact, they give strongly biased estimates of spatial trend, because most species are substantially more detectable in open habitats such as farmland than in closed habitats such as woodland. Temporal trends are also compromised in the presence of habitat succession, or if observers of varying ability are used in different years.

Point counts to a fixed distance w are another form of quadrat sampling if it can be assumed that all objects within w of each point are detected. Thus if the design comprises k points, each comprising a circular plot of radius w, the sampled area is $k\pi w^2$. If n objects are detected within these circular plots, density D is estimated by

$$\hat{D} = \frac{n}{k\pi w^2} \tag{1.5}$$

Point counts have the same disadvantages relative to point transects (below) as strip transects have relative to line transects.

1.2.5 *Point transect sampling*

The term point transect was coined because it may be considered as a line transect of zero length (i.e. a point). This analogy is of little conceptual use because line and point transect theory differ. Point transects are often termed variable circular plots in the ornithological literature, where the points are often placed at intervals along straight line transects (Fig. 1.5). We consider a series of k points positioned randomly, or a grid of k equally spaced points randomly superimposed on the study area. An observer measures the sighting (radial) distance r_i from the random point to each of the objects detected (Fig. 1.6). Upon completion of the survey of the k points, there are n distance measurements to the detected objects. In circular plot sampling, an area of πw^2 around each point is censused, whereas in point transect sampling, only the area close to the point must be fully censused; a proportion of objects away from the point but within the sampled area remains undetected. Conceptually, we can use the distances $r_i, i = 1, \ldots, n$, to estimate the effective radius ρ, for which as many objects beyond ρ are

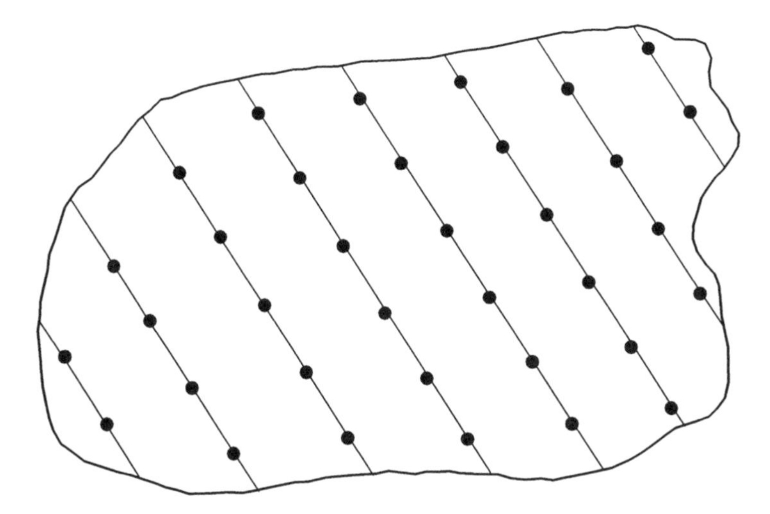

Fig. 1.5. Point transect surveys are often based on points laid out systematically along parallel lines. Alternatively, the points could be placed completely at random or in a stratified design. Variance estimation is dependent upon the point placement design.

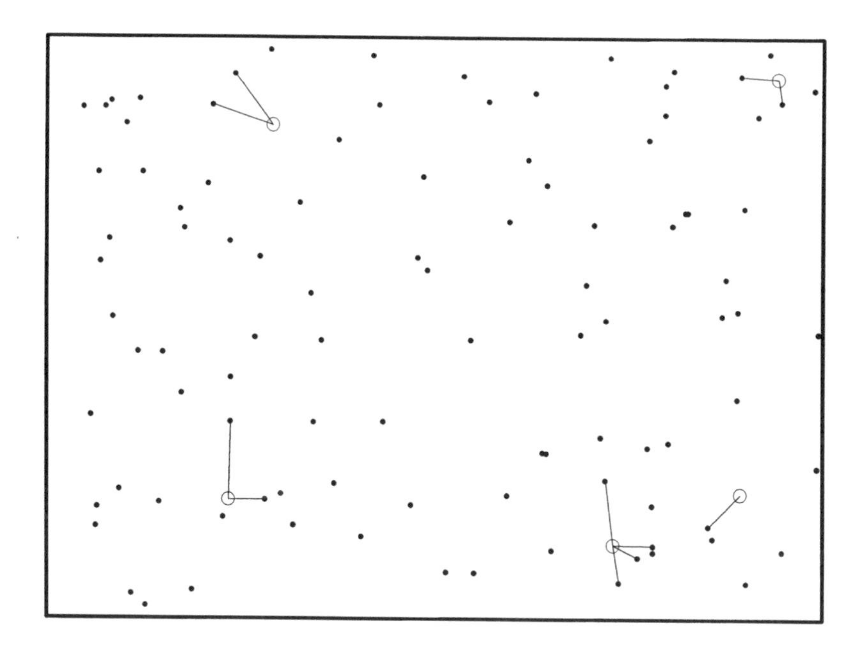

Fig. 1.6. Point transect sampling approach with five randomly placed points ($k = 5$), denoted by the open circles. Eleven objects were detected and the 11 sighting distances $r_1, r_2, \ldots, r_{11}$ are shown.

detected as are missed within ρ. Density D is then estimated by

$$\hat{D} = \frac{n}{k\pi\hat{\rho}^2} \tag{1.6}$$

This is analogous to eqn (1.5), with the known radius w replaced by the estimated effective radius, $\hat{\rho}$.

If instead we define P_a to be the expected proportion of objects detected within a radius w, estimated by $\hat{P}_a$, then

$$\hat{D} = \frac{n}{k\pi w^2 \hat{P}_a} \tag{1.7}$$

Again we formulate the estimation problem into a more standard statistical framework later.

1.2.6 *Trapping webs*

Trapping webs (Anderson *et al.* 1983; Wilson and Anderson 1985b) provide an alternative to traditional capture–recapture sampling for estimating animal density. They represent the only application of distance sampling in which trapping is an integral part. Traps are placed along lines radiating from randomly chosen points (Fig. 1.7); the traditionally used rectangular trapping grid cannot be used as a trapping web. Here 'detection' by an observer is replaced by animals being caught in traps at a known distance from the centre of a trapping web. Trapping continues for t occasions and data from either the initial capture of each animal or all captures and recaptures are analysed. To estimate density over a wider area, several randomly located webs are required.

1.2.7 *Cue counting*

Cue counting (Hiby 1985) was developed as an alternative to line transect sampling for estimating whale abundance from sighting surveys. Observers on a ship or aircraft record all sighting cues within a sector ahead of the platform and their distance from the platform. The cue used depends on species, but might be the blow of a whale at the surface. The sighting distances are converted into the estimated number of cues per unit time per unit area using a point transect modelling framework. The cue rate (usually corresponding to blow rate) is estimated from separate experiments, in which individual animals or pods are monitored over a period of time.

1.2.8 *Dung counts*

Dung counts are surveys in which the density of dung is estimated, and then converted into an estimate of abundance of the animals that produce the dung. In fact, the objects can be anything produced by the animals of

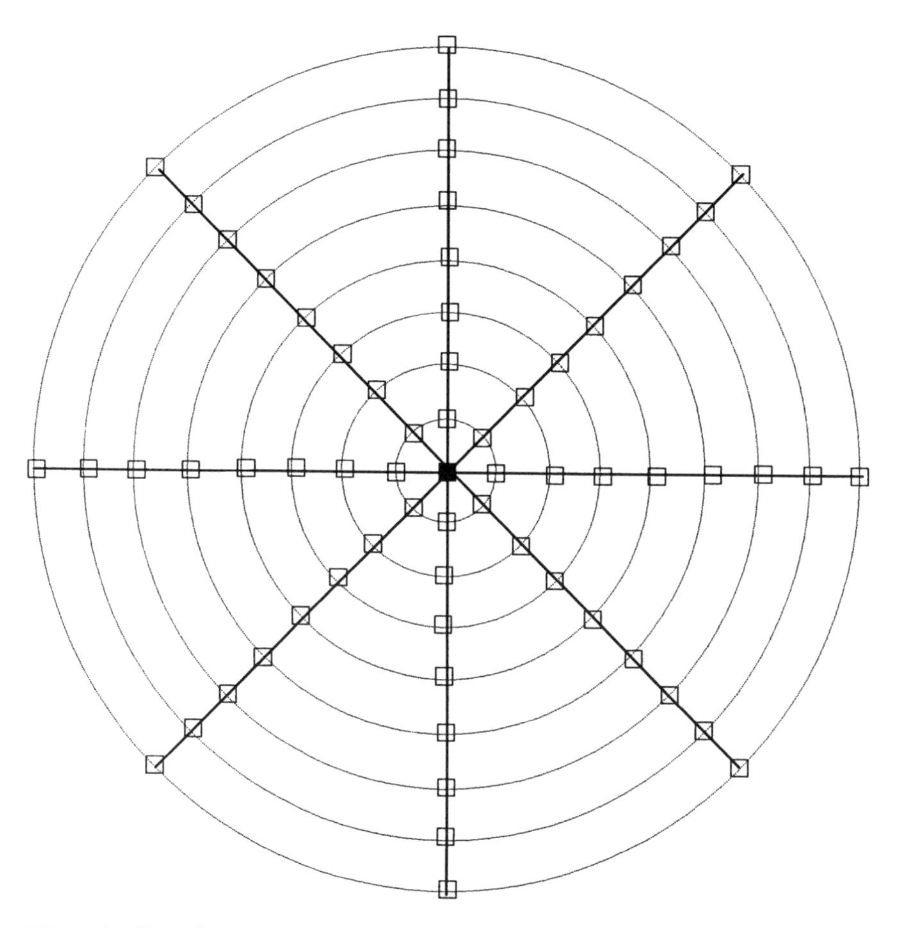

Fig. 1.7. Use of a trapping web to sample small mammal populations is an extension of point transect theory. Traps (e.g. live traps, snap traps or pitfall traps), represented by squares, are placed at the intersections of the radial lines with the concentric circles.

interest, provided that production is linearly related to animal abundance. In the case of apes for example, the objects are usually nests. The surveys themselves are typically quadrat sampling, strip transect sampling or line transect sampling.

In order to convert estimated dung density to animal abundance, it is necessary to estimate two further parameters: the rate at which dung is deposited and the disappearance rate of dung. Sometimes when animal density is high, it is feasible to clear the quadrats or strips of old dung. The survey is then carried out a known time period after clearance, before any new dung has disappeared. This circumvents the need to estimate disappearance rate, which may vary appreciably by habitat, season, weather conditions, etc. In areas of low density, a large area must be sampled, in which case it is impractical to clear the sampled area. In such circumstances, line transect sampling is likely to be more cost-effective than either quadrat or strip transect sampling.

1.2.9 *Related techniques*

In point transect sampling, the distance of each detected object from the point is recorded. In point-to-object methods, the distance of the nearest object from the point is recorded (Clark and Evans 1954; Eberhardt 1967). The method may be extended, so that the distances of the n nearest objects to the point are recorded (Holgate 1964; Diggle 1983). Thus the number of detected objects from a point is predetermined, and the area around the point must be searched exhaustively to ensure that no objects are missed closer to the point than the farthest of the n identified objects. Generally the method is inefficient for estimating density, and estimators are prone to bias.

Nearest neighbour methods are closely similar to point-to-object methods, but distances are measured from a random object, not a random point (Diggle 1983). If objects are randomly distributed, the methods are equivalent, whereas if objects are aggregated, distances under this method will be smaller on average. Diggle (1983) summarizes *ad hoc* estimators that improve robustness by combining data from both methods; if the assumption that objects are randomly distributed is violated, biases in the point-to-object and nearest neighbour density estimates tend to be in opposite directions.

1.3 The detection function

Central to the concept of distance sampling is the detection function $g(y)$:

$$\begin{aligned} g(y) &= \text{the probability of detecting an object, given that it is at} \\ &\quad\ \text{distance } y \text{ from the random line or point} \\ &= \text{pr}\{\text{detection} \mid \text{distance } y\} \end{aligned}$$

The distance y refers to either the perpendicular distance x for line transects or the sighting (radial) distance r for point transects. Generally, the detection function decreases with increasing distance, but $0 \leq g(y) \leq 1$ always. In the development to follow we usually assume that $g(0) = 1$, that is, objects on the line or point are detected with certainty. Typical graphs of $g(y)$ are shown in Fig. 1.8. Often, only a small percentage of the objects of interest are detected in field surveys. However, a proper analysis of the associated distances allows reliable estimates of true density to be made. The detection function $g(y)$ could be written as $g(y \mid v)$, where v is the collection of variables other than distance affecting detection, such as

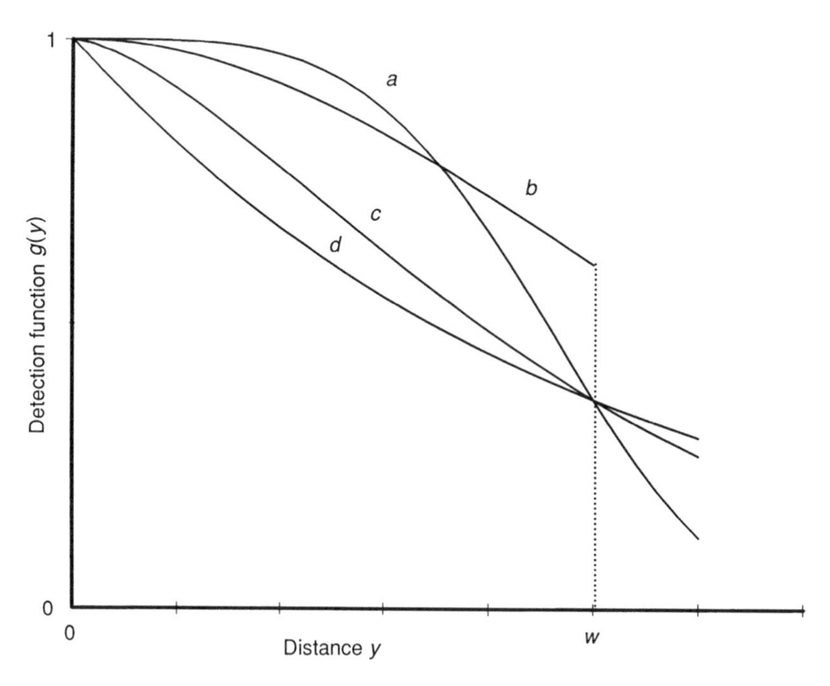

Fig. 1.8. Some examples of the detection function $g(y)$. Function b is truncated at w and thus takes the value zero for all $y > w$. Functions with shapes similar to a, b and c are common in distance sampling. Function d usually results from poor survey design and conduct, and is problematic.

object size. We will not use this explicit notation, but it is understood. Methods that model such effects will be addressed in Buckland *et al.* (in preparation), as will methods for studies in which detection on the line is not certain ($g(0) < 1$).

1.4 Range of applications

1.4.1 *Objects of interest*

Studies of birds represent a major use of both point and line transect sampling. Birds are often conspicuous by their bright colouration or distinctive song or call, thus making detection possible even in dense habitats. Surveys in open habitats often use line transects, whereas surveys in more closed habitats with high canopies often use point transects. Distance sampling methods have seen use in studying populations of many species of gamebirds, raptors, passerines and shorebirds.

Many terrestrial mammals have been successfully surveyed using distance sampling methods (e.g. pronghorn antelope, feral pigs, fruit bats, mice and several species of deer, rabbits, hares, kangaroos, primates and African ungulates). Marine mammals (several species of dolphin, porpoise, seal and whale) have been the subject of many surveys reported in the literature. Reptiles, amphibians, beetles and wolf spiders have all been the subject of distance sampling surveys, and fish (in coral reefs) and red crab densities have been estimated from underwater survey data.

Many inanimate or dead objects have been surveyed using distance sampling, including birds' nests, mammal burrows, and dead deer and pigs. Plant populations and even plant diseases are candidates for density estimation using distance sampling theory. One military application is estimation of the number of mines anchored to the seabed in minefields.

1.4.2 *Method of transect coverage*

Distance sampling methods have found use in many situations. Specific applications are still being developed from the existing theory. The versatility of the method is partially due to the variety of ways in which the transect line can be traversed. Historically, line transects were traversed on foot by a trained observer. In recent years, terrestrial studies have used trail bikes, all-terrain vehicles or horses. Transect surveys have been conducted using fixed wing aircraft and helicopters; 'ultralight' aircraft are also appropriate in some instances.

Transect surveys in aquatic environments can be conducted by divers with snorkels or SCUBA gear, or from surface vessels ranging in size from small boats to large ships, or various aircraft, or by sleds with mounted video units pulled underwater by vessels on the surface. Small submarines may have utility in line or point transect surveys if proper visibility can be achieved. Remote sensing may find extensive use as the technology develops (e.g. acoustic instruments, radar, remotely controlled cameras, multispectral scanners).

In general, the observer can traverse a line transect at a variable speed, travelling more slowly to search heavy cover. The observer may leave the line and walk an irregular path, keeping within w on each side of the line. However, the investigator must ensure that all objects on the line are detected, and that the recorded effort L is the length of the line, not the total distance the observer travels. Point transects are usually surveyed for a fixed time (e.g. 10 min per sample point). To avoid bias however, a 'snapshot' method should be adopted, in which object locations are 'frozen' at a point in time. Without this approach, random object movement can generate substantial upward bias in point transect estimates of density.

1.4.3 *Clustered populations*

Distance sampling is best explained in terms of 'objects of interest', rather than a particular species of bird or mammal. Objects of interest might be dead deer, bird nests, jackrabbits, etc. Often, however, interest lies in populations whose members are naturally aggregated into clusters. Here we will take clusters as a generic term to indicate herds of mammals, flocks of birds, coveys of quail, pods of whales, prides of lions, schools of fish, etc. A cluster is a relatively tight aggregation of objects of interest, as opposed to a loosely clumped spatial distribution of objects. More commonly, 'group' is used, but we prefer 'cluster' to avoid confusion with the term 'grouped data', defined below.

Surveying clustered populations differs in a subtle but important way between strip transect sampling and line transect sampling. In strip transect sampling, all individuals inside the strip are censused; one ignores the fact that the objects occur in clusters. In contrast, in distance sampling with a fixed w, one records all clusters detected if the centre of the cluster is inside the strip (i.e. 0 to w). If the centre of the cluster is inside the strip, then the count of the size of the cluster must include all individuals in the cluster, even if some individuals are beyond w. On the other hand, if the centre of the cluster is outside the strip, then no observation is recorded, even though some individuals in the cluster are inside the strip.

In distance sampling theory, the clusters must be considered to be the object of interest and distances should be measured from the line or point to the geometric centre of the cluster. Then, estimation of the density of clusters is straightforward. The sample size n is the number of clusters detected during the survey. If a count is also made of the number of individuals (s) in each observed cluster, one can estimate the average cluster size in the population, $E(s)$. The density of individuals D can be expressed as a product of the density of clusters D_s times the average cluster size:

$$D = D_s \cdot E(s) \tag{1.8}$$

The simplest estimate of $E(s)$ is $\bar{s}$, the mean size of the n detected clusters. However, detection may be a function of cluster size. Sometimes, those clusters detected at some distance from the line are substantially larger on average than those detected close to the line, because smaller clusters are less detectable at large distances. In this case, the estimator of D_s is still unbiased, but the mean cluster size $\bar{s}$ is a positively biased estimate of $E(s)$, resulting in upward bias in estimated density.

A well developed general theory exists for the analysis of distance data from clustered populations. Here the detection probability is dependent on both distance from the line or point and cluster size (this phenomenon is called size-biased sampling). Several approaches are possible: (1) stratify by cluster size and apply the usual methods within each stratum, then sum the estimated densities of individuals; (2) treat cluster size as a covariate and use parametric models for the bivariate distance–cluster size data (Drummer and McDonald 1987); (3) truncate the distance data to reduce the correlation between detection distance and cluster size and then apply robust semiparametric line transect analysis methods; (4) first estimate cluster density, then regress cluster size on $\hat{g}(y)$ to estimate mean cluster size where detection is certain ($\hat{g}(y) = 1$); (5) attempt an analysis by individual object rather than cluster, and use robust inference methods to allow for failure of the assumption of independent detections. Strategy (3) is straightforward and generally quite robust; appropriate data truncation after data collection can greatly reduce the dependence of detection probability on cluster size, and more severe truncation can be used for mean cluster size estimation than for fitting the line transect model, thus reducing the bias in $\bar{s}$ further. Strategy (4) is also effective. More advanced methods are addressed in Buckland *et al.* (in preparation).

1.5 Types of data

Distance data can be recorded accurately for each individual detected, or grouped ('binned') into distance categories, so that the observations are frequencies by category. Rounding errors in measurements often cause the data to be grouped to some degree, but they must then be analysed as if they had been recorded accurately, or grouped further, in an attempt to reduce the effects of rounding on bias. Distances are often assigned to predetermined distance intervals, and must then be analysed using methods developed for the analysis of frequency data.

1.5.1 *Ungrouped data*

Two types of ungrouped data can be taken in line transect surveys: perpendicular distances x_i or sighting distances r_i and angles θ_i. If sighting distances and angles are taken, they should be transformed to perpendicular distances for analysis. Only sighting distances r_i are used in the estimation of density in point transects. Trapping webs use the same type of measurement r_i, which is then the distance from the centre of the web to the trap containing animal i. The cue counting method also requires sighting distances r_i, although only those within a sector ahead of the observer are recorded. Angles (0° to 360° from some arbitrary baseline) are potentially useful in testing assumptions in point transects and trapping webs, but have not usually been taken. In cue counting also, angles (sighting angles θ_i) are not usually recorded, except to ensure that they fall between $\pm\phi$, where 2ϕ is the sector angle. In all cases we will assume that n distances $\{y_1, y_2, \ldots, y_n\}$ are measured corresponding to the n detected objects. Of course, n itself is usually a random variable, although one could design a survey in which searching continues until a pre-specified number of objects n is detected; L is then random and the theory is modified slightly (Rao 1984).

Sample size n should generally be at least 60–80, although for some purposes, as few as 40 might be adequate. Formulae are available to determine the sample size that one expects to achieve a given level of precision (measured, for example, by the coefficient of variation). A pilot survey is valuable in predicting sample sizes required, and will usually show that a sample as small as 40 objects for an entire study is unlikely to achieve the desired precision. Often, sample sizes of several hundred are required for effective management.

1.5.2 *Grouped data*

Data grouping arises in distance sampling in two ways. First, ungrouped data $y_i, i = 1, \dots, n$, may be taken in the field, but analysed after deliberate grouping into frequency counts n_j, $j = 1, \dots, u$, where u is the number of groups. Such grouping into distance intervals is often done to achieve robustness in the analysis of data showing systematic errors such as heaping (i.e. rounding errors). Grouping the r_i and θ_i data by intervals in the field or for analysis in line transect surveys is strongly discouraged because it complicates calculation of perpendicular distances.

Second, the data might be taken in the field only by distance intervals. For example, in aerial surveys it may only be practical to count the number of objects detected in the following distance intervals: 0–20, 20–50, 50–100, 100–200 and 200–500 m. Thus, the exact distance of an object detected anywhere from 0 to 20 m from the line or point would not be recorded, but only that the object was in the first distance interval. The resulting data are a set of frequency counts n_j by specified distance categories rather than the set of exact distances, and total sample size is equal to $n = \sum n_j$.

Distance intervals are defined by boundaries called cutpoints c_j. For u such boundaries, we have cutpoints $0 < c_1 < c_2 < \cdots < c_u$. By convention, let $c_0 = 0$ and $c_u = w$, where w can be finite or infinite (i.e. unbounded). Typically in line transect sampling the intervals defined by the cutpoints will be wider near w and narrower near the centreline. However, in point transect sampling, the first interval may be quite wide because the area corresponding to it is relatively small. The sum of the counts in each distance interval equals the total number of detections n, which is the sample size. In the example above, $u = 5$, and the cutpoints are $0 < c_1 = 20 < c_2 = 50 < c_3 = 100 < c_4 = 200 < c_5 = w = 500$. Suppose the frequency counts n_j are 80, 72, 60, 45 and 25, respectively. Then $n = \sum n_j = 282$ detections.

1.5.3 *Data truncation*

In designing a line transect survey, a distance $y = w$ might be specified beyond which detected objects are ignored. In this case, the width of the transect to be searched is $2w$, and the area searched is of size $2wL$. In point transects, a radius w can similarly be established, giving the total area searched as $k\pi w^2$. In the general theory, w may be assumed

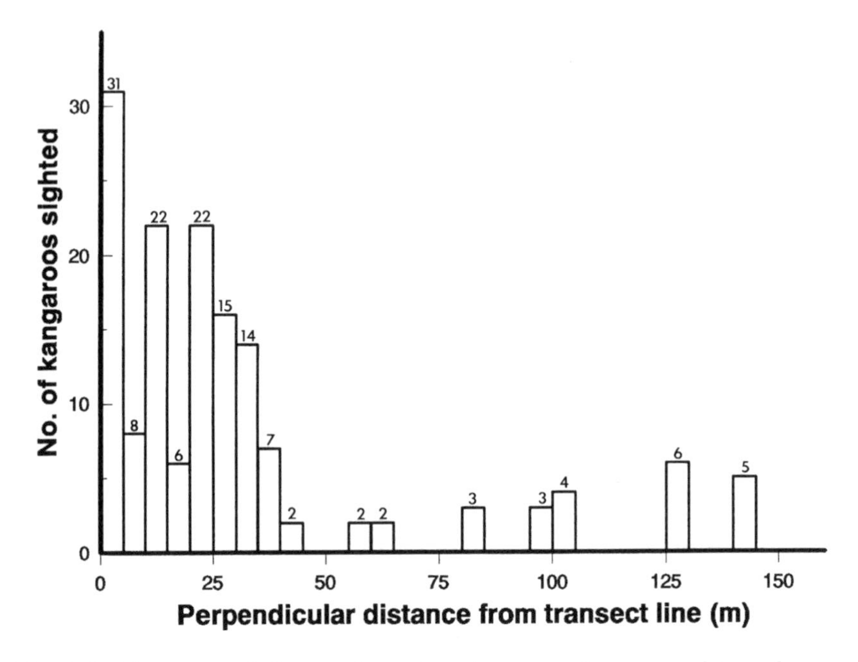

Fig. 1.9. Histogram of the number of eastern grey kangaroos detected as a function of distance from a line transect survey on Rotamah Island, Australia (redrawn from Coulson and Raines 1985). These data illustrate some heaping in the first, third and fifth distance classes, and the need to truncate observations beyond about 50 m.

to be infinite so that objects may be detected at quite large distances. In such cases, the width of the transect or radius around the point is unbounded.

Distance data can be truncated (i.e. larger distances discarded) prior to analysis. We use w to denote the distance beyond which detections are discarded. Such truncation allows us to delete outliers that make modelling of the detection function $g(y)$ difficult (Fig. 1.9). For example, w might be chosen such that $\hat{g}(w) \approx 0.15$. Such a rule might eliminate many detections in some point transect surveys, but only relatively few detections in line transect surveys. A simpler (but often unsatisfactory) rule might be to truncate 5–10% of the objects detected at the largest distances. If data are truncated in the field, further truncation may be carried out at the analysis stage if this seems useful.

General methodology is available for 'left-truncation' (Alldredge and Gates 1985). This theory is potentially useful in aerial surveys if visibility directly below the aircraft is limited and, thus, $g(0) < 1$. Quang and Lanctot (1991) provide an alternative solution to this problem. Selection of a model for the distance data is critical under left-truncation because estimation may be very model-dependent, especially if truncation extends beyond the width of the shoulder. A simpler solution for line transect sampling is to offset the line to a distance at which detection is thought to be certain.

1.5.4 *Units of measurement*

The derivation of the theory assumes that the units of y_i, L and D are all on the same measurement scale. Thus, if the distances y_i are measured in metres, then L should be in metres and density will be in numbers per square metre. In practice it is a simple but important matter to convert the y_i, l_i or D from any unit of measure into any other; in fact, computer software facilitates such conversions (e.g. feet to metres or acres to square kilometres or numbers/m^2 to numbers/km^2).

1.5.5 *Ancillary data*

In some cases, there is interest in age or sex ratios of animals detected, in which case these ancillary data must be recorded. Cluster size is a type of ancillary data. Size of the animal, its reproductive state (e.g. kangaroos carrying young in the pouch), or presence of a marker or radio transmitter are other examples of ancillary data collected during a survey. Such ancillary information can be incorporated in a variety of ways. If probability of detection is a function of the ancillary variable, then it might be used to stratify the data, or it might enter the analysis as a covariate, to improve estimation.

1.6 Known constants and parameters

1.6.1 *Known constants*

Several known constants are used in this book and their notation is given below:

A = size of the area occupied by the population of interest;
k = number of lines or points surveyed;
l_i = length of the ith transect line, $i = 1, \ldots, k$;
L = total line length = $\sum l_i$; and
w = the width of the area searched on each side of the line transect, or the radius searched around a point transect, or the truncation point beyond which data are not used in the analysis.

1.6.2 *Parameters*

In line and point transect surveys there are only a few unknown parameters of interest. These are defined below:

D = density (number per unit area);
N = population size in the study area;

$E(s)$ = mean cluster size in the population (not the same as, but often estimated by, the sample mean $\bar{s}$ of detected objects);

$f(0)$ = the probability density function of detected distances from the line, evaluated at zero distance;

$h(0)$ = the slope of the probability density function of detected distances from the point, evaluated at zero distance; and

$g(0)$ = probability of detection on the line or point, usually assumed to be 1. For some applications (e.g. species of whale which spend substantial periods underwater and thus avoid detection, even on the line or point), this parameter must be estimated from other types of information.

Density D may be used in preference to population size N in cases where the size of the area is not well-defined. Often an encounter rate n/L is computed as an index for sample size considerations or even as a crude relative density index.

1.7 Assumptions

Statistical inference in distance sampling rests on the validity of several assumptions. First, the survey must be competently designed and conducted. No analysis or inference theory can make up for fundamental flaws in survey procedure. Second, the physical setting is idealized:

1. Objects are spatially distributed in the area to be sampled according to some stochastic process with rate parameter D (= number per unit area).
2. Randomly placed lines or points are surveyed and a sample of n objects is detected, measured and recorded.

It is not necessary that the objects be randomly (i.e. Poisson) distributed. Rather, it is critical that the line or point be placed randomly with respect to the distribution of objects. Random line or point placement ensures a representative sample of the relevant distances and hence a valid density estimate. It also ensures that the surveyed lines or points are representative of the whole study area, so that the estimated density $\hat{D}$ applies to the entire area, and not just to the surveyed strips or circles associated with those lines or points. The use of transects along trails or roads does not constitute a random sample and represents poor survey practice in most circumstances. In practice, a systematic grid of lines or points, randomly placed in the study area, suffices and may often be advantageous.

Three assumptions are essential for reliable estimation of density from line or point transect sampling. These assumptions are given in order from most to least critical:

1. Objects directly on the line or point are always detected (i.e. they are detected with probability 1, or $g(0) = 1$).
2. Objects are detected at their initial location, prior to any movement in response to the observer.
3. Distances (and angles where relevant) are measured accurately (ungrouped data) or objects are correctly counted in the proper distance interval (grouped data).

Some investigators include the assumption that one must be able to identify the object of interest correctly. In rich communities of songbirds, this problem is often substantial. Marine mammals often occur in mixed schools, so it is necessary both to identify all species present and to count the number of each species separately. In rigorous theoretical developments, assumption (2) is taken to be that objects are immobile. However, slow movement relative to the speed of the observer causes few problems in line transects. In contrast, responsive movement of animals to the approaching observer can create serious problems. In point transects, undetected movement of animals is always problematic because the observer is stationary. The problem is compounded because a moving animal is more likely to be detected when it is close to the point, biasing detection distances down and abundance estimates up.

Careful attention must be given to survey design and field protocol, to ensure that key assumptions are met, at least to a good approximation. Equally important, observers should have adequate training and relevant experience.

The effects of partial failure of these assumptions will be covered at length in later chapters. Methodology for when $g(0) < 1$ is covered in Buckland *et al.* (in preparation). All of the above assumptions can be relaxed under certain circumstances. We note that no assumption is made regarding symmetry of $g(y)$ on the two sides of the line or around the point, although extreme asymmetry would be problematic. Generally, we believe that asymmetry near the line or point will seldom be large, and good design will ensure that it is not, although topography may sometimes cause difficulty. If data are pooled to give a reasonable sample size, such problems can usually be ignored.

Assumptions are discussed further in Chapter 2.

1.8 Fundamental concept

It may seem counterintuitive that observers can fail to detect over 50% of the objects of interest in the sampled area (strips of dimension L by

$2w$ or circular plots of size πw^2), and still obtain accurate estimates of population density. The following two sections provide insights into how distances are the key to the estimation of density when some of the objects remain undetected. We will illustrate the intuitive ideas for the case of line transect sampling; those for point transects are similar.

Consider an arbitrary area of size A with objects of interest distributed according to some random process. Assume a randomly placed line and grouped data taken in each of eight one-foot distance intervals from the line on either side, so that $w = 8$. If all objects were detected, we would expect, on average, a histogram of the observations to be uniform as in Fig. 1.10a. In other words, on average, one would not expect many more or fewer observations to fall, say, within the 7th interval than the first interval, or any other interval.

In contrast, distance data from a survey of duck (*Anas* and *Aythya* spp.) nests at the Monte Vista National Wildlife Refuge in Colorado, USA (Anderson and Pospahala 1970) are shown in Fig. 1.10b as a histogram. Approximately 10 000 acres of the refuge were surveyed using $L = 1600$ miles of transect, and an area $w = 8$ feet on each side of the transect was searched. A total of 534 nests was found during 1967 and 1968 and the distance data were grouped for analysis into one-foot intervals. Clearly there is evidence from this large survey that some nests went undetected in the outer three feet of the area searched. Visual inspection might suggest that about 10% of the nests were missed during the survey. Note that the intuitive evidence that nests were missed is contained in the distances, here plotted as a histogram.

Examination of such a histogram suggests that a 'correction factor', based on the distance data, is needed to correct for undetected objects. Note that such a correction factor would be impossible if the distances (or some other ancillary information) were not recorded. Anderson and Pospahala (1970) fitted a simple quadratic equation to the midpoints of each histogram class to obtain an objective estimate of the number of nests not detected (Fig. 1.10c). Their equation, fitted by least squares, was

$$\text{frequency} = 77.05 - 0.4039x^2$$

The proportion (P_a) of nests detected was estimated as the unshaded area under the curve in Fig. 1.10c divided by the total area under the horizontal line (shaded + unshaded). The estimated proportion of nests detected from 0 to 8 feet can be computed to be $\hat{P}_a = 0.888$, suggesting that a correction factor of 1.126 (= 1/0.888) be applied to the total count of $n = 534$. Thus, the estimated number of nests within eight feet of the sample transects was $n/\hat{P}_a = 601$, and because the transects sampled 5.5% of the refuge, the estimate of the total number of nests on the refuge during the two-year period was $601/0.055 = 10\,927$. This procedure provides the intuition that distances are central to reliable density

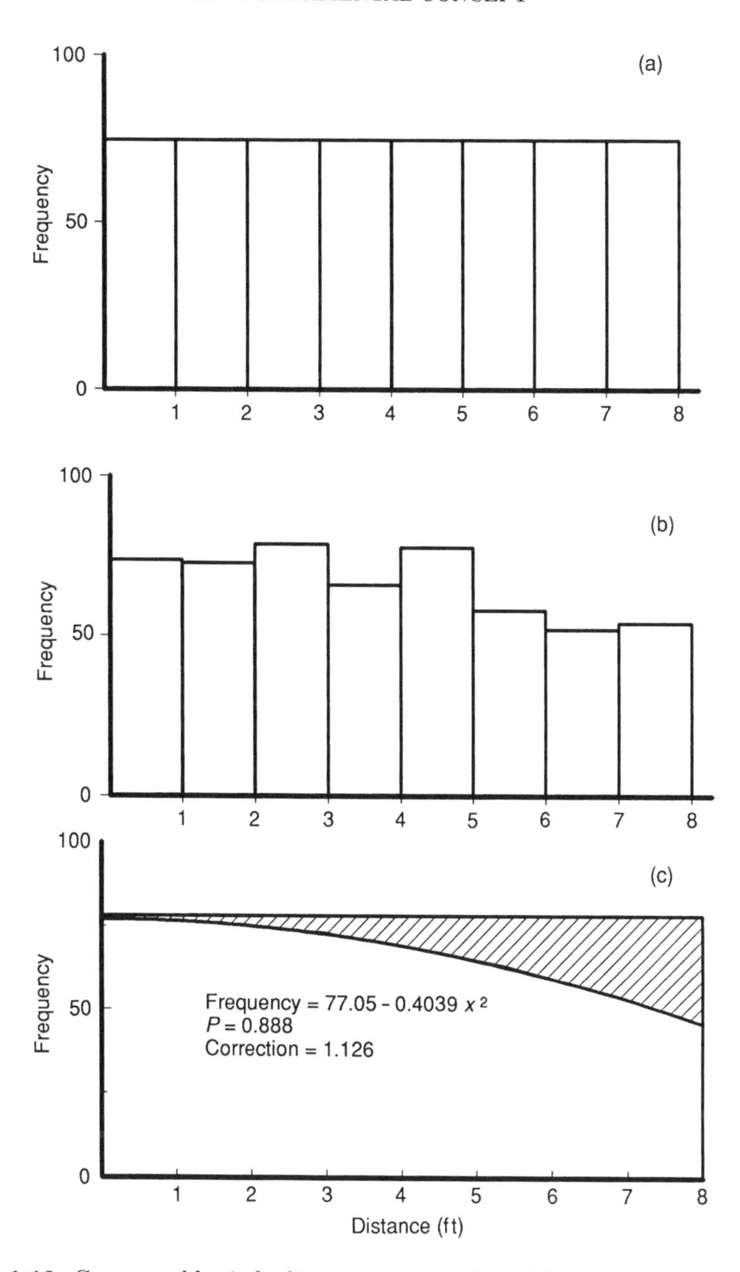

Fig. 1.10. Conceptual basis for line transect sampling: (a) the expected number of objects detected in eight distance classes if no objects were left undetected; (b) real data where a tendency to detect fewer objects at greater distances from the line can be noticed; (c) simple methods can be used to estimate the proportion of the objects left undetected (shaded area). The proportion detected, P_a, can be estimated from the distance data.

estimation. The method performs well even if most of the objects in the surveyed strips are not detected, provided field protocol ensures that the three key assumptions are well approximated. The Anderson–Pospahala method is no longer recommended since superior analysis methods are now available, but it illustrates the principle underlying the theory. The next two chapters will put this intuitive argument on a more formal basis.

1.9 Detection of objects

When a survey has been conducted, n objects would have been detected. Considerable confusion regarding the meaning of n exists in the literature. Here an attempt is made to factor n into its fundamental components. Burnham *et al.* (1981), Dawson (1981) and Morgan (1986: 9–25) give a discussion of these issues. Ramsey (1981) provides an informative example.

The number detected n is a confounding of true density and probability of detection. The latter is a function of many factors, including cue production by, or characteristics of, the object of interest, observer effectiveness and the environment. Of these factors, one could hope that only the first, density, influences the count. While this might be ideal, it is rarely true.

1.9.1 *Cue production*

The object of interest often provides cues that lead to its detection by the survey observer. Obvious cues may be a loud or distinctive song or call. A splash made by a marine mammal, or a flock of seabirds above a school of dolphins, are other examples of cues. Large size, bright or contrasting colouring, movement or other behaviour may be causes for detection. These cues are frequently species-specific and may vary by age or sex of the animal, time of day, season of the year or whether the animal is mated. Thus, the total count n can vary for reasons unrelated to density (Mayfield 1981; Richards 1981; Bollinger *et al.* 1988). Most often, the probability of detection of objects based on some cue diminishes as distance from the observer increases.

1.9.2 *Observer effectiveness*

Observer variability is well known in the literature on biological surveys. Interest in the survey, training and experience are among the dominant reasons why observers vary widely in their ability to detect objects of interest. However, both vision and hearing acuity may be major variables which are often age-specific (Ramsey and Scott 1981a; Scott *et al.* 1981b). Fatigue is a factor on long or difficult surveys. Even differing heights of observers may be important for surveys carried out on foot, with tall observers detecting objects at a higher rate. Generally, the detection of objects decreases with increasing distance, and the rate of decrease varies by observer. For example, in Laake's stake surveys (Burnham *et al.* 1980), between 27% and 67% of the stakes present were detected and recorded by different observers, yet accurate estimates of stake density were obtained using distance sampling theory.

1.9.3 *Environment*

Environmental variables often influence the number of objects detected (Best 1981; Ralph 1981; Verner 1985). The habitat type and its phenology are clearly important (Bibby and Buckland 1987). Physical conditions often inhibit detection: wind, precipitation, darkness, sun angle, etc. Cue production varies by time of day, which can have a tenfold effect in the detectability of some avian species (Robbins 1981; Skirvin 1981). Often, these variables interact to cause further variability in detection and the count n.

Distance sampling provides a general and comprehensive approach to the estimation of population density. The distances y_i allow reliable estimates of density in the face of variability in detection due to factors such as cue production, observer effectiveness and environmental differences. The specific reasons why an object was not detected are unimportant. Furthermore, it seems unnecessary to research the influence of these environmental variables or to standardize survey protocol for them, if distances are taken properly and appropriate analysis carried out. Distance sampling methods fully allow for the fact that many objects will remain undetected, as long as they are not on the line or point.

1.10 History of methods

1.10.1 *Line transects*

We draw here on the reviews by Hayes (1977), Gates (1979) and Buckland *et al.* (2000). Strip transects appear to date back to bird surveys in Illinois conducted in 1906 by J.E. Gross (Forbes 1907; Forbes and Gross 1921). Nice and Nice (1921) introduced roadside counts, which were assumed to yield total counts out to a predetermined distance from a road. Kelker (1945) refined the concept of a strip count a little, by recording distances from the line, and subsequently estimating a distance up to which it is reasonable to assume that all objects are detected. The width of this strip was judged subjectively from a histogram of the data. The so-called 'Kelker strip' has periodically been re-invented by other authors. Hahn (1949) modified the approach by estimating at sample locations the distance to the 'vegetation line', beyond which it was assumed that objects are not detectable. The width of the strip was then estimated as the average of these distances.

The first attempt to estimate abundance in the spirit of line transect sampling, in which a proportion of animals within the strip remain undetected, appears to date back to R.T. King in the 1930s (Leopold 1933; Gates 1979). He used the average of the radial (sighting) distances r as an estimate of the effective half-width of the surveyed strip.

Hayne (1949) developed a rigorous estimator based on radial distances, which remained the only line transect estimator with a solid mathematical basis for nearly 20 years. During that period, line transect sampling was used on a variety of species. The assumptions were sharpened in the wildlife literature and some evaluations of the method were presented (e.g. Robinette *et al.* 1956). Hayne's estimator continued to be used into the 1980s, but it was not robust when the mean sighting angle differed appreciably from the value of 32.7° implied by the model, and Hayes and Buckland (1983) showed that the model was a poor representation of the detection process.

In 1968, two important papers were published in which some of the key ideas and conceptual approaches to line transect sampling finally appeared (Eberhardt 1968; Gates *et al.* 1968). Gates *et al.* (1968) published the first truly rigorous statistical development of a line transect estimator, applicable only to untruncated and ungrouped perpendicular distance data. They proposed that the detection function $g(x)$ be a negative exponential form, $g(x) = \exp(-ax)$, where a is an unknown parameter to be estimated. Under that model, the effective half-width of search is $\mu = 1/a$. Gates *et al.* (1968) obtained the maximum likelihood estimator of a based on a sample of perpendicular distances ($\hat{a} = 1/\bar{x}$) and provided an estimator of the sampling variance. For the first time, rigorous consideration was given to questions such as optimal estimation under the model, construction of confidence intervals and tests of assumptions. The one weakness was that the assumed exponential shape for the detection function has been found to be unsatisfactory, both because it is inflexible and because it assumes that the detection function is 'spiked', with probability of detection dropping quickly close to the line. Good survey protocol seeks to ensure that this does not occur.

In contrast, Eberhardt (1968) conceptualized a fairly general model in which the probabilities of detection decreased with increasing perpendicular distance. He reflected on the shape of the detection function $g(x)$, and suggested both that there was a lack of information about the appropriate shape and that the shape might change from survey to survey. Consequently, he suggested that the appropriate approach would be to adopt a family of curves to model $g(x)$. He suggested two such families, a power series and a modified logistic, both of which are fairly flexible parametric functions. His statistical development of these models was limited, but important considerations had been advanced.

In the years following 1968, line transect sampling was developed along rigorous statistical inference principles. Parametric approaches to modelling $g(x)$ were predominant, with the exception of Anderson and Pospahala (1970), who introduced some of the basic ideas that underlie a nonparametric or semiparametric approach to the analysis of line transect data. Emlen (1971) proposed an *ad hoc* method that was popular in avian studies.

A general model structure for line transect sampling based on perpendicular distances was presented by Seber (1973: 28–30). For an arbitrary detection function, Seber gave the probability distribution of the distances $x_1, \ldots, x_n$ and the general form of the estimator of animal density D. This development was left at the conceptual stage and not pursued to the final step of a workable general approach for deriving line transect estimators.

More work on sighting distance estimators appeared (Gates 1969; Overton and Davis 1969). There was a tendency to think of approaches based on perpendicular distances as appropriate for inanimate or non-responsive objects, whereas methods for flushing animals were to be based on sighting distances and angles (Eberhardt 1968, 1978a). That artificial distinction tended to prevent the development of a unified theory for line transect sampling. By the mid-1970s, line transect sampling remained a relatively unexplored methodology for the estimation of animal density. Robinette *et al.* (1974) reported on a series of field evaluations of various line transect methods. Their field results were influential in the development of the general theory.

Burnham and Anderson (1976) pursued the general formulation of line transect sampling and gave a basis for the general construction of line transect estimators. They developed the general result $\hat{D} = n \cdot \hat{f}(0)/2L$, in which $\hat{f}(0)$ is the estimated probability density function of perpendicular distances, evaluated at zero distance. The key problem of line transect data analysis was seen to be the modelling of the detection function $g(x)$ (possessing the property that $g(0) = 1$), or equivalently, the fitting of the probability density function $f(x)$ and the subsequent estimation of $f(0)$. The nature of the specific data (grouped or ungrouped, truncated or untruncated) is irrelevant to the basic estimation problem. Consequently, their formulation is applicable for the development of any parametric or semiparametric line transect estimator, using the wide statistical literature on fitting density functions. Further, the general theory is applicable to point transect sampling with some modification (Buckland 1987a).

Burnham and Anderson's (1976) paper heralded a period of new statistical theory. Major contributions published during the 1976–80 period include Schweder (1977), Crain *et al.* (1978, 1979), Pollock (1978), Patil *et al.* (1979b), Quinn (1979), Ramsey (1979), Seber (1979) and Quinn and Gallucci (1980). Other papers developing methodology during this short period include Anderson *et al.* (1978, 1979a, 1980), Eberhardt (1978a,b, 1979), Sen *et al.* (1978), Burnham *et al.* (1979), Patil *et al.* (1979a) and Smith (1979). Anderson *et al.* (1979b) provided guidelines for field sampling, including practical considerations. Burdick (1979) produced an advanced method to estimate spatial patterns of abundance from line transect sampling where there are major gradients in population density. Laake *et al.* (1979) and Gates (1980) produced computer software packages, TRANSECT and LINETRAN, respectively, for the analysis of line transect data.

Gates (1979) gave a readable summary of developments in line transect sampling theory up to that date, and Ramsey's (1979) paper provided the theoretical framework for parametric models. Hayes (1977) gave an excellent summary of methodology and provided many useful insights at that time.

Burnham *et al.* (1980) published a major monograph on line transect sampling theory and application. This work provided a review of previous methods, gave guidelines for field use, and identified a small class of estimators that seemed generally useful. Usefulness was based on four criteria: model robustness, pooling robustness, a shape criterion and estimator efficiency. Theoretical and Monte Carlo studies led them to suggest the use of estimators based on the Fourier series (Crain *et al.* 1978, 1979), the exponential power series (Pollock 1978), and the exponential quadratic model.

Since 1980, more theory has been developed on a wide variety of issues. Seber (1986) and Ramsey *et al.* (1988) give brief reviews. Major contributions during the 1980s include Butterworth (1982a,b), Patil *et al.* (1982), Hayes and Buckland (1983), Buckland (1985), Burnham *et al.* (1985), Johnson and Routledge (1985), Quinn (1985), Drummer and McDonald (1987), Ramsey *et al.* (1987), Thompson and Ramsey (1987) and Zahl (1989). Other papers during the decade include Buckland (1982), Stoyan (1982), Burnham and Anderson (1984), Anderson *et al.* (1985a,b) and Gates *et al.* (1985). Several interesting field evaluations where density was known have appeared since 1980, including Burnham *et al.* (1981), Hone (1986, 1988), White *et al.* (1989), Bergstedt and Anderson (1990) and Otto and Pollock (1990). In addition, other field evaluations where the true density was not known have been published, but these results are difficult to interpret.

The approaches of this book find their roots in the work of Burnham and Anderson (1976) and Burnham *et al.* (1980). Buckland (1992a,b) developed the unified 'key plus series adjustment' formulation, which includes as special cases the Fourier series model of Crain *et al.* (1978, 1979) and the Hermite polynomial model of Buckland (1985). In parallel with the first edition of this book, DISTANCE, the DOS version of the distance sampling software, was developed (Laake *et al.* 1993).

1.10.2 *Point transects*

Point transect sampling has a much shorter history. Point counts or 'circular plots' have been around for some time. For example, the North American Breeding Bird Survey, based on 50 roadside point counts along each of a large number of 24.5-mile routes, was piloted in 1965 and began officially in 1966 (Link and Sauer 1997). In point counts, all detected birds are counted, either out to a given radius or without truncation. The approach is still in wide use, despite its shortcomings. In the 1970s, work started on developing variable circular plots, which we term point transects. The initial work was not published until 1980 (Reynolds *et al.* 1980), but the methods were already in fairly wide use by then, and several papers preceded it into print. Notable amongst these is Wiens and Nussbaum (1975), which draws on the methods developed by Emlen (1971) for line transect sampling, and which remain popular despite the availability of improved methods. Ramsey and Scott (1979) recognized early both the potential and the pitfalls of point transects, and provided a similar theoretical basis as that for line transects given by Burnham and Anderson (1976) and Ramsey (1979). Ramsey and Scott (1979) noted that, by squaring distances, the estimation problem became essentially the same as for line transect sampling, and Burnham *et al.* (1980) endorsed this approach. However, Buckland (1987a) noted the effect that transformation has on the shape of the detection function, and favoured an analysis of the untransformed distances, so that the same models for the detection function could be used for both line and point transects. Software Distance is based on this approach.

1.11 Program Distance

The computation for most estimators is arduous and prone to errors if done by hand. Estimators of sampling variances are similarly tedious. Data should be plotted and estimates of $f(y)$ should be graphed for visual comparison with the observed distance data. All of these tasks are far better done by computer, and Distance (Thomas *et al.* 1998) provides comprehensive PC software, available free from http://www.ruwpa.st-and.ac.uk/distance/.

Versions 3.5 and later of Distance are fully windows-based and require Microsoft Windows 95 or later to run. The minimum hardware requirement for Distance 3.5 is a Pentium-class processor with 32MB RAM and 800×600 SVGA monitor with 32 768 colours. An older DOS-based version, DISTANCE, with more modest hardware requirements, is also available from the internet site.

The program allows researchers to focus on the results and interpretation, rather than on computational details. Almost all the examples presented in this book were analysed using Distance; project files containing the distance data and associated analyses for some of the examples are available as an aid to researchers. The program is useful both for data analysis and as a research tool. Only occasional references to Distance are made throughout this book because the software comes with extensive online documentation.

2

Assumptions and modelling philosophy

Before we address the assumptions, we note that a good design and survey protocol, as outlined in Chapter 7, are essential.

2.1 Assumptions

This section provides material for a deeper understanding of the assumptions required for the successful application of distance sampling theory. The validity of the assumptions allows the investigator assurance that valid inference can be made concerning the density of the population sampled. Before we address the assumptions, we note that the design and conduct of the survey must pay due regard to good survey practice, as outlined in Chapter 7. If the survey is poorly designed or executed, the estimates may be of little value. Sound theory and analysis procedures cannot make up for poor sampling design or inadequate field protocol.

It is assumed that a population comprises objects of interest that are distributed in the area to be sampled according to some stochastic process with rate parameter D (= expected number per unit area). In particular, it is not necessary that the objects have a Poisson distribution, although this is mistakenly given in several places in the literature. Rather, it is critical that the lines or points be placed randomly with respect to the distribution of objects. Random line or point placement justifies the extrapolation of the sample statistics to the population of interest.

Given the random line or point placement, we can usually safely assume that objects are uniformly distributed with respect to perpendicular distance from the line or with respect to distance in any given direction from the point. However, if the study area is not large relative to typical detection distances, or if it is very fragmented, a large proportion of the strips of half-width w or circles of radius w might fall outside the study area, and this proportion will tend to increase with distance from the line or point. Consequently, objects within the study area will not be uniformly distributed with respect to distance from the line or point. In such circumstances, estimation may be appreciably biased if allowance is not made for the effect. The more general methods of Buckland *et al.* (in preparation) allow estimation in these circumstances. Fortunately, the bias from this source is negligible for the large majority of studies.

Three assumptions are critical to achieving reliable estimates of density from line or point transect sampling. The effects of partial failure of these assumptions and corresponding theoretical extensions are covered at length in later sections or in Buckland *et al.* (in preparation). All three assumptions can be relaxed under certain circumstances.

2.1.1 *Assumption 1: objects on the line or point are detected with certainty*

It is assumed that all objects at zero distance are detected, that is $g(0) = 1$. In practice, detection on or near the line or point should be nearly certain. Design of surveys must fully consider ways to assure that this assumption is met; its importance cannot be overemphasized.

The theory can be generalized such that density can be computed if the value of $g(y)$ is known for some value of y. However, this result is of little practical significance in biological sampling unless an assumption that $g(y) = 1$ for some $y > 0$ is made. This is potentially useful for aerial surveys, in which the observer cannot see the line directly beneath the aircraft but can detect all animals a short distance away from the centreline (Quang and Lanctot 1991).

If objects on or near the line or point are missed, the standard estimate will be biased low (i.e. $E(\hat{D}) < D$). The bias is a simple function of $g(0)$: $E(\hat{D}) - D = -[1 - g(0)] \cdot D$, which is zero (unbiased) when $g(0) = 1$. Many things can be done in the field to help ensure that $g(0) = 1$. For example, video cameras have been used in aerial and underwater surveys to allow a check of objects on or very near the line; the video can be monitored after completion of the field survey. Trained dogs have been used in ground surveys to aid in detection of grouse close to the line.

Although we stress that every effort should be made to ensure $g(0) = 1$, the practice of 'guarding the centreline' during shipboard or aerial line transect surveys can be counterproductive. For example, suppose that most search effort is carried out using $20\times$ or $25\times$ tripod-mounted binoculars on a ship, but an observer is assigned to search with the naked eye, to ensure animals very close to the ship are not missed. If $g(0)$ in the absence of this observer is appreciably below 1, then the detection function may be as illustrated in Fig. 2.1. This function may violate the shape criterion described later, and cannot be reliably modelled unless sample size is very large. The problem may be exacerbated if animals are attracted to the ship; the observer guarding the centreline may only detect animals as they move in towards the bow. Thus, field procedures should ensure both that $g(0) = 1$ and that the detection function does not fall steeply with distance from the line or point.

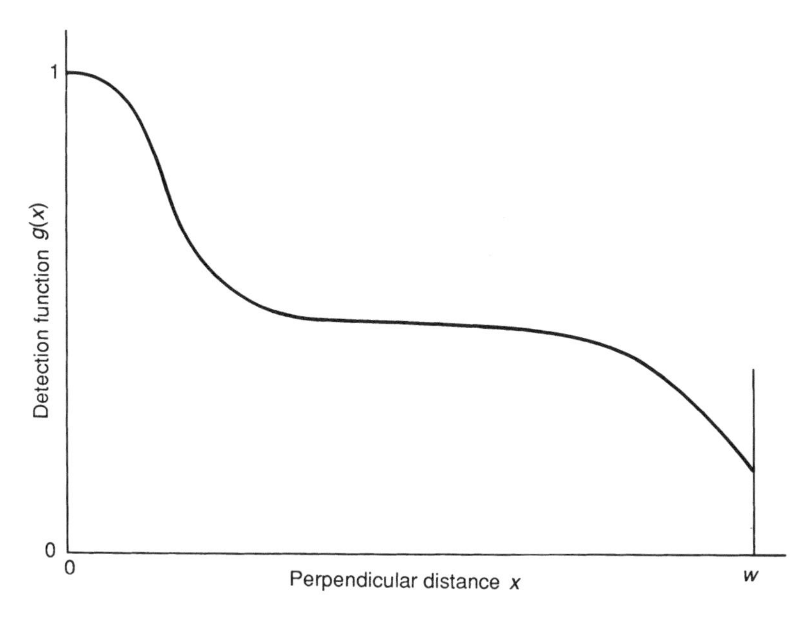

Fig. 2.1. Hypothetical detection function illustrating the danger of assigning an observer to 'guard the centreline'. This problem is most common in shipboard and aircraft surveys involving more than one observer.

Where it is not possible to design a survey so that $g(0) = 1$ at least approximately, then double or multiple platform surveys should be considered. In their simplest form, observers search independently from two platforms, usually a shipboard survey with two platforms that are shielded visually and aurally from each other. For each animal detected from one platform, an assessment is made of whether it is also detected from the other (a duplicate detection). These surveys require more advanced analysis methods, which are addressed in Buckland *et al.* (in preparation).

2.1.2 *Assumption 2: objects are detected at their initial location*

The theory of distance sampling and analysis is idealized in terms of dimensionless points or 'objects of interests'. Conceptually, a 'snapshot' is taken at the time of the survey, and distances measured to the line or point. Surveys of dead deer, plants or duck nests are easily handled in this framework. More generally, movement independent of the observer causes no problems, unless the object is counted more than once on the same unit of transect sampling effort (usually the line or point) or if it is moving at roughly half the speed of the observer or faster. Animals such as jackrabbits or pheasants may flush suddenly as an observer approaches. The measurement must be taken to the animal's original location. In these cases, the flush is often the cue that leads to detection. Animal movement after detection is not a problem, as long as the original location can be established accurately and the appropriate distance measured. Similarly, it is of no concern if an animal is detected more than once on different occasions of sampling the same transect. Animals that move to the vicinity of the next transect in response to disturbance by the observer are problematic. However, chance movement that leads to detection of the same animal from different transects is not a serious concern; if movement is random, or at least not systematically in a single direction, then animals moving in one direction will tend to be compensated by animals moving the other way. By contrast, if the observer unknowingly records the same animal several times while traversing a single transect, due to undetected movement ahead of him (as occurs with some songbirds), bias can be large.

In studies of mobile animals, it is possible that an animal moves from its original location for some distance prior to being detected. The measured distance is then from the random line or point to the location of the detection, not the animal's original location. If such undetected movements prior to detection were random (see Yapp 1956), no serious problem would result, provided that the animal's movement is slow relative to the speed of the observer. If movement is not slow, its effect must be modelled (Schweder 1977), or field procedures must be modified (Section 6.5). However, movement may be in response to the observer. Animals may take evasive movement prior to detection. A jackrabbit might hop several metres away from the observer into heavy cover and wait. As the observer moves closer, the rabbit might eventually flush. If the new location is thought to be the original location and this distance is measured and recorded, then the assumption is violated. Similarly, some species of partridge will tend to run away from the line, then as the observer continues to approach, they flush. If a substantial portion of the population moves further from the line prior to detection, this movement will often be apparent from examination of the histogram of the distance data (Fig. 2.2). However, if some animals move a considerable perpendicular distance and others remain in their original location, then the effect may not be detectable from the data. If evasive movement occurs prior to detection, the estimator will be biased low ($E(\hat{D}) < D$) (Fig. 2.2b or c). Less frequently, animals (e.g. some songbirds and marine mammals) may be attracted to the observer (Bollinger *et al.* 1988; Buckland and Turnock 1992). If animals move toward the observer prior to being detected, a positive bias in estimated density can be expected ($E(\hat{D}) > D$). However, in this case, the movement is unlikely to be detected in the histogram, even if it is severe.

Ideally, the observer on a line transect survey would try to minimize such movement by looking well ahead as the area is searched. Field procedures should try to ensure that most detections occur beyond the likely range of the effect of the observer on the animals. In point transect surveys, one must be careful not to disturb animals as the sample point is

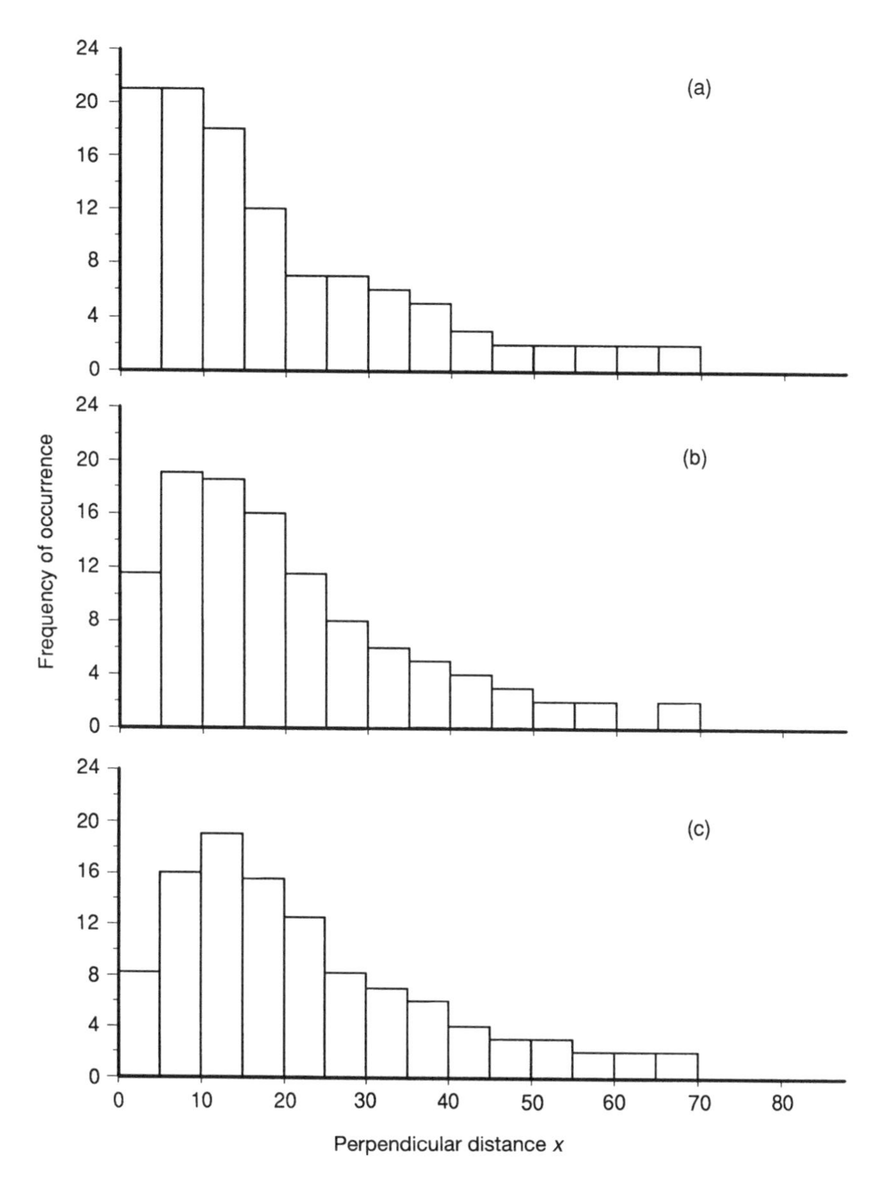

Fig. 2.2. Three histograms of perpendicular distance data, for equally spaced cutpoints, illustrating the effect of evasive movement prior to detection. Expected values are shown for the case where relatively little movement away from the observer was experienced prior to detection (a), while (b) and (c) illustrate cases where movement prior to detection was more pronounced. Data taken from Laake (1978).

approached, or perhaps wait a while upon reaching the point. It seems unlikely that methods will be developed for the reliable estimation of density for cases where a high proportion of the objects moves in response to the observer prior to detection without making some very critical and untestable assumptions (e.g. Smith 1979), unless relevant and reliable ancillary data can be gathered (Turnock and Quinn 1991; Buckland and Turnock 1992).

2.1.3 *Assumption 3: measurements are exact*

Ideally, recorded distances (and angles, where relevant) are exact, without measurement errors, recording errors or heaping. For grouped data, detected objects are assumed to be correctly assigned to distance categories. Reliable estimates of density may be possible even if the assumption is violated. Although the effect of inaccurate measurements of distances or angles can often be reduced by careful analysis (e.g. grouping), it is better to gather good data in the field, rather than to rely on analytical methods. It is important that measurements near the line or point are made accurately. Rounding errors in measuring angles near zero are problematic, especially in the analysis of ungrouped data, and for shipboard surveys. If errors in distance measurements are random and not too large, then reliable density estimates are still likely (Gates *et al.* 1985). Biased measurements pose a larger problem (e.g. a strong tendency to underestimate the distances using ocular judgements), which is usually avoidable by careful consideration of field methods, good observer training programmes and/or use of modern technical aids such as survey lasers.

For duck nests and other stationary objects, distances can be measured with a steel tape or similar device, and for most terrestrial surveys (especially those in more open habitats), distances between 20 m and a few hundred metres can be measured accurately by survey lasers or laser binoculars. When distances cannot be measured with good precision, a useful alternative is to take grouped data in, say, 5–7 distance intervals, such that the width of the intervals increases toward w. We denote the cutpoints that define these intervals by c_0 (usually $c_0 = 0$), $c_1, c_2, c_3, \ldots$, and for line transects, we would typically have $[c_1 - 0] \leq [c_2 - c_1] \leq [c_3 - c_2] \leq \cdots$. Thus, careful measurement is required only near the cutpoints c_i.

2.1.3.1 *Heaping*

Often, when distances are estimated (e.g. ocular estimates, 'eyeballing'), the observer may 'round' to convenient values (e.g. 5, 10, 50 or 100) when recording the result. Thus, a review of the n distance values will frequently reveal many 'heaped' values such as 0, 10, 25, 50 or 100, and relatively few numbers such as 1, 4, 7, 19 or 31. Heaping is common in sighting angles, which are often strongly heaped at 0, 15, 30, 45, 60 and 90 degrees. A histogram of the data will often reveal evidence of heaping. Goodness of fit tests are very sensitive to heaping, and care is needed in choosing suitable distance intervals when grouping the data for such a test. In the presence of heaping, analysis might be improved by using the same groups together with an analysis of grouped data, rather than analysing the ungrouped data. Cutpoints for grouping distances from the line or point should be selected so that large 'heaps' fall approximately at the midpoints of the groups. For line transects, sighting distances and angles should not be grouped prior to conversion into perpendicular distances. Where practical, heaping can be avoided in the field by measuring distances, rather than merely estimating them. The effects of heaping can be reduced during the analysis by smearing (Butterworth 1982b). Heaping at perpendicular distance zero can result in serious overestimation of density. This problem is sometimes reduced if a model is used that always satisfies the shape criterion (Section 2.3.2), although accurate measurement is the most effective solution.

2.1.3.2 *Systematic bias*

When distances are estimated, it is common to find that the error has a systematic component as well as a random component. For example, there is sometimes a strong tendency to underestimate distances at sea. Each distance may tend to be over- or underestimated. In surveys where only grouped data are taken, the counts may be in error because the cutpoints c_i are in effect $c_i + \delta_i$ where δ_i is some systematic increment. Thus, n_1 is not the count of objects detected between perpendicular distances 0 and c_1, it is the count of objects detected between 0 and $c_1 + \delta_1$. Little can be done to reduce the effect of these biased measurements in the analysis of the data unless experiments are carried out to estimate the bias; a calibration equation then allows the biased measurements to be corrected. Again, careful measurements are preferable to rough estimates of distances.

2.1.3.3 *Outliers*

If data are collected with no fixed width w, it is possible that a few extreme outliers will be recorded. A histogram of the data will reveal outliers. These data values contain little information about the density and will frequently be difficult to fit (Fig. 1.9). Generally, such extreme values will not be useful in the final analysis of density, and should be truncated. If all detected animals are recorded in the field, irrespective of their distance from the line, we would typically expect to truncate the largest 5–10% of those distances prior to analysis.

2.1.4 *Other assumptions*

Other aspects of the theory can be considered as assumptions. The assumption that detections are (statistically) independent events is often mentioned. If detections are somewhat dependent (e.g. 'string' flushes of quail), then the theoretical variances will be underestimated. However, we recommend that empirically based estimates of sampling variances be made, thus alleviating the need for this assumption. That is, if $var(n)$ is estimated from variation in encounter rate between independent replicate lines or points, then estimation is remarkably robust to extreme violations of the assumption of within line or point independence, even though strictly it is required for forming the likelihood from which we estimate the model parameters. Independence of detection of individual animals is clearly violated in clustered populations. This is usually handled by defining the cluster as the object of interest and measuring the ancillary variable, cluster size. This solution can be unsatisfactory for objects that occur in loose, poorly defined clusters, so that the location and size of the cluster may be difficult to determine or estimate without bias. In this case, because the methods are largely unaffected by dependence between detections, the distance to each individual detected object can be recorded and analysed.

Statistical inference methods used here (e.g. maximum likelihood estimators of parameters, theoretical sampling variance estimators and goodness of fit tests) assume independence among detections. Failure of the assumption of independence has little effect on the point estimators, and the robust variance estimators we recommend here, but causes underestimation of theoretical variance estimates (Cox and Snell 1989). The goodness of fit test is sensitive to non-independence.

It is of little concern if detection on either side of the line or around the point is not symmetric, provided that the asymmetry is not extreme, in which case it may prove difficult to model $g(y)$. In the case of line transect sampling, it may be better to analyse the data to the left of the line separately from those to the right, and to average the two density estimates, perhaps weighting by the inverse variance.

A more practically important consideration relates to the shape of the detection function near zero distance. This shape can often be judged by examining histograms of the distance data using different groupings. Distance sampling theory performs well when a 'shoulder' in detectability exists near the line or around the point. That is, detectability is certain near the line or point and stays certain or nearly certain for some distance. This will be defined as the 'shape criterion' in Section 2.3. If detectability falls sharply just off the line or point, then estimation tends to be poor. Thus if data are to be analysed reliably, the detection function from which they come should possess a shoulder; to this extent, the shape criterion is an assumption.

Some papers imply that an object should not be counted on more than one line or point. This, by itself, is not true as no such assumption is required. In surveys with $w = \infty$, an object of interest can be detected from two different lines without violating any assumptions. As noted above, if in line transect sampling an animal moves ahead of the observer and is counted repeatedly, abundance will be overestimated. This is undetected movement in response to the observer; double counting, by itself, is not a cause of bias if such counts correspond to different units of counting effort. Bias is likely to be small unless repeated counting is common during a survey. Detections made behind the observer in line transect sampling may be utilized, unless the object is located before the start of a transect leg, in which case it is outside the rectangular strip being surveyed.

These assumptions, their importance, models robust to partial violations of assumptions, and field methods to meet assumptions adequately will be addressed in the material that follows.

2.2 Fundamental models

This section provides a glimpse of the theory underlying line and point transect sampling. This material is an extension of Section 1.8.

2.2.1 *Line transects*

In strip transect sampling, if strips of width $2w$ and total length L are surveyed, an area of size $a = 2wL$ is censused. All n objects within the strips are enumerated, and estimated density is the number of objects per unit area:

$$\hat{D} = \frac{n}{2wL} \tag{2.1}$$

In line transect sampling, only a proportion of the objects in the area a surveyed is detected. Let this unknown proportion be P_a. If P_a were known, then we would estimate density by

$$\hat{D} = \frac{n}{2wLP_a} \tag{2.2}$$

Thus if P_a can be estimated from the distance data, the estimate of density becomes

$$\hat{D} = \frac{n}{2wL\hat{P}_a} \tag{2.3}$$

Now, some formalism is needed for the estimation of P_a from the distances. The unconditional probability of detecting an object in the strip (of area $a = 2wL$) is

$$P_a = \frac{\int_0^w g(x)\,dx}{w} \tag{2.4}$$

Thus we need to estimate the function $g(x)$ from the distance data. In the duck nest example of Chapter 1, $g(x)$ was found by dividing the estimated quadratic equation by the intercept (77.05), to give

$$\hat{g}(x) = 1 - 0.0052x^2 \tag{2.5}$$

Note that $\hat{g}(8) = 1 - 0.0052(8^2) = 0.66$, indicating that approximately one-third of the nests near the edges of the transect were never detected. Then

$$\hat{P}_a = \frac{\int_0^8 (1 - 0.0052x^2)\,dx}{8} = 0.888 \tag{2.6}$$

Substituting the estimator of P_a from eqn (2.4) into $\hat{D}$ from eqn (2.3) gives

$$\hat{D} = \frac{n}{2L\int_0^w \hat{g}(x)\,dx} \tag{2.7}$$

because the w and $1/w$ cancel out. Then the integral $\int_0^w g(x)\,dx$ becomes the critical quantity and is denoted as μ for simplicity. Thus,

$$\hat{D} = n/2L\hat{\mu} \tag{2.8}$$

There is a very convenient way to estimate the quantity $1/\mu$. The derivation begins by noting that the probability density function (pdf) of the perpendicular distance data, conditional on the object being detected, is merely

$$f(x) = \frac{g(x)}{\int_0^w g(x)\,dx} \tag{2.9}$$

This result follows because the expected number of objects (including those that are not detected) at distance x from the line is independent of x. This implies that the density function is identical in shape to the detection function; it can thus be obtained by rescaling, so that the function integrates to unity.

By assumption, $g(0) = 1$, so that the pdf, evaluated at zero distance, is

$$f(0) = \frac{1}{\int_0^w g(x)\,dx} = \frac{1}{\mu} \tag{2.10}$$

The parameter $\mu = \int_0^w g(x)\,dx$ is a function of the measured distances. Therefore, we will often write the general estimator of density for line transect sampling simply as

$$\hat{D} = \frac{n \cdot \hat{f}(0)}{2L} = \frac{n}{2L\hat{\mu}} \tag{2.11}$$

This estimator can be further generalized, but the conceptual approach remains the same. $\hat{D}$ is valid whether w is bounded or unbounded (infinite) and when the data are grouped or ungrouped. Note that either form of eqn (2.11) is equivalent to eqn (2.3).

For the duck nest example, an estimate of the effective strip half-width is $\hat{\mu} = w\hat{P}_a = 8(0.888) = 7.10\,\text{ft}$, $\hat{f}(0) = 1/7.10$ and $\hat{D} = 534/(2 \times 1600 \times 7.10)$ nests/mile/ft $= 124$ nests/square mile.

The density estimator expressed in terms of an estimated pdf, evaluated at zero, is convenient, as a large statistical literature exists on the subject of estimating a pdf. Thus, a large body of general knowledge can be brought to bear on this specific problem.

2.2.2 *Point transects*

In traditional circular plot sampling, k areas each of size πw^2 are censused and all n objects within the k plots are enumerated. By definition, density is the number per unit area, thus

$$\hat{D} = \frac{n}{k\pi w^2} \tag{2.12}$$

In point transect sampling, only a proportion of the objects in each sampled area is detected. Again, let this proportion be P_a. Then the estimator of density is

$$\hat{D} = \frac{n}{k\pi w^2 \hat{P}_a} \tag{2.13}$$

Consider an annulus of width dr at distance r from a point (Fig. 2.3). The area of this annulus is approximately $2\pi r\, dr$. The proportion of the circle of radius w that falls within this annulus is therefore

$$\frac{2\pi r\, dr}{\pi w^2} \tag{2.14}$$

which is also the expected proportion of objects in the circle that lie within the annulus. Thus the expected proportion of objects in the circle that are both in the annulus and detected is

$$\frac{2\pi r g(r)\, dr}{\pi w^2} \tag{2.15}$$

The unconditional probability of detecting an object that is in one of the k circular plots is therefore

$$P_a = \int_0^w \frac{2\pi r g(r)\, dr}{\pi w^2} = \frac{2}{w^2} \int_0^w r g(r)\, dr \tag{2.16}$$

Substituting eqn (2.16) into eqn (2.13) and cancelling the w^2 terms, the estimator of density is

$$\hat{D} = \frac{n}{2k\pi \int_0^w r\hat{g}(r)\, dr} \tag{2.17}$$

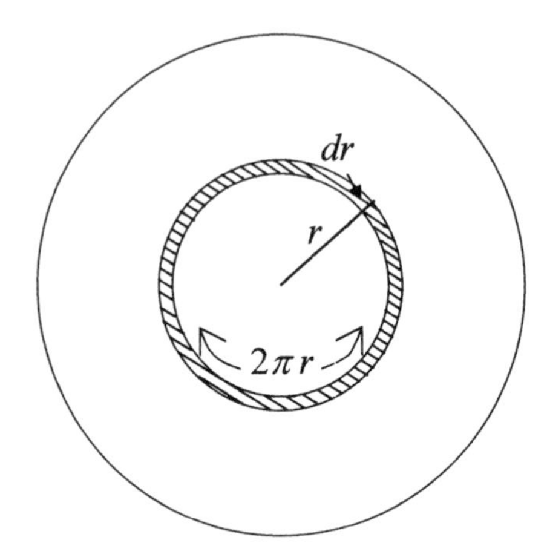

Fig. 2.3. An annulus of width dr and distance r from the point has approximate area $2\pi r\,dr$. Thus the expected number of objects in this annulus is $2\pi r\,dr\,D$, of which we expect to detect $2\pi r\,dr\,Dg(r)$.

Defining

$$\nu = 2\pi \int_0^w rg(r)\,dr \tag{2.18}$$

then

$$\hat{D} = \frac{n}{k\hat{\nu}} \tag{2.19}$$

Clearly, ν is the critical quantity to be estimated from the distance data for a point transect survey.

2.2.3 *Summary*

The statistical problem in the estimation of density of objects is the estimation of μ or ν, using the measured distances x_i (line transects) or r_i (point transects). Then the estimator of density for line transect sampling is

$$\hat{D} = \frac{n}{2L\hat{\mu}} \tag{2.20}$$

where

$$\mu = \int_0^w g(x)\,dx \tag{2.21}$$

The estimator of density for point transect surveys can be given in a similar form:

$$\hat{D} = \frac{n}{k\hat{\nu}} \tag{2.22}$$

where

$$\nu = 2\pi \int_0^w r g(r)\, dr \qquad (2.23)$$

This, then, entails careful modelling and estimation of $g(y)$. Good statistical theory exists for these general problems.

2.3 Philosophy and strategy

The true detection function $g(y)$ is not known. Furthermore, it varies due to numerous factors (Section 1.9). Therefore, it is important that strong assumptions about the shape of the detection function are avoided. In particular, a flexible or 'robust' model for $g(y)$ is essential.

The strategy used here is to select a few models for $g(y)$ that have desirable properties. These models are the ones shown to provide a sound basis for modelling the detection function $g(y)$. This class of models excludes those that are not robust, have restricted shapes, or have inefficient estimators. Because the estimator of density is closely linked to $g(y)$, it is of critical importance to select models for the detection function carefully. Three properties desired for a model for $g(y)$ are, in order of importance, model robustness, a shape criterion and efficiency.

2.3.1 *Model robustness*

The most important property of a model for the detection function is model robustness. This means that the model is a general, flexible function that can take a variety of plausible shapes for the detection function. Often, this property excludes single parameter models, as experience has shown that models with two or three parameters are frequently required. Most of the models recommended have a variable number of parameters, depending on how many are required to fit the specific data set. These are sometimes called semiparametric models.

The concept of pooling robustness (Burnham *et al.* 1980) is included here under model robustness. Models of $g(y)$ are pooling robust if the data can be pooled over many factors that affect detection probability (Section 1.9) and still yield a reliable estimate of density. Consider two approaches: stratified estimation $\hat{D}_{\mathrm{st}}$ and pooled estimation $\hat{D}_{\mathrm{p}}$. In the first case, the data could be stratified by factors affecting detectability (e.g. three observers and four habitat types) and an estimate of density made for each stratum. These separate estimates could be combined into an estimate of average density $\hat{D}_{\mathrm{st}}$. In the second case, all data could be pooled, regardless of any stratification (e.g. the data for the three observers and four habitat types would be pooled) and a single estimate of density computed, $\hat{D}_{\mathrm{p}}$. A model is pooling robust if $\hat{D}_{\mathrm{st}} \approx \hat{D}_{\mathrm{p}}$. Pooling robustness is a desirable property. Only models that are linear in the parameters satisfy the condition with strict equality, although general models that are model robust, such as those recommended in this book, approximately satisfy the pooling robust property.

2.3.2 *Shape criterion*

Theoretical considerations and the examination of empirical data suggest that the detection function should have a 'shoulder' near the line or point. That is, detection remains nearly certain at small distances from the line or point. Mathematically, the derivative $g'(0)$ should be zero. This shape criterion excludes functions that are spiked near zero distance. Frequently, a histogram of the distance data will not reveal the presence of a shoulder, particularly if the histogram classes are large (Fig. 2.4), or if the data include several large values (a long tail). Generally, good models for $g(y)$ will satisfy the shape criterion near zero distance. The shape criterion is especially important in the analysis of data where some heaping at zero distance is suspected. This occurs most frequently when small sighting angles are rounded to zero, or when the centreline is not located accurately, and may give rise to histograms that show no evidence of a shoulder, even when the true detection function has a substantial shoulder.

2.3.3 *Efficiency*

Other things being equal, it is desirable to select a model that provides estimates that are relatively precise (i.e. have small variance). We recommend maximum likelihood methods, which have many good statistical properties, including that of asymptotic minimum variance. Efficient estimation is of benefit only for models that are model robust and have a shoulder near zero distance; otherwise, estimation might be precise but biased.

2.3.4 *Model fit*

Ideally, there would be powerful statistical tests of the fit of the model for $g(y)$ to the distance data. The only simple omnibus test available is the χ^2 goodness of fit test based on grouping the distance data. This test compares the observed frequencies n_i (based on the grouping selected) with the estimated expected frequencies under the model, $\hat{E}(n_i)$, in the usual way:

$$\chi^2 = \sum_{i=1}^{u} \frac{[n_i - \hat{E}(n_i)]^2}{\hat{E}(n_i)} \tag{2.24}$$

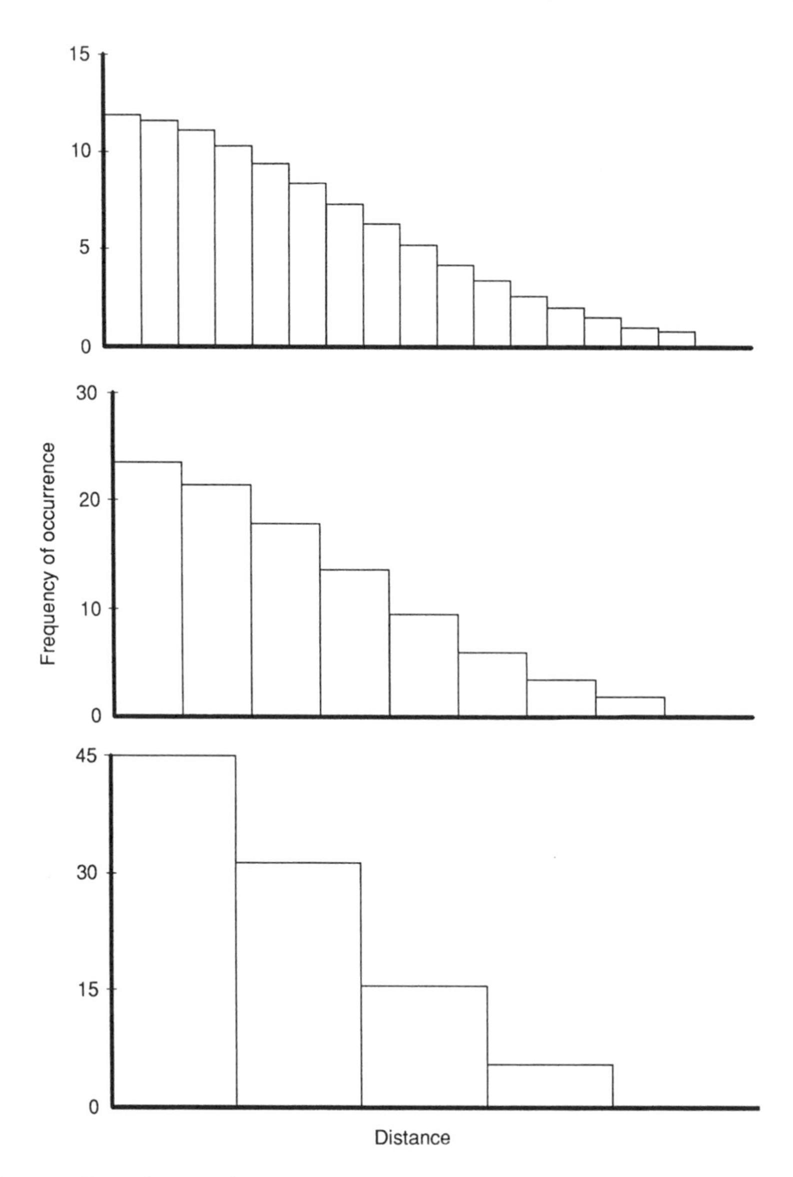

Fig. 2.4. Data ($n = 100$) from the half-normal model with $\sigma = 33.3$ and $w = 100$ shown with three different sets of group interval. As the group interval increases, the data appear to become more spiked. Adapted from Burnham *et al.* (1980).

is approximately χ^2 with $u - m - 1$ degrees of freedom (df), where u is the number of groups and m is the number of parameters estimated. In isolation, this approach has severe limitations for choosing a model for $g(y)$, given a single data set (Fig. 2.5).

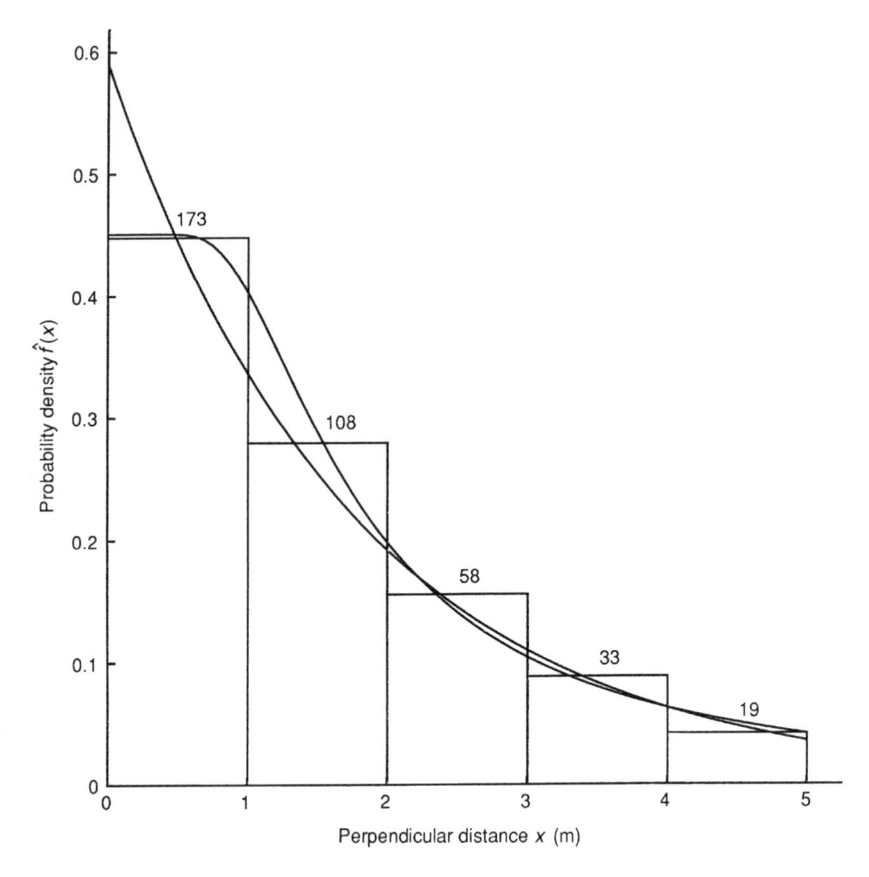

Fig. 2.5. The distance data are often of little help in testing the relative fit among models. Here, fits of the negative exponential model and the hazard-rate model to a line transect data set are shown. Both models provide an excellent fit ($\chi^2 = 0.49$, 3 df, $p = 0.92$ and $\chi^2 = 0.33$, 2 df, $p = 0.85$, respectively), even though the estimates of $f(0)$ are quite different ($\hat{f}(0) = 0.589$ and 0.450, respectively).

Generally, as the number of parameters in a model increases, the bias decreases but the sampling variance increases. A proper model should be supported by the particular data set and thus have enough parameters to avoid large bias but not so many that precision is lost (the Principle of Parsimony). The relative fit of alternative models may be evaluated using Akaike's Information Criterion, AIC (Akaike 1973; Sakamoto *et al.* 1986; Burnham and Anderson 1998), or AICc, which incorporates a small-sample bias correction (Hurvich and Tsai 1989, 1995). These technical subjects are presented in the following chapters.

2.3.5 *Test power*

The power of the goodness of fit test is quite low and, therefore, of limited use in selecting a good model of $g(y)$ for the analysis of a particular data set. In particular, this test is incapable of discriminating between quite different models near the line or point, the most critical region (Fig. 2.5). In addition, grouping data into fewer groups frequently diminishes the power of the test still further and may give the visual impression that the data arise from a spiked distribution such as the negative exponential, when the true detection function has a shoulder (Fig. 2.4).

While goodness of fit test results should be considered in the analysis of distance data, they will be of limited value in selecting a model. Thus, a class of reliable models is recommended here, based on the three properties: model robustness, the shape criterion and estimator efficiency.

2.4 Robust models

Several models of $g(y)$ are recommended for the analysis of line or point transect data. These models, as implemented in program Distance, have the three desired properties of model robustness, shape criterion and estimator efficiency. Following Buckland (1992a), the modelling process can be conceptualized in two steps. First, a 'key function' is selected as a starting point, possibly based on visual inspection of the histogram of distances, after truncation of obvious outliers. Often, a simple key function is adequate as a model for $g(y)$, especially if the data have been properly truncated. The uniform and the half-normal key functions should probably receive initial consideration (Fig. 2.6a). The uniform key function has no parameters, whereas the half-normal key has one unknown parameter to be estimated from the distance data. The hazard-rate model (Fig. 2.6b) might also be considered as a key function, although it requires that two key parameters be estimated.

Second, a flexible form, called a 'series expansion', is used to adjust the key function, using perhaps one or two more parameters, to improve the fit of the model to the distance data. Conceptually, the detection function is modelled in the following general form:

$$g(y) \propto \text{key}(y)[1 + \text{series}(y)] \tag{2.25}$$

The key function alone may be adequate for modelling $g(y)$, especially if sample size is small or the distance data are easily described by a simple model. Theoretical considerations often suggest a series expansion appropriate for a given key. Three series expansions are considered here: (1) the cosine series, (2) simple polynomials and (3) Hermite polynomials (Stuart and Ord 1987: 220–7). All three expansions are linear in their parameters.

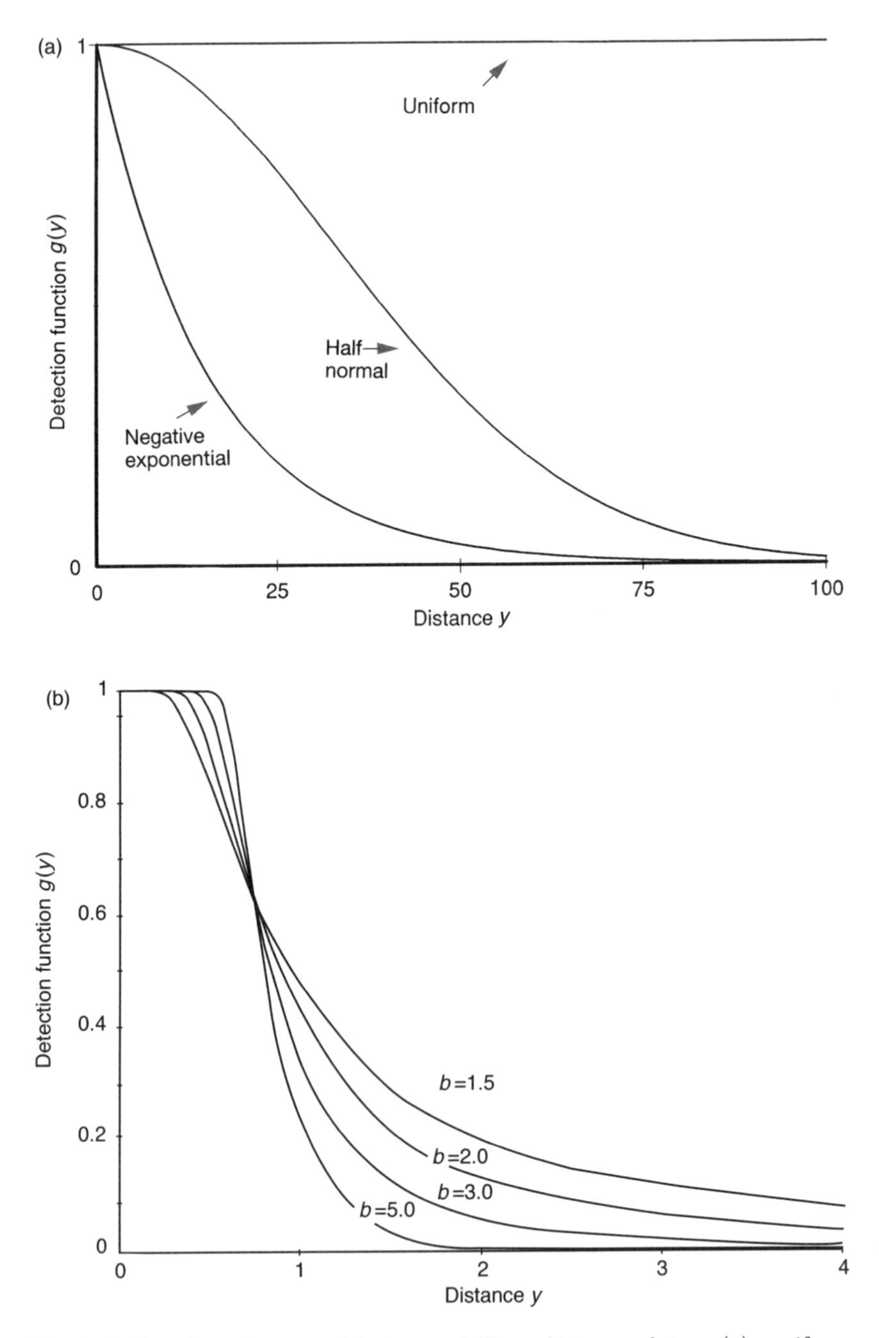

Fig. 2.6. Key functions useful in modelling distance data: (a) uniform, half-normal and negative exponential and (b) hazard-rate model for four different values of the shape parameter b.

Thus, some generally useful models of $g(y)$ are as follows:

Key function	Series expansion
Uniform, $1/w$	Cosine, $\sum_{j=1}^{m} a_j \cos(j\pi y/w)$
Uniform, $1/w$	Simple polynomial, $\sum_{j=1}^{m} a_j (y/w)^{2j}$
Half-normal, $\exp(-y^2/2\sigma^2)$	Cosine, $\sum_{j=2}^{m} a_j \cos(j\pi y/w)$
Half-normal, $\exp(-y^2/2\sigma^2)$	Hermite polynomial, $\sum_{j=2}^{m} a_j H_{2j}(y_s)$, where $y_s = y/\sigma$.
Hazard-rate, $1 - \exp(-(y/\sigma)^{-b})$	Cosine, $\sum_{j=2}^{m} a_j \cos(j\pi y/w)$
Hazard-rate, $1 - \exp(-(y/\sigma)^{-b})$	Simple polynomial, $\sum_{j=2}^{m} a_j (y/w)^{2j}$

The uniform + cosine expression is the Fourier series model of Crain *et al.* (1979) and Burnham *et al.* (1980). This is an excellent omnibus model and has been shown to perform well in a variety of situations. The uniform + simple polynomial model includes the models of Anderson and Pospahala (1970), Anderson *et al.* (1980) and Gates and Smith (1980).

It may be desirable to use the half-normal key function with either a cosine expansion or Hermite polynomials. Because histograms of distance data often decline markedly with distance from the line, the half-normal may often represent a good choice as a key function. Similarly, the uniform key and one cosine term will often provide a good standard for possible further fitting with series adjustment terms. Theoretical reasons suggest the use of Hermite polynomials in conjunction with the half-normal key, especially for the untruncated case. This is a minor point, and the reader should think of this as only an alternative form of a polynomial.

The final two models listed above use Buckland's (1985) hazard-rate as a two-parameter key function and use cosine and simple polynomial expansions for additional fitting, if required. The hazard-rate model is a derived model in contrast to the others, which are proposed shapes. That is, the shape of this family of models is the result of *a priori* assumptions about the detection process. The hazard-rate model has been shown to possess good properties. It can have a marked shoulder that can be nearly flat for some distance from the line or point. Even for data appearing to be spiked, this model can fit a flat shoulder, yet provide a good fit. However, if the data are very spiked, the hazard-rate can give density estimates with large positive bias, especially when the spike is an artefact of rounding in the data.

These series-expansion models are nonparametric in the sense that the number of parameters used is data-dependent. The estimation theory for these models, including rules to select the number of parameters to use, is given in the following chapter. Typically, given suitable distance truncation, an adequate model for $g(y)$ will include only one or two parameters, sometimes three. Sometimes the key function by itself will be adequate, with no terms in the series expansion. We emphasize that truncation will often be required as part of the modelling, especially if the data are ungrouped. Outlier observations provide relatively little information about density, but are often difficult to model, so that proper truncation should always be considered in modelling $g(y)$. Program Distance allows the combination of any of the key functions with any of the series expansions as a model for $g(y)$. Some models have appeared in the literature that assume $g(y) = 1$ for some considerable distance from the line or point; the models suggested above do not impose this assumption.

Program Distance allows any key function to be used with any series expansion; however, the combinations listed above should be satisfactory for general use. Further effort directed at model evaluation and development might now be better directed at survey design and data collection techniques to meet critical assumptions. More advanced, general methods of modelling the distance data are covered in Buckland *et al.* (in preparation).

The exponential + simple polynomial is available in program Distance for the salvage analysis of poorly collected data where there is strong reason to believe that the distance data are truly spiked. It has the form:

$$\exp(-y/\lambda) \cdot \left[1 + \sum_{j=1}^{m} a_j (y/\lambda)^{2j}\right] \tag{2.26}$$

Use of this approach should be accompanied by adequate justification and we recommend its use only in unusual circumstances. Every consideration should be given to the use of the hazard-rate model for distance data that appear spiked because this model enforces the shape criterion, offers greater flexibility in fitting a spike and gives a more realistic (larger) variance when the data are inadequate for reliable modelling.

2.5 Some analysis guidelines

Distance sampling represents a broad area and includes many types of application and degree of complexity of design and data. Thus, specific 'cookbook' procedures for data analysis cannot be safely given. Instead, we will suggest a useful strategy that could be considered when planning the analysis of a data set. In this section we will consider only a simple survey and will not address stratification and other complications given in later chapters.

2.5.1 *Exploratory phase*

The exploratory phase of the analysis involves the preparation of histograms of the distance data under several groupings. Sometimes it is effective to partition the data into 10–20 groups to get a fine-grained picture of the distance data. Examination of such histograms can provide insight into the presence of heaping, evasive movement, outliers and the occasional gross error. Prominent heaps can be mitigated by judicious grouping or splitting prior to further analysis. Evasive movement is problematic, but it is important to know that movement is present. (Movement towards the line or point generally cannot be detected from the distance data alone.) Some truncation of the distance data is nearly always suggested, even if no obvious outlier is noticed. If there is no truncation of larger distances in the field, we recommend that 5–10% of the largest observations be truncated. A more refined rule of thumb is to truncate the data when $g(x) \approx 0.15$ for line transects or 0.10 for point transects. If the data were taken as grouped data in the field, then options for further truncation are more limited.

Empirical estimates of $var(n)$ can be computed and compared with the variance under the Poisson assumption (i.e. $\widehat{var}(n) = n$). One can examine the stability of the ratio $\widehat{var}(n)/n$ over various design features. If the data are from a clustered population, plots of s or $\log_e(s)$ against x or r should be made and examined. Of course, data entry errors and other anomalies should be screened and corrected. This analysis phase is open-ended but the analyst is encouraged to begin to understand the data and possible violations of the assumptions. Chatfield (1988, 1991) offers some general practical advice relevant here.

2.5.2 *Model selection*

Model selection cannot proceed until proper truncation and, where relevant, grouping have been tentatively addressed. Thus, this phase begins once a data set has been properly prepared. Several robust models should be considered (e.g. those in Section 2.4). The following chapters will introduce and demonstrate the use of likelihood ratio tests, goodness of fit tests and AIC as aids in objective model selection. Here it might be appropriate to remind the analyst that it is the fit of the model to the distance data near the line or point that is most important (unless there is thought to be heaping at zero distance). Usually analysis will suggest additional exploratory work, so that the process is iterative. For example, it may become apparent that the fit of one or more models could be substantially improved by selecting a different truncation point w, or by grouping ungrouped data, or by changing the choice of group intervals for grouped data. Further, if data are available over several years, taken in the same habitat type by the same observer, then it might be prudent to pool the data for the estimation of $f(0)$, but to use the year-specific sample sizes n_i, where i is year, to estimate annual abundance. The validity of this approach must then be assessed, for example, by comparing AIC values to determine whether a common value for $f(0)$ can be assumed.

2.5.3 *Final analysis and inference*

At some point the analyst selects a model believed to be the best for the data set under consideration. In some cases, there may be several competing models that seem equally good, and we will later consider how inference can reflect this. In many cases, there will be a subset of models that can be excluded from final consideration because they perform poorly relative to other models. Often, if two or three models seem to fit equally well to a data set, estimation of density D and mean cluster size $E(s)$ under these models will be quite similar (see examples in Chapter 8).

Once a single model (or a subset of models) has been selected, the analyst can address further issues. Thus, one might consider bootstrapping to obtain improved estimates of precision, or carry out a Monte Carlo study to understand further the effect of some assumption failure (e.g. overestimation of a significant proportion of detection distances in an aerial survey, due to the aircraft flying too low at times). We might include a component of variance to reflect model selection uncertainty (Buckland *et al.* 1997; Burnham and Anderson 1998). Finally, estimates of density or abundance and their precision are made, and qualifying statements presented, such as discussion of the effects of failures of assumptions.

The above guidelines give a broad indication of how the analyst might proceed. They will be developed in the following chapters, both to give substance to the theory required at each step, and to show how the philosophy for analysis is implemented in real examples.

3
Statistical theory

3.1 General formula

3.1.1 *Standard distance sampling*

The analysis methods for distance sampling described here model measured or estimated distances from a line or point so that density of objects in a study area may be estimated. Conceptually, object density varies spatially, and lines or points are placed at random or systematically in the study area to allow mean density to be estimated.

Suppose in a given survey that objects do not occur in clusters and that distances are only recorded out to a distance w from the line or point, or equivalently that recorded distances are truncated at distance w. Suppose further that the true density is D objects per unit area, distributed within a study area of size A, so that population size is $N = D \cdot A$. Let the area covered by the survey within distance w of the line or point be a, so that $a = 2wL$ for line transects, where L is the total line length, and $a = k\pi w^2$ for point transects, where k is the number of points. Let the probability of detection for an object within this area, unconditional on its actual position, be P_a. Then the expected number of objects detected within area a (or equivalently, within distance w of a line or point), $E(n)$, is equal to the expected number of animals in the surveyed area, $D \cdot a$, multiplied by the probability of detection, P_a, so that

$$D = \frac{E(n)}{a \cdot P_a} \tag{3.1}$$

If objects occur in clusters, so that $E(n)$ is the expected number of clusters, then the above equation should be multiplied by $E(s)$, the expectation of cluster size for the population, to obtain the density of individuals:

$$D = \frac{E(n) \cdot E(s)}{a \cdot P_a} \tag{3.2}$$

Replacing parameters in eqn (3.2) by their estimators gives

$$\hat{D} = \frac{n \cdot \hat{E}(s)}{a \cdot \hat{P}_a} \tag{3.3}$$

The variance of $\hat{D}$ may be approximated using the delta method (Seber 1982: 7–9). Assuming correlations between the three estimation components are zero, the variance estimate is then:

$$\widehat{var}(\hat{D}) = \hat{D}^2 \cdot \left\{ \frac{\widehat{var}(n)}{n^2} + \frac{\widehat{var}[\hat{E}(s)]}{[\hat{E}(s)]^2} + \frac{\widehat{var}(a \cdot \hat{P}_a)}{(a \cdot \hat{P}_a)^2} \right\} \tag{3.4}$$

Equivalently, if we define coefficient of variation of an estimator, cv, to be its standard error divided by itself, we can write this equation as

$$[cv(\hat{D})]^2 = [cv(n)]^2 + [cv\{\hat{E}(s)\}]^2 + [cv(a \cdot \hat{P}_a)]^2 \tag{3.5}$$

Although area $a \to \infty$ as $w \to \infty$, the product $a \cdot \hat{P}_a$ remains finite, so that all three equations hold when there is no truncation. To estimate $a \cdot P_a$, a form must be specified, explicitly or implicitly, for the detection function $g(y)$, which represents the probability of detection of an object or object cluster at a distance y from the line or point. The simplest form is that of the Kelker strip: the truncation point w is selected such that it is reasonable to assume that $g(y) = 1.0$ for $0 \leq y \leq w$. More generally it seems desirable that the detection function has a 'shoulder'; that is, $g'(0)$ should be zero, so that the detection function is flat at zero. This is the shape criterion defined by Burnham *et al.* (1980). The detection function should also be non-increasing, and have a tail that goes asymptotically to zero.

The relationship between the detection function, $g(y)$, and the probability density function (pdf) of distances, $f(y)$, is different for line and for point transects. We use the notation y to represent either x, the perpendicular distance of an object from the centreline in line transect sampling, or r, the distance of an object from the observer in point transect sampling.

3.1.2 *Line transect sampling*

For line transect sampling, the relationship between $g(x)$ and $f(x)$ is particularly simple. We start with two concepts: (1) the area of a strip of incremental width dx at distance x from the line is independent of x and (2) random line placement ensures that objects are uniformly distributed with respect to distance from the line. Thus it seems reasonable that the density function should be identical in shape to the detection function, but rescaled so that it integrates to unity. This result may be proven as follows. Suppose for the moment that w is finite.

$$\begin{aligned} f(x)\,dx &= \text{pr}\{\text{object in } (x,\, x+dx) \mid \text{object detected}\} \\ &= \frac{\text{pr}\{\text{object in } (x,\, x+dx) \text{ and object detected}\}}{\text{pr}\{\text{object detected}\}} \end{aligned}$$

$$= [\text{pr}\{\text{object detected} \,|\, \text{object in } (x,\, x + dx)\}$$
$$\cdot \text{ pr}\{\text{object in } (x,\, x + dx)\}]/P_a$$
$$= \frac{g(x) \cdot (dx \cdot L)/(w \cdot L)}{P_a}$$

Thus $f(x) = g(x)/(w \cdot P_a)$. It is convenient to define $\mu = w \cdot P_a$, so that

$$f(x) = \frac{g(x)}{\mu} \tag{3.6}$$

Because $\int_0^w f(x)\,dx = 1$, it follows that $\mu = \int_0^w g(x)\,dx$. Figure 3.1 illustrates the result that P_a, the probability of detecting an object given that it is within w of the centreline, is $\mu = \int_0^w g(x)\,dx$ (the area under the curve) divided by $1.0w$ (the area of the rectangle); that is, $w \cdot P_a = \mu$, which is well-defined even when w is infinite.

The parameter μ is often termed the effective strip width, or more strictly, the effective strip half-width; if all objects were detected out to a distance μ on either side of the transect, and none beyond, then the expected number of objects detected would be the same as for the actual survey.

Let the total length of transect be L. Then the area surveyed is $a = 2wL$, and $a \cdot P_a = 2\mu L$. Since $\mu = g(x)/f(x)$ and $g(0) = 1.0$, then $\mu = 1/f(0)$,

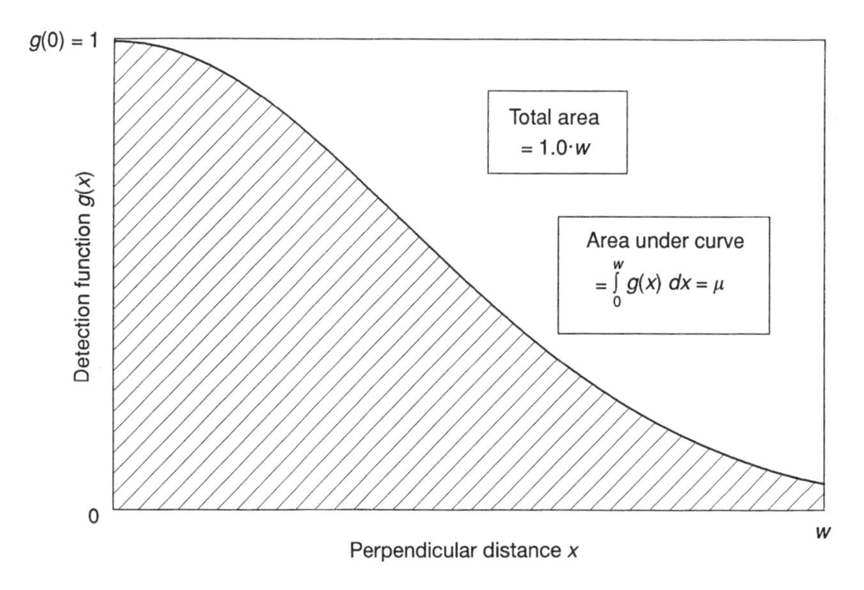

Fig. 3.1. The unconditional probability that an animal within distance w of the line is detected is the area under the detection function μ divided by the area of the rectangle $1.0w$.

and so for line transects, eqn (3.2) becomes:

$$D = \frac{E(n) \cdot f(0) \cdot E(s)}{2L} \tag{3.7}$$

The parameter $f(0)$ is statistically well-defined and is estimable from the perpendicular distances $x_1, \ldots, x_n$ in a variety of ways.

It is worth noting the effect of truncation of perpendicular distances on eqn (3.7). Truncation is increased by decreasing w. Expected sample size $E(n)$ decreases, and $f(0)$ increases to compensate, so that the product $E(n) \cdot f(0)$ remains constant, and density D is still defined by eqn (3.7).

3.1.3 *Point transect sampling*

The derivation for point transects is similar, but the relationship between $g(r)$ and $f(r)$ is less simple. The area of a ring of incremental width dr at distance r from the observer is proportional to r (Fig. 2.3). Thus we might expect that $f(r)$ is proportional to $r \cdot g(r)$; using the constraint that $f(r)$ integrates to unity,

$$f(r) = \frac{rg(r)}{\int_0^w rg(r)\,dr} \tag{3.8}$$

A more rigorous proof follows:

$$\begin{aligned} f(r)\,dr &= \text{pr}\{\text{object in annulus}(r, r+dr) \mid \text{object detected}\} \\ &= \frac{\text{pr}\{\text{object in } (r, r+dr) \text{ and object detected}\}}{\text{pr}\{\text{object is detected}\}} \\ &= [\text{pr}\{\text{object detected} \mid \text{object in } (r, r+dr)\} \\ &\quad \cdot \text{pr}\{\text{object in } (r, r+dr)\}]/P_a \\ &= \frac{g(r) \cdot (2\pi r dr/\pi w^2)}{P_a} \end{aligned}$$

so that

$$f(r) = \frac{2\pi r \cdot g(r)}{\pi w^2 \cdot P_a} \tag{3.9}$$

To be a valid density function, $\int_0^w f(r)\,dr = 1$, so that $f(r) = (2\pi r \cdot g(r))/\nu$, with $\nu = 2\pi \int_0^w rg(r)\,dr = \pi w^2 \cdot P_a$. This result also holds for infinite w. Analogous to μ, ν is sometimes called the effective area of detection.

Another way to derive P_a is to note that the cumulative distribution function of distances from a point to objects, given that they are within a circle of radius w centred on that point, and irrespective of whether they are detected, is

$$\text{pr}\{R \leq r\} = \frac{\pi r^2}{\pi w^2}, \quad 0 \leq r \leq w \tag{3.10}$$

Hence the pdf of such distances is obtained by differentiation:

$$u(r) = \frac{2\pi r}{\pi w^2} \tag{3.11}$$

We now note that P_a is the expected value over r of the probability of detection, $g(r)$ (i.e. the long-term average of the detection probabilities):

$$P_a = E_r[g(r)] = \int_0^w g(r)\,u(r)\,dr = \int_0^w g(r)\frac{2\pi r}{\pi w^2}\,dr = \frac{2}{w^2}\int_0^w rg(r)\,dr \tag{3.12}$$

It follows from eqn (3.9) that

$$f(r) = \frac{2\pi r g(r)}{\nu} \tag{3.13}$$

with ν defined as above.

Let there be k points, so that the surveyed area is $a = k\pi w^2$. The probability of detection of an object, given that it is within a distance w of the observer, is now $P_a = \nu/(\pi w^2)$, so that $a \cdot P_a$ becomes $k\nu$; again this holds as $w \to \infty$. Since $\nu = 2\pi r g(r)/f(r)$ and $g(0) = 1.0$, then for point transects eqn (3.2) becomes:

$$D = \frac{E(n) \cdot h(0) \cdot E(s)}{2\pi k} \tag{3.14}$$

where $h(0) = \lim_{r \to 0} f(r)/r = 2\pi/\nu$.

Note that $h(0)$ is merely the derivative of the probability density $f(r)$, evaluated at $r = 0$; alternative notation would be $f'(0)$. Thus $h(0)$ is estimable from the detection distances $r_1, \ldots, r_n$. Whereas $f(x)$ and $g(x)$ have identical shapes in line transect sampling, for point transects, $g(r)$ is proportional to $f(r)/r$. The constant of proportionality is $1/h(0)$.

Results equivalent to eqns (3.3) and (3.4) follow in the obvious way. Note that for both line and point transects, behaviour of the pdf at zero distance is critical to object density estimation.

A special case that is sometimes of interest is when the full circle is not necessarily surveyed. This can happen for example because an obstruction obscures visibility in part of the circle, a particular problem with spotlight surveys. A simple solution is to define a sector of angle ψ that spans the obstructed region (Fig. 3.2), and discard any detections in front of the obstruction that fall within this sector. The contribution to the surveyed area from this circle is now $(\pi - \psi/2)w^2$ instead of πw^2, and the total

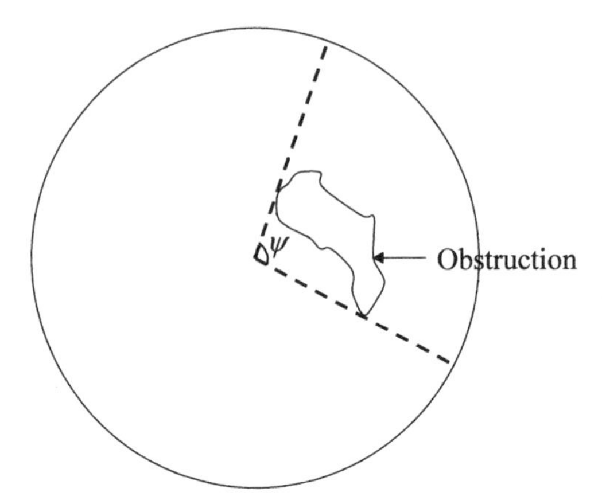

Fig. 3.2. In point transect sampling, if the observer is dependent on visual detection cues, obstructions can reduce the surveyed area around a point. By excluding detections within the sector shown here, and calculating the surveyed area about the point as $(\pi - \psi/2)w^2$, density may still be estimated.

surveyed area is modified accordingly in the above derivation, by summing the surveyed area corresponding to each circle. This is readily carried out in Distance, by specifying the effort associated with a point to be the proportion, $1 - \psi/2\pi$, of the corresponding circle that is surveyed (multiplied by the number of times the point is surveyed, if more than once). If the same fraction of each circle is surveyed, then this may be included as a single multiplier to be applied to the density estimate (Section 3.1.4). A more sophisticated solution to the problem of obstructions is given in Section 6.7. If object density is correlated with the presence of obstructions, systematic exclusion of sectors in which obstructions occur generates some bias in the estimated density of objects.

Burnham *et al.* (1980: 195) recommended that distances in point transect sampling should be transformed to 'areas' before analysis. Thus, the ith recorded area would be $u_i = \pi r_i^2, i = 1, \ldots, n$. If $f_u(u)$ denotes the pdf of areas u_i, it may be shown that $f_u(u) = f(r)/(2\pi r) = g(r)/\nu$. The advantage of this transformation should now be apparent; the new density is identical in form to that for perpendicular distances in line transect sampling (where $f(x) = g(x)/\mu$), so line transect software may be used to analyse the data. Further, if r is allowed to tend to zero, then $f_v(0) = h(0)/(2\pi)$, and the development based on areas is therefore equivalent to that based on distances. This seems to suggest that modelling of areas rather than distances is preferable. However, as noted by Buckland (1987a), the detection function expressed as a function of area differs in shape from that expressed as a function of distance, and it is no longer clear that a model for area should satisfy the shape criterion. For example the half-normal model for distances, which satisfies the shape criterion, transforms to the negative exponential model for areas, which does not satisfy the shape criterion, whereas the hazard-rate model of Hayes and Buckland (1983) retains both its parametric form and a shoulder under the transformation, although the shoulder becomes narrower. We now recommend modelling the untransformed distance data, because line transect detection functions may then be more safely carried across to point transects, thus allowing the focus of analysis to be the detection function in both cases.

3.1.4 *Distance sampling with multipliers*

Frequently, we need to include multipliers in eqn (3.3). We can express this equation as

$$\hat{D} = \frac{n \cdot \hat{E}(s)}{c \cdot a \cdot \hat{P}_a} \tag{3.15}$$

where c might be a known constant, an estimated parameter, or the product of more than one parameter estimate. For example, if just one side of a line is surveyed in line transect sampling, we must include $c = 0.5$ in the divisor, reflecting the fraction of the strip that is surveyed. In point transect sampling, if just one quarter of the circle were to be recorded, we would have $c = 0.25$. Similarly, if each point were counted three times, we would take $c = 3$. Cue counting (Section 6.4) uses point transect sampling theory, but only a sector of the circle is covered, and so a multiplier is needed to reflect this. If for line or point transects $g(0) < 1$ (i.e. detection on the line or at the point is not certain), but we have an independent estimate of $g(0)$, then this estimate may be substituted for c in the above equation. (Better ways of handling this problem are addressed in Buckland *et al.* in preparation.)

When line transect surveys of dung are carried out, two multipliers are required to convert the estimated abundance of dung to estimated abundance of animals (Section 6.2). First we need to include the mean time for dung to decay in the divisor, to estimate the number of droppings produced by the population per day. Second, we also need to include an estimate of the number of droppings produced per animal per day in the divisor, to convert this estimate to an estimate of the number of animals. Suppose we wish to apply two multipliers to eqn (3.3): γ, estimated by $\hat{\gamma}$, to the numerator, and η, estimated by $\hat{\eta}$, to the denominator. Then eqn (3.3) becomes

$$\hat{D} = \frac{n \cdot \hat{E}(s) \cdot \hat{\gamma}}{a \cdot \hat{P}_a \cdot \hat{\eta}} \tag{3.16}$$

and, assuming independence between components, the delta method yields a generalization of eqn (3.4):

$$\widehat{var}(\hat{D}) = \hat{D}^2 \cdot \left\{ \frac{\widehat{var}(n)}{n^2} + \frac{\widehat{var}[\hat{E}(s)]}{[\hat{E}(s)]^2} + \frac{\widehat{var}(a \cdot \hat{P}_a)}{(a \cdot \hat{P}_a)^2} + \frac{\widehat{var}(\hat{\gamma})}{\hat{\gamma}^2} + \frac{\widehat{var}(\hat{\eta})}{\hat{\eta}^2} \right\} \tag{3.17}$$

This can be expressed equivalently, but more simply, in terms of squared coefficients of variation:

$$[cv(\hat{D})]^2 = [cv(n)]^2 + [cv\{\hat{E}(s)\}]^2 + [cv(a \cdot \hat{P}_a)]^2 + [cv(\hat{\gamma})]^2 + [cv(\hat{\eta})]^2 \tag{3.18}$$

Note that if a multiplier is a known constant, such as the fraction of the strip or circle surveyed, then its variance is zero, and so its contribution to the above variance is also zero.

3.2 The key function formulation for distance data

Most formulations proposed for the probability density of distance data from line or point transects may be categorized into one of two groups. If there are theoretical reasons for supposing that the density has a given parametric form, then parametric modelling may be carried out. Otherwise, robust or nonparametric procedures such as Fourier series, splines, kernel methods or polynomials might be preferred. In practice it may be reasonable to assume that the true density function is close to a known parametric form, yet systematic departures can occur in some data sets. In this instance, a parametric procedure may not always give an adequate fit, yet a nonparametric method may be too flexible, perhaps giving very different fits to two related data sets from a single study, due to small random fluctuations in the data. An example of the latter occurs when a one-term Fourier series model is selected for one data set and a two-term model for a second. The second data set might be slightly larger, or show a slightly smaller shoulder; both increase the likelihood of rejecting the one-term fit. Bias in estimation of $f(0)$ can be a strong function of the number of Fourier series terms selected (Buckland 1985), so that comparisons across data sets may be misleading. The technique described by Buckland (1992a,b) and summarized below specifies a key function to incorporate knowledge of the likely shape of the density function, whether theoretical or from past experience, and allows polynomial or cosine series adjustments to be made, to ensure a good fit to the data. Barabesi and Fattorini (1993) developed and investigated a very similar approach.

Simple polynomials were used for fitting line transect data by Anderson and Pospahala (1970). However, low order simple polynomials may have unsuitable shapes. By taking the best available parametric form for the density, $\alpha(y)$, and multiplying it by a simple polynomial, this shortcoming is removed. We call $\alpha(y)$ the key function. If it is a good fit, it needs no adjustment; the worse the fit, the greater the adjustments required.

When the key function is the untruncated normal (or half-normal) density, Hermite polynomials (Stuart and Ord 1987: 220–7) are orthogonal with respect to the key, and may therefore be preferred to simple polynomials. Hermite polynomials are traditionally fitted by the method of moments, leading to unstable behaviour when the number of observations is not large or when high order terms are included. Buckland (1985) overcame these difficulties by using numerical techniques to obtain maximum likelihood estimates of the polynomial coefficients. These procedures have the further advantage that the fitted model is identical whether the Gram-Charlier type A or the Edgeworth formulation is adopted; for the method of moments, this is not so (Stuart and Ord 1987: 222–5).

Let the density function be expressed as

$$f(y) = \frac{\alpha(y)}{\beta} \cdot \left[1 + \sum_{j=1}^{m} a_j \cdot p_j(y_s)\right] \tag{3.19}$$

where $\alpha(y)$ is a parametric key, containing k parameters (usually 0, 1 or 2);

$$p_j(y_s) = \begin{cases} y_s^j, & \text{if a simple polynomial is required, or} \\ H_j(y_s), & \text{the } j\text{th Hermite polynomial, } j = 1, \ldots, m, \text{ or} \\ \cos(j\pi y_s), & \text{if a cosine series is required;} \end{cases}$$

y_s is a standardized y value (see below);
a_j is zero if term j of $p_j(y_s)$ is not used in the model, or is estimated by maximum likelihood;
β is a normalizing function of the parameters (key parameters and series coefficients) alone.

It is necessary to scale the observed distances. For the simple polynomial formulation, estimation is invariant to choice of scale, but the operation is generally necessary to avoid numeric problems when fitting the model. If the key function is parameterized such that a single key parameter, σ say, is a scale parameter, y_s may be found as y/σ for each observation. If the parameters of the key function are fully integrated into the estimation routine, σ can be estimated by maximum likelihood (see below). Otherwise the key function may be fitted by maximum likelihood in the absence of polynomial adjustments, and subsequent fitting of polynomial terms can be carried out conditional on those key parameter estimates. For the Fourier series formulation, analyses are conditional on w, the truncation point, and $y_s = y/w$. In practice, it is simpler to use this standardization for all models, a strategy used in Distance.

For line transect sampling, the standard form of the Fourier series model is obtained by setting the key function equal to the uniform distribution, so that $\alpha(y) = 1/w$. Used in conjunction with simple polynomials, this key gives the method of Anderson and Pospahala (1970). The standard form of the Hermite polynomial model arises when the key function is the half-normal. Point transect keys are found by multiplying their line transect counterparts by y (or, equivalently, $2\pi y$). The key need not be a valid density function, provided that it is always non-negative, has a finite integral, and for practical purposes, is non-increasing with finite $f(0)$. For the half-normal line transect key, define $\alpha(x) = \exp[-(x/\sigma)^2/2]$, and absorb the denominator of the half-normal density, $\sqrt{(\pi\sigma^2/2)}$, into β. Similarly, if a half-normal detection function is assumed in point transect sampling, the key may be defined as $\alpha(r) = r\exp[-(r/\sigma)^2/2]$. In general, absorb any part of the key that is a function of the parameters alone into β.

For line transect sampling, the detection function is generally assumed to be symmetric about the line. Similarly for point transect sampling, detection probability is assumed to be independent of angle. Estimation is robust to failures of these assumptions. The detection function may be envisaged as a continuous function on $(-w, +w)$; for line transects, negative distances would correspond say to sightings to the left of the line and positive to the right, and for point transects, this function can be thought of as a section through the detection 'dome', passing through the central point. Since the function is assumed to be symmetric about zero, only cosine terms are used for the Fourier series model, and only polynomials of even order for polynomial models, so that the detection function is an even function. In the case of the Hermite polynomial model, the parameter of the half-normal key corresponds to the second moment term, so that the first polynomial term to be considered is of order four (corresponding to an adjustment for kurtosis) if terms are assessed for inclusion sequentially. Other options for identifying a subset of potential terms (e.g. polynomials of order 4, 6, 8 and 10) to include are forwards stepping, in which the term giving the greatest improvement in fit, as judged by say Akaike's Information Criterion (AIC, Section 3.4.2) or likelihood ratio testing (Section 3.4.3), is entered first; and an all subsets approach, in which AIC is evaluated for every possible combination of terms (including the case of no adjustment terms), and the model with the lowest overall AIC is selected.

When the key function is not normal and terms are selected sequentially, it is less clear which polynomial term should be considered first. Any key will contain a parameter, or a function of parameters, that corresponds to scale, so a possible rule is to start with the term of order four, whatever the key. An alternative rule is to start with the term of order $2 \cdot (k+1)$, where k is the number of key parameters. We advise against the use of keys with more than two parameters. For stepwise and all subsets approaches, all even terms down to order two and up to an arbitrary limit can be included.

3.3 Maximum likelihood methods

We concentrate here on the likelihood function for the detection distances, $y_i, i = 1, \ldots, n$, conditional on n. If the full data set was to be modelled in a comprehensive way, then the probability of realizing the data $\{n, y_1, \ldots, y_n, s_1, \ldots, s_n\}$ might be expressed as

$$\begin{aligned}\mathrm{pr}(n, y_1, \ldots, y_n, s_1, \ldots, s_n) &= \mathrm{pr}(n) \cdot \mathrm{pr}(y_1, \ldots, y_n, s_1, \ldots, s_n \mid n) \\ &= \mathrm{pr}(n) \cdot \mathrm{pr}(y_1, \ldots, y_n \mid n) \\ &\quad \cdot \mathrm{pr}(s_1, \ldots, s_n \mid n, y_1, \ldots, y_n)\end{aligned}$$

Thus, inference on the distances y_i can be made conditional on n, and inference on the cluster sizes s_i can be conditional on n and the y_i. This provides the justification for treating estimation of D as a series of three univariate problems.

Rao (1973) and Burnham *et al.* (1980) present maximum likelihood estimation methods for both grouped and ungrouped distance data. Applying those techniques to the key formulation of Section 3.2 yields the following useful results.

3.3.1 *Ungrouped data*

Define

$$\mathcal{L}(\underline{\theta}) = \prod_{i=1}^{n} f(y_i)$$

where y_i is the ith recorded distance, $i = 1, \ldots, n$, $\theta_1, \ldots, \theta_k$ are the parameters of the key function and $\theta_{k+j} = a_j, j = 1, \ldots, m$, are the parameters (coefficients) of the adjustment terms. Then

$$\log_e[\mathcal{L}(\underline{\theta})] = l = \sum_{i=1}^{n} \log_e[f(y_i)] = \sum_{i=1}^{n} \log_e[f(y_i) \cdot \beta] - n \cdot \log_e \beta \qquad (3.20)$$

Hence

$$\begin{aligned}\frac{\partial l}{\partial \theta_j} &= \sum_{i=1}^{n} \frac{\partial \log_e[f(y_i)]}{\partial \theta_j} \qquad (3.21) \\ &= \sum_{i=1}^{n} \left\{ \frac{1}{f(y_i) \cdot \beta} \cdot \frac{\partial [f(y_i) \cdot \beta]}{\partial \theta_j} \right\} - \frac{n}{\beta} \cdot \frac{\partial \beta}{\partial \theta_j}, \quad j = 1, \ldots, k + m\end{aligned}$$

where

$$\frac{\partial[f(y_i)\cdot\beta]}{\partial\theta_j} = \begin{cases} \alpha(y_i)\cdot\left[\sum_{j'=1}^{m} a_{j'}\cdot\frac{\partial p_{j'}(y_{is})}{\partial y_{is}}\right]\cdot\frac{\partial y_{is}}{\partial\theta_j} \\ \quad +\left[1+\sum_{j'=1}^{m} a_{j'}\cdot p_{j'}(y_{is})\right]\cdot\frac{\partial\alpha(y_i)}{\partial\theta_j}, \quad 1\le j\le k \\ \alpha(y_i)\cdot p_{j-k}(y_{is}), \quad \text{for all } j>k \\ \qquad\qquad \text{for which } a_{j-k} \text{ is non-zero} \end{cases} \tag{3.22}$$

$$\frac{\partial p_{j'}(y_{is})}{\partial y_{is}} = \begin{cases} j\cdot p_{j-1}(y_{is}), \text{ with } p_0(y_{is})=1, \text{for simple and} \\ \quad \text{Hermite polynomials} \\ -j\pi\cdot\sin(j\pi y_s), \text{ for the Fourier series model} \end{cases} \tag{3.23}$$

When $k=1$ and $y_{is}=y_i/\theta_1$, $\partial y_{is}/\partial\theta_1 = -y_i/\theta_1^2$; when $k=1$ and $y_{is} = y_i/w$, $\partial y_{is}/\partial\theta_1 = 0$.

The eqns $\partial l/\partial\theta_j = 0$, $j=1,\ldots,k+m$, may be solved using, for example, Newton-Raphson or a simplex procedure. To change between simple and Hermite polynomials, it is merely necessary to redefine $p_j(y_s)$, $j=1,\ldots,m$; to change between polynomial and cosine series adjustments, the derivative of $p_j(y_s)$ with respect to y_s must also be redefined. If a different key $\alpha(y)$ is required, the only additional algebra needed to implement the method is to find $\partial\alpha(y)/\partial\theta_j$ and $\partial y_s/\partial\theta_j$, $1\le j\le k$; β and $\partial\beta/\partial\theta_j$ are evaluated by numerical integration.

The Fisher information matrix per observation may be estimated by the Hessian matrix $H(\hat{\underline{\theta}})$, with jhth element

$$H_{jh}(\hat{\underline{\theta}}) = \frac{1}{n}\cdot\left[\sum_{i=1}^{n}\frac{\partial\log_e[f(y_i)]}{\partial\hat{\theta}_j}\cdot\frac{\partial\log_e[f(y_i)]}{\partial\hat{\theta}_h}\right] \tag{3.24}$$

This may be formed from quantities already calculated. If a function of the parameters, $g(\underline{\theta})$, is to be estimated by $g(\hat{\underline{\theta}})$, then

$$\widehat{var}\{g(\hat{\underline{\theta}})\} = \frac{1}{n}\cdot\left[\frac{\partial g(\hat{\underline{\theta}})}{\partial\hat{\underline{\theta}}}\right]'[H(\hat{\underline{\theta}})]^{-1}\left[\frac{\partial g(\hat{\underline{\theta}})}{\partial\hat{\underline{\theta}}}\right] \tag{3.25}$$

3.3.2 *Grouped data*

Suppose the observations y are grouped, the ith group spanning the interval (c_{i1}, c_{i2}), $i=1,\ldots,u$. In general, the data may be truncated at either or both ends. For line and point transects, it is usual that $c_{i1}=0$ (no left truncation) and $c_{i2}=c_{i+1,1}$, $i=1,\ldots,u-1$. The likelihood function is now multinomial. Let the group frequencies be $n_1,\ldots,n_u$, with cell probabilities

$$\pi_i = \int_{c_{i1}}^{c_{i2}} f(y)\,dy \tag{3.26}$$

Then

$$\mathcal{L}(\underline{\theta}) = \frac{n!}{n_1! \cdots n_u!} \prod_{i=1}^{u} \pi_i^{n_i}, \qquad \text{with } n = \sum_{i=1}^{u} n_i \tag{3.27}$$

$$\log_e[\mathcal{L}(\underline{\theta})] = l = \sum_{i=1}^{u} n_i \cdot \log_e(\pi_i) + \text{ a constant} \tag{3.28}$$

$$\frac{\partial l}{\partial \theta_j} = \sum_{i=1}^{u} \frac{n_i}{\pi_i} \cdot \frac{\partial \pi_i}{\partial \theta_j}, \quad j = 1, \ldots, k+m \tag{3.29}$$

Define $P_i = \pi_i \cdot \beta$, so that

$$\frac{\partial \pi_i}{\partial \theta_j} = \frac{1}{\beta} \cdot \left[\frac{\partial P_i}{\partial \theta_j} - \frac{\partial \beta}{\partial \theta_j} \cdot \pi_i \right] \tag{3.30}$$

Then if P_i and $\partial P_i / \partial \theta_j$, $j = 1, \ldots, k+m$, $i = 1, \ldots, u$, can be found,

$$\beta = \sum_{i=1}^{u} P_i \quad \text{and} \quad \frac{\partial \beta}{\partial \theta_j} = \sum_{i=1}^{u} \frac{\partial P_i}{\partial \theta_j}$$

Given parameter estimates, the P_i may be evaluated by numerically integrating the numerator, $f(y) \cdot \beta$, of the density function:

$$P_i = \pi_i \cdot \beta = \int_{c_{i1}}^{c_{i2}} f(y) \cdot \beta \, dy \tag{3.31}$$

Similarly,

$$\frac{\partial P_i}{\partial \theta_j} = \int_{c_{i1}}^{c_{i2}} \frac{\partial \{f(y) \cdot \beta\}}{\partial \theta_j} \, dy \tag{3.32}$$

and may be found using numerical integration on

$$\frac{\partial [f(y) \cdot \beta]}{\partial \theta_j} = \begin{cases} \alpha(y) \cdot \left[\sum\limits_{j'=1}^{m} a_{j'} \cdot \dfrac{\partial p_{j'}(y_s)}{\partial y_s} \right] \cdot \dfrac{\partial y_s}{\partial \theta_j} \\ \quad + \left[1 + \sum\limits_{j'=1}^{m} a_{j'} \cdot p_{j'}(y_s) \right] \cdot \dfrac{\partial \alpha(y)}{\partial \theta_j}, \quad 1 \leq j \leq k \\ \alpha(y) \cdot p_{j-k}(y_s), \quad \text{for all } j > k \text{ for which} \\ \qquad\qquad\qquad a_{j-k} \text{ is non-zero} \end{cases} \tag{3.33}$$

The implications of changing between simple and Hermite polynomials or between polynomial and cosine series adjustments, and of changing the key function, are identical to the case of ungrouped data.

Again, a robust iterative procedure is required to maximize the likelihood. Variances follow as for ungrouped data, except that the information matrix per observation, $I(\underline{\theta})$, now has jhth element

$$I_{jh}(\underline{\theta}) = \sum_{i=1}^{u} \frac{1}{\pi_i} \cdot \frac{\partial \pi_i}{\partial \theta_j} \cdot \frac{\partial \pi_i}{\partial \theta_h}, \quad j, h = 1, \ldots, k+m \tag{3.34}$$

All of these quantities are now available, and so a function of the parameters $g(\underline{\theta})$ is estimated by $g(\hat{\underline{\theta}})$ with variance

$$\widehat{var}\{g(\hat{\underline{\theta}})\} = \frac{1}{n} \cdot \left[\frac{\partial g(\hat{\underline{\theta}})}{\partial \underline{\theta}}\right]' [I(\hat{\underline{\theta}})]^{-1} \left[\frac{\partial g(\hat{\underline{\theta}})}{\partial \hat{\underline{\theta}}}\right] \tag{3.35}$$

If data are analysed both grouped and ungrouped, and the respective maxima of the likelihood functions are compared, the constant combinatorial term in the likelihood for grouped data should be omitted. As the number of groups tends to infinity and interval width tends to zero, the likelihood for grouped data tends to that for ungrouped data, provided the constant is ignored.

3.3.3 *Special cases*

Suppose no polynomial or cosine adjustments are required. The method then reduces to a straightforward fit of a parametric density. The above results hold, except the range of j is now from 1 to k, and for ungrouped data eqn (3.22) reduces to

$$\frac{\partial[f(y_i) \cdot \beta]}{\partial \theta_j} = \frac{\partial \alpha(y_i)}{\partial \theta_j}, \quad j = 1, \ldots, k \tag{3.36}$$

For grouped data, eqn (3.33) reduces to the above, with the suffix i deleted from y.

For the Hermite polynomial model, it is sometimes convenient to fit the half-normal model as described in the previous paragraph, and then to condition on that fit when making polynomial adjustments. For the standard Fourier series model, the key is a uniform distribution on $(0, w)$, where w is the truncation point, specified before analysis. In each of these cases, the adjustment terms are estimated conditional on the parameters of the key. Thus eqn (3.22) reduces to

$$\frac{\partial[f(y_i) \cdot \beta]}{\partial \theta_j} = \alpha(y_i) \cdot p_{j-k}(y_{is}),$$

$$\text{for non-zero } a_{j-k} \text{ and } k < j \leq k+m$$

Equation (3.33) reduces similarly, but with suffix i deleted; otherwise results follow through exactly as before, but with j restricted to the range $k+1$ to $k+m$. This procedure is necessary whenever the uniform key is selected. For keys that have at least one parameter estimated from the data, the conditional maximization is useful only if simultaneous maximization across all parameters fails to converge.

A third option that is sometimes useful is to refit the key, conditional on polynomial or cosine adjustments. Equation (3.22) then becomes

$$\frac{\partial[f(y_i)\cdot\beta]}{\partial\theta_j} = \alpha(y_i)\cdot\left[\sum_{j'=1}^{m} a_{j'}\cdot\frac{\partial p_{j'}(y_{is})}{\partial y_{is}}\right]\cdot\frac{\partial y_{is}}{\partial\theta_j} + \left[1+\sum_{j'=1}^{m} a_{j'}\cdot p_{j'}(y_{is})\right]\cdot\frac{\partial\alpha(y_i)}{\partial\theta_j},\quad 1\le j\le k \tag{3.37}$$

and similarly for eqn (3.33), but minus the suffix i throughout. The range of j is from 1 to k; otherwise results follow exactly as before.

3.3.4 *The half-normal detection function*

If the detection function is assumed to be half-normal, and the data are both ungrouped and untruncated, the above approach leads to closed form estimators and a particularly simple analysis for both line and point transect sampling. Suppose the detection function is given by $g(y) = \exp(-y^2/2\sigma^2)$, $0 \le y < \infty$. We consider the derivation for line transects $(y = x)$ and point transects $(y = r)$ separately.

3.3.4.1 *Line transects*

With no truncation, the density function of detection distances is $f(x) = g(x)/\mu$, where

$$\mu = \int_0^\infty g(x)\,dx = \int_0^\infty \exp(-x^2/2\sigma^2)\,dx = \sqrt{\frac{\pi\sigma^2}{2}} \tag{3.38}$$

Given n detections, the likelihood function is

$$\mathcal{L} = \prod_{i=1}^{n}\left\{\frac{g(x_i)}{\mu}\right\} = \left\{\prod_{i=1}^{n}\exp(-x_i^2/2\sigma^2)\right\}\Big/\mu^n \tag{3.39}$$

so that

$$l = \log_e(\mathcal{L}) = -\sum_{i=1}^{n}\left\{\frac{x_i^2}{2\sigma^2}\right\} - n\cdot\log_e\left\{\sqrt{\frac{\pi\sigma^2}{2}}\right\} \tag{3.40}$$

Differentiating l with respect to σ^2 (i.e. $k = 1$, $\theta_1 = \sigma^2$ and $m = 0$ in terms of the general notation) and setting the result equal to zero gives:

$$\frac{dl}{d\sigma^2} = \sum_{i=1}^{n}\frac{x_i^2}{2\sigma^4} - \frac{n}{2\sigma^2} = 0 \tag{3.41}$$

so that $\hat{\sigma}^2 = \sum_{i=1}^{n} x_i^2/n$. Then $\hat{f}(0) = 1/\hat{\mu} = \sqrt{2/(\pi\hat{\sigma}^2)}$.

By evaluating the Fisher information matrix, we get $var(\hat{\sigma}^2) = 2\sigma^4/n$ from which

$$var\{\hat{f}(0)\} = \frac{1}{n\pi\sigma^2} = \frac{\{f(0)\}^2}{2n} \tag{3.42}$$

Equation (3.7), with $E(s)$ set equal to one, yields

$$D = \frac{E(n) \cdot f(0)}{2L} \tag{3.43}$$

from which

$$\hat{D} = \frac{n \cdot \hat{f}(0)}{2L} = \left[2\pi L^2 \sum_{i=1}^{n} \frac{x_i^2}{n^3}\right]^{-0.5} \tag{3.44}$$

The methods of Section 3.6 yield an estimated variance for $\hat{D}$.

Quinn (1977) investigated the half-normal model, and derived an unbiased estimator for $f(0)$.

3.3.4.2 *Point transects*

The density function of detection distances is given by $f(r) = 2\pi r \cdot g(r)/\nu$. For the half-normal detection function,

$$\begin{aligned}\nu &= 2\pi \int_0^w r \cdot g(r)\, dr = 2\pi \int_0^w r \cdot \exp\left(\frac{-r^2}{2\sigma^2}\right) dr \\ &= \left[-2\pi\sigma^2 \cdot \exp\left(\frac{-r^2}{2\sigma^2}\right)\right]_0^w \\ &= 2\pi\sigma^2 \left\{1 - \exp\left(\frac{-w^2}{2\sigma^2}\right)\right\} \end{aligned} \tag{3.45}$$

Because there is no truncation, $w = \infty$, so that $\nu = 2\pi\sigma^2$. Note that if we substitute $w = \sigma$ into this equation, then the expected proportion of sightings within σ of the point transect is $2\pi\sigma^2\{1 - \exp(-0.5)\}/\nu =$ 39%. This compares with 68% for line transects; thus for the half-normal model, nearly 70% of detections occur within one standard deviation of the observer's line of travel for line transects, whereas less than 40% occur within this distance from the observer for point transects. This highlights the fact that recorded distances tend to be greater for point transects than for line transects, a difference which is even more marked if the detection function is long-tailed.

If n independent detections are made, the likelihood function is given by

$$\mathcal{L} = \prod_{i=1}^{n} \{2\pi r_i \cdot g(r_i)/\nu\} = \left\{\prod_{i=1}^{n} r_i \cdot \exp\left(\frac{-r_i^2}{2\sigma^2}\right)\right\} \Big/ (\sigma^{2n}) \tag{3.46}$$

so that

$$l = \log_e(\mathcal{L}) = \sum_{i=1}^{n} \left\{ \log_e(r_i) - \frac{r_i^2}{2\sigma^2} \right\} - n \cdot \log_e(\sigma^2) \tag{3.47}$$

This is maximized by differentiating with respect to σ^2 and setting equal to zero:

$$\frac{dl}{d\sigma^2} = \sum_{i=1}^{n} \frac{r_i^2}{2\sigma^4} - \frac{n}{\sigma^2} = 0 \tag{3.48}$$

so that $\hat{\sigma}^2 = \sum_{i=1}^{n} r_i^2/2n$. It follows that $\hat{h}(0) = 2\pi/\hat{\nu} = 1/\hat{\sigma}^2$. Equation (3.14), with $E(s)$ set equal to one, gives

$$D = \frac{E(n) \cdot h(0)}{2\pi k} \tag{3.49}$$

so that

$$\hat{D} = \frac{n \cdot \hat{h}(0)}{2\pi k} = \frac{n^2}{\pi k \sum_{i=1}^{n} r_i^2} \tag{3.50}$$

The maximum likelihood method yields $var[\hat{h}(0)]$. The half-normal detection function has just one parameter (σ^2), so that the information matrix is a scalar. It yields

$$var(\hat{\sigma}^2) = \frac{\sigma^4}{n} \tag{3.51}$$

and

$$var\{\hat{h}(0)\} = \frac{1}{n\sigma^4} = \frac{\{h(0)\}^2}{n} \tag{3.52}$$

Estimation of $var(n)$ and $var(\hat{D})$ is covered in Section 3.6.

3.3.5 *Constrained maximum likelihood estimation*

The maximization routine used by Distance allows constraints to be placed on the fitted detection function. In all analyses, the constraint $\hat{g}(y) \geq 0$ is imposed. In addition, $\hat{g}(y)$ is evaluated at ten equally spaced y values, y_1^* to y_{10}^*, and the non-linear constraint $\hat{g}(y_i^*) \geq \hat{g}(y_{i+1}^*)$, $i = 1, \ldots, 9$, is enforced. The user may override this constraint, or replace it by the weaker constraint that $\hat{g}(0) \geq \hat{g}(y_i^*)$, $i = 1, \ldots, 10$. If the same data set is analysed by Distance and by TRANSECT (Laake *et al.* 1979), different estimates may be obtained; TRANSECT does not impose constraints, and in addition does not fit the Fourier series model by maximum likelihood. Distance warns the user when a constraint has caused estimates to be modified. In these instances, the estimates are no longer the true (unconstrained) maximum likelihood estimates, so the analytic variance of $\hat{f}(0)$ or $\hat{h}(0)$ may be unreliable, and we recommend that the bootstrap option for variance estimation (Section 3.6.4) is selected.

3.4 Choice of model

The key + adjustment formulation for line and point transect models outlined above has been implemented in Distance, so that a large number of models are available to the user. Although this gives great flexibility, it also creates a problem of how to choose an appropriate model. We consider here criteria that models for the detection function should satisfy, and methods that allow selection between contending models. Burnham and Anderson (1998) give a much more detailed discussion of model selection issues.

3.4.1 *Criteria for robust estimation*

Burnham *et al.* (1979, 1980: 44) identified four criteria that relate to properties of the assumed model for the detection function. In order of importance, they were model robustness, pooling robustness, the shape criterion and estimator efficiency.

3.4.1.1 *Model robustness*

Given that the true form of the detection function is not known except in the case of computer simulations, models are required that are sufficiently flexible to fit a wide variety of shapes for the detection function. An estimator based on such a model is termed model robust. The adoption of the key + series expansion formulation means that any parametric model can yield model robust estimation, by allowing its fit to be adjusted when the data dictate. A model of this type is sometimes called 'semiparametric'.

3.4.1.2 *Pooling robustness*

Probability of detection is a function of many factors other than distance from the observer or line. Weather, time of day, observer, habitat, behaviour of the object, its size and many other factors influence the probability that the observer will detect it. Conditions will vary during the course of a survey, and different objects will have different intrinsic detectabilities. Thus the recorded data are realizations from a heterogeneous assortment of detection functions. A model is pooling robust if it is robust to variation in detection probability for any given distance y. A fuller definition of this concept is given by Burnham *et al.* (1980: 45).

3.4.1.3 *Shape criterion*

The shape criterion can be stated mathematically as $g'(0) = 0$. In words, it states that a model for the detection function should have a shoulder;

that is, detection is assumed to be certain not only on the line or at the point itself, but also for a small distance away from the line or point. The restriction is reasonable given the nature of the sighting process. Hazard-rate modelling of the detection process (see Buckland *et al.* in preparation) gives rise to detection functions which possess a shoulder for all parameter values, even when sharply spiked hazards with infinite slope at zero distance are assumed. If the shape criterion is violated, robust estimation of object density is problematic if not impossible.

3.4.1.4 *Estimator efficiency*

Estimators that have poor statistical efficiency (i.e. that have large variances) should be ruled out. However, an estimator that is highly efficient should be considered only if it satisfies the first three criteria. High estimator efficiency is easy to achieve at the expense of bias, and the analyst should be satisfied that an estimator is roughly unbiased, or at least that there is no reason to suppose it might be more biased than other robust estimators, before selecting on the basis of efficiency.

3.4.2 *Akaike's Information Criterion*

A fundamental basis for model selection is Kullback–Leibler information (Kullback and Leibler 1951). This quantity, often denoted as $I(f, g)$, is the information (I) lost when a particular model (g) is used to approximate full reality (f). Kullback–Leibler information cannot be computed directly as neither f nor the parameter values for the approximating model are known. Akaike (1973) found a simple relationship between the maximized log-likelihood function and the estimated mean value of Kullback–Leibler information. This led to AIC, which provides a quantitative method for model selection, whether models are nested or not. AIC is simple to compute and treats model selection within an optimization rather than a hypothesis testing framework. Burnham and Anderson (1998) review the theory and application of AIC, provide a number of extensions, and reference the key papers by Akaike and others. AIC is defined as

$$\text{AIC} = -2 \cdot \log_{\text{e}}(\mathcal{L}) + 2q \tag{3.53}$$

where $\log_{\text{e}}(\mathcal{L})$ is the log-likelihood function evaluated at the maximum likelihood estimates of the model parameters and q is the number of estimated parameters in the model. We can interpret the first term, $-2 \cdot \log_{\text{e}}(\mathcal{L})$, as a measure of how well the model fits the data, while the second term is a 'penalty' for the addition of parameters. Model fit can always be improved, and bias reduced, by adding parameters, but at a cost of adding model complexity and increasing variance. AIC provides a satisfactory trade-off between bias and variance.

For a given data set, AIC is computed for each candidate model and the model with the lowest AIC is selected for inference. Thus, AIC attempts to identify a model that fits the data well but does not have too many parameters (the principle of parsimony). Assuming normally distributed data, Hurvich and Tsai (1989, 1995) developed a second order criterion, called AICc, based on the work of Sugiura (1978):

$$\text{AICc} = -2 \cdot \log_e(\mathcal{L}) + 2q + \frac{2q(q+1)}{n-q-1} \tag{3.54}$$

where n is the sample size (number of detections). As n/q increases, AIC $\rightarrow$ AICc. Thus, for large samples, either may be used, but for smaller samples (say $n < 20q$), AICc might be considered. Because AICc is not necessarily better than AIC when the data are not normally distributed, software Distance uses AIC as the default model selection option.

For analyses of grouped data, Distance omits the constant term from the multinomial likelihood when it calculates AIC. This ensures that AIC tends to the value obtained from analysis of ungrouped data as the number of groups tends to infinity, with each interval length tending to zero.

Often, one model will be very much better than other models in the candidate set. In this case, it is often best to base inference on this selected model. However, it is common to see several models that are somewhat similar. Buckland *et al.* (1997) and Burnham and Anderson (1998) provide theory to allow inference to be based on all the models in the set. This procedure, termed model averaging, is both simple and effective and is implemented in Distance.

3.4.3 *The likelihood ratio test*

We include the likelihood ratio test for completeness, although we recommend use of AIC or AICc. The likelihood ratio test is only useful for testing between hierarchical models, so its potential relevance here is for choosing the number of adjustment terms to include. For the special case of testing for inclusion of a single extra adjustment term, AIC is algebraically equivalent to a likelihood ratio test of size 15.7%. We prefer the conceptual basis of AIC (see previous section), and AIC allows comparison of non-nested models, making it of greater practical use.

Suppose that a fitted model has m_1 adjustment terms (Model 1). A likelihood ratio test allows an assessment of whether the addition of another m_2 terms improves the adequacy of a model significantly. The null hypothesis is that Model 1, with m_1 adjustment terms, is the true model, whereas the alternative hypothesis is that Model 2 with all $m_1 + m_2$ adjustment terms is the true model. The test statistic is

$$\chi^2 = -2\log_e(\mathcal{L}_1/\mathcal{L}_2) = -2[\log_e(\mathcal{L}_1) - \log_e(\mathcal{L}_2)] \tag{3.55}$$

where $\mathcal{L}_1$ and $\mathcal{L}_2$ are the maximum values of the likelihood functions for Models 1 and 2, respectively. If Model 1 is the true model, the test statistic follows a χ^2 distribution with m_2 degrees of freedom.

The usual way to use likelihood ratio tests for line and point transect series expansion models is to fit the key function, and then fit a low order adjustment term. If it provides no significant improvement as judged by the above test, the fit of the key alone is taken. If the adjustment term does improve the fit, the next term is added, and a likelihood ratio test is again carried out. The process is repeated sequentially until the test is not significant, or until a maximum number of terms has been attained. Forward or backward stepping procedures can also be used.

3.4.4 *Goodness of fit*

Goodness of fit can be useful in assessing the quality of the distance data and understanding the general shape of the detection function. Suppose the n distance data from line or point transects are split into u groups, with sample sizes $n_1, n_2, \ldots, n_u$. Let the cutpoints between groups be defined by $c_0, c_1, \ldots, c_u = w$ ($c_0 > 0$ corresponds to left-truncation of the data). Suppose that a pdf $f(y)$ with q parameters is fitted to the data, yielding the estimated function, $\hat{f}(y)$. The area under this function between cutpoints c_{i-1} and c_i is

$$\hat{\pi}_i = \int_{c_{i-1}}^{c_i} \hat{f}(y)\, dy \tag{3.56}$$

Then

$$\chi^2 = \sum_{i=1}^{u} \frac{(n_i - n \cdot \hat{\pi}_i)^2}{n \cdot \hat{\pi}_i} \tag{3.57}$$

has a χ^2 distribution with $u - q - 1$ degrees of freedom if the fitted model is the true model.

Although a significantly poor fit need not be of great concern, it provides a warning of a problem in the data or the selected detection model structure, which should be investigated through closer examination of the data or by exploring other models and fitting options. Note that it is the fit of the model near zero distance that is most critical; the model selection criteria considered here do not give special emphasis to this region.

3.5 Estimation for clustered populations

Although the general formula of Section 3.1 incorporates the case in which the detections are clusters of objects, estimation of the expected cluster size $E(s)$ is often problematic. The obvious estimator, the average size of detected clusters, may be subject to size bias; if large clusters are detectable at greater distances than small clusters, mean size of detected clusters will be a positively biased estimator of $E(s)$, the mean cluster size in the population. Solutions to this problem are addressed below.

3.5.1 *Truncation*

The simplest solution is to truncate clusters that are detected far from the line or point. The truncation distance need not be the same as that used for fitting the detection function to the distance data; if size bias is potentially severe, truncation should be greater. To be certain of eliminating the effects of size bias, the truncation distance should correspond roughly to the width of the shoulder of the detection function. Then $E(s)$ is estimated by $\bar{s}$, the mean size of the n clusters detected within the truncation distance. Generally, a truncation distance v corresponding to an estimated probability of detection $\hat{g}(v)$ in the range 0.6–0.8 ensures that bias in this estimate is small for line transects; truncation should be rather more severe for point transects. Variance of $\bar{s}$ is then estimated by:

$$\widehat{var}(\bar{s}) = \frac{\sum_{i=1}^{n_v}(s_i - \bar{s})^2}{n_v(n_v - 1)} \tag{3.58}$$

where s_i denotes the size of cluster i and n_v is the number of observations within truncation distance v of the line or point. This estimator remains unbiased when the individual s_i have different variances.

3.5.2 *Stratification by cluster size*

Truncation may prove unsatisfactory if sample size is small. The second option is to stratify by cluster size. Suppose that objects in a population occur either singly or in pairs. We can first select all the data on single objects, and analyse them, yielding an abundance estimate. In a separate analysis, we can analyse the pairs of objects, and again obtain an abundance estimate of individuals (which will be double the abundance of clusters). We can now add these estimates together, to yield an overall abundance estimate.

In practice, many different cluster sizes might be recorded, and there may be too few data to carry out an independent analysis for each recorded cluster size. Strata can then be formed by pooling similar cluster sizes. For example, the first stratum might correspond to clusters of size 1–2, the second to cluster sizes 3–5, the third to cluster sizes 6–10 and so on.

Quinn (1979) showed that under stratification, mean cluster size in the population is estimated by

$$\hat{E}(s) = \frac{\sum_v n_v s_v \hat{f}_v(0 \,|\, s = s_v)}{\sum_v n_v \hat{f}_v(0 \,|\, s = s_v)} \tag{3.59}$$

where summation is over the recorded cluster sizes. Thus there are n_v detections of clusters of size s_v, and the effective strip half-width for these clusters is $1/f_v(0 \mid s = s_v)$. The estimate is therefore the average size of detected clusters, weighted by the inverse of the estimated effective strip half-width at each cluster size. For point transect sampling, $\hat{h}_v(0 \mid s = s_v)$ would replace $\hat{f}_v(0 \mid s = s_v)$.

3.5.3 *Weighted average of cluster sizes*

Stratification still requires a large sample size so that the $f_v(0 \mid s = s_v)$ can be individually estimated. For line transect sampling, Quinn (1979) noted that data can be pooled with respect to cluster size if we assume that the effective strip half-width is proportional to the logarithm of cluster size, since then

$$\hat{E}(s) = \frac{\sum_v n_v s_v / \log_e(s_v)}{\sum_v n_v / \log_e(s_v)} \tag{3.60}$$

This method is used in the procedures developed by Holt and Powers (1982) for estimating dolphin abundance in the eastern tropical Pacific. If used, mean perpendicular distance should be plotted as a function of cluster size to assess the functional relationship between cluster size and effective strip half-width (Quinn 1985). The method should not be used in conjunction with truncation of clusters at larger distances, because cluster size is then underestimated. The purpose of truncation is to restrict the mean cluster size calculation to those clusters that are relatively unaffected by size bias, so effective strip half-width of the retained clusters is not then proportional to the logarithm of cluster size. Clusters beyond the truncation distance are larger than average when size bias is present, so that the above weighted mean, if applied after truncating distant clusters, corrects for the effects of size bias twice.

Quinn (1985) showed that the stratification method necessarily yields a higher coefficient of variation for abundance of clusters than the above method in which data are pooled across cluster size, but found that the result does not extend to estimates of object abundance. For his example, he concludes that the method of pooling is superior for estimating cluster abundance, and the method of stratification for estimating object abundance. This conclusion is likely to be true more generally. Stratification strategies relevant to this issue are discussed in more detail in Section 3.7.

3.5.4 *Regression estimators*

The solution of plotting mean perpendicular distance y against cluster size in line treansect sampling was proposed by Best and Butterworth (1980), who predicted mean cluster size at zero distance, using a weighted linear regression of cluster size on distance. This suffers from the difficulty that, if the detection function has a shoulder, mean cluster size does not increase with distance until detection distance exceeds the width of the shoulder. Sample size is seldom sufficient to determine that a straight line fit is inadequate, so that estimated mean cluster size at zero distance is biased downwards, and population abundance is underestimated. A solution to this problem is to replace perpendicular distance (line transects) or detection distance (point transects) y_i for the ith detection by $\hat{g}(y_i)$ in the regression, where $\hat{g}(y)$ is the estimated detection function from the fit of the selected model to the distances from the line or point to detected clusters. Mean cluster size in the population is then estimated by the predicted mean size of detected clusters in the region around the line or point for which detection is estimated to be certain ($\hat{g}(y) = 1.0$). Thus if there are n detections, at distances y_i and of sizes s_i , if $E_d(s \mid y)$ denotes the expected size of detected clusters at distance y, and $E(s)$ denotes the expected size of all clusters, whether detected or not (assumed independent of y), we have

$$\hat{E}_d(s \mid y) = a + b \cdot \hat{g}(y) \tag{3.61}$$

where a and b are the intercept and slope respectively of the regression of s on $\hat{g}(y)$. Then

$$\hat{E}(s) = \hat{E}_d(s \mid y = 0) = a + b \tag{3.62}$$

and

$$\widehat{var}[\hat{E}(s)] = \left[\frac{1}{n} + \frac{(1 - \bar{g})^2}{\sum_{i=1}^{n}\{\hat{g}(y_i) - \bar{g}\}^2}\right] \cdot \hat{\sigma}^2 \tag{3.63}$$

with $\hat{\sigma}^2$ = residual mean square and $\bar{g} = \sum_{i=1}^{n} \hat{g}(y_i)/n$

A further problem of the regression method occurs when cluster size is highly variable, so that one or two large clusters might have large influence on the fit of the regression line. Their influence may be reduced by transformation, for example to $z_i = \log_e(s_i)$. Suppose a regression of z_i on $\hat{g}(y_i)$ yields the equation $\hat{z} = a + b \cdot \hat{g}(y)$. Thus at $\hat{g}(y) = 1.0$, mean log cluster size is estimated by $a + b$, and $E(s)$ is estimated by

$$\hat{E}(s) = \exp(a + b + \widehat{var}(\hat{z})/2) \tag{3.64}$$

where

$$\widehat{var}(\hat{z}) = \left[1 + \frac{1}{n} + \frac{(1 - \bar{g})^2}{\sum_{i=1}^{n}\{\hat{g}(y_i) - \bar{g}\}^2}\right] \cdot \hat{\sigma}^2 \tag{3.65}$$

$\hat{\sigma}^2$ is the residual mean square, and $\bar{g}$ is as above. Further,

$$\widehat{var}\{\hat{E}(s)\} = \exp\{2(a + b) + \widehat{var}(\hat{z})\} \cdot \{1 + \widehat{var}(\hat{z})/2\} \cdot \widehat{var}(\hat{z})/n \tag{3.66}$$

Software Distance allows the user to regress either y_i or $\hat{g}(y_i)$ on either s_i or $\log(s_i)$. The default is a regression of $\hat{g}(y_i)$ on $\log(s_i)$.

Detected clusters far from the line or point tend to be larger than average. Hence a regression of cluster size (or log cluster size) on distance is expected to have a positive slope, and a regression on estimated probability of detection a negative slope. Sometimes, the regression yields a slope that is significantly different from zero, but with the wrong sign. This occurs when observers tend to underestimate the size of detected clusters, and the degree of underestimation is greater for clusters further away, as often happens in woodland surveys of songbirds that are typically recorded in small flocks, or in shipboard surveys of marine mammals. Provided the sizes of clusters that are on or very close to the line or point are estimated without bias, the regression method still provides valid estimation of mean cluster size in the population. Thus, if the regression method is used, it corrects both for size-biased detection and for underestimation of size of detected clusters, provided neither of these effects occur at the line or point.

3.5.5 *Use of covariates*

The pooling method, with calculation of a weighted average cluster size, may be improved upon theoretically by incorporating cluster size as a covariate in the model for the detection function. Drummer and McDonald (1987) considered replacing detection distance y in a parametric model for the detection function by y/s^{γ}, where s is size of the cluster recorded at distance y and γ is a parameter to be estimated. Although their method was developed for line transect sampling, it can also be implemented for point transects. Ramsey *et al.* (1987) included covariates by relating the logarithm of effective area searched to a linear function of covariates, one of which could be cluster size; this is in the spirit of generalized linear models. Quang (1991) developed a method of modelling the bivariate detection function $g(y, s)$ using Fourier series. In Buckland *et al.* (in preparation), we extend the approach of Ramsey *et al.* (1987) to provide a general framework for modelling covariates in distance sampling.

3.5.6 *Replacing clusters by individual objects*

The problems of estimating mean cluster size can sometimes be avoided by taking the sampling unit to be the object, not the cluster. Even when detected clusters show extreme selection for large clusters, this approach can yield an unbiased estimate of object abundance, provided all clusters on or near the line are detected. The assumption of independence between sampling units is clearly violated, so robust methods of variance estimation that are insensitive to failures of this assumption should be adopted. Use of resampling methods such as the bootstrap (Section 3.6.4) allows the individual line or point to be the sampling unit instead of the object, so that valid variance estimation is possible. Under this approach, results from goodness of fit tests, likelihood ratio tests and AIC should not be used for model selection, since they will all favour over-parameterized models, due to overdispersion in the data. One solution is to select a model based on an analysis of clusters, then to refit the model, with the same number of adjustment terms, to the data recorded by object.

If the number of clusters detected is small, if cluster size is highly variable, or if mean cluster size is large, this approach may perform poorly.

3.6 Density, variance and interval estimation

3.6.1 *Basic formulae*

We start by noting the relationship between the *variance* (var), *standard error* (se) and *coefficient of variation* (cv) of an estimator. The variance is a measure of precision of an estimator, whose units are the square of those for the estimator. Thus for an abundance estimator, with units of animals, its variance will have units of (animals)^2. The standard error of the estimator is the square root of the variance, and has the same units as the estimator. We define the coefficient of variation of an estimator to be its standard error divided by itself.

Substituting estimates into eqn (3.7), the general formula for estimating object density from line transect data is

$$\hat{D} = \frac{n \cdot \hat{f}(0) \cdot \hat{E}(s)}{2L} \tag{3.67}$$

From eqn (3.4), the variance of $\hat{D}$ is approximately

$$\widehat{var}(\hat{D}) = \hat{D}^2 \cdot \left\{ \frac{\widehat{var}(n)}{n^2} + \frac{\widehat{var}[\hat{f}(0)]}{[\hat{f}(0)]^2} + \frac{\widehat{var}[\hat{E}(s)]}{[\hat{E}(s)]^2} \right\} \tag{3.68}$$

Equivalent expressions for point transect sampling are

$$\hat{D} = \frac{n \cdot \hat{h}(0) \cdot \hat{E}(s)}{2\pi k} \tag{3.69}$$

and

$$\widehat{var}(\hat{D}) = \hat{D}^2 \cdot \left\{ \frac{\widehat{var}(n)}{n^2} + \frac{\widehat{var}[\hat{h}(0)]}{[\hat{h}(0)]^2} + \frac{\widehat{var}[\hat{E}(s)]}{[\hat{E}(s)]^2} \right\} \tag{3.70}$$

If all objects occur singly, then $E(s) = 1$ and the term involving $\hat{E}(s)$ is zero in the above variance formulae.

To estimate the precision of $\hat{D}$, the precision of each component in the estimation equation must be estimated. Alternatively, resampling or empirical methods can be used to estimate $var(\hat{D})$ directly; some options are described in Sections 3.6.2–3.6.4. If precision is estimated component by component, then methods should be adopted for estimating mean cluster size that provide a variance estimate, $\widehat{var}[\hat{E}(s)]$. Estimates of $f(0)$ or $h(0)$ and corresponding variance estimates are obtained using the information matrix from maximum likelihood theory (Section 3.3). If objects are distributed randomly, then sample size n has a Poisson distribution, and $\widehat{var}(n) = n$. Generally, biological populations show some degree of aggregation, and Burnham *et al.* (1980: 55) suggested multiplication of the Poisson variance by two if no other approach for estimating $var(n)$ was available. If data are recorded by replicate lines or points, then a better method is to estimate $var(n)$ from the observed variation between lines and points. This method is described in the next section.

Having obtained $\hat{D}$ and $\widehat{var}(\hat{D})$, an approximate $100(1 - 2\alpha)\%$ confidence interval is given by

$$\hat{D} \pm z_\alpha \cdot \sqrt{\widehat{var}(\hat{D})} \tag{3.71}$$

where z_α is the upper α point of the $N(0,1)$ distribution ($z_\alpha = z_{0.025} = 1.96$ for a 95% confidence interval). However, the distribution of $\hat{D}$ is positively skewed, and an interval with better coverage is obtained by assuming that $\hat{D}$ is log-normally distributed. Following the derivation of Burnham *et al.* (1987: 212), a $100(1 - 2\alpha)\%$ confidence interval is given by

$$(\hat{D}/C, \hat{D} \cdot C) \tag{3.72}$$

where

$$C = \exp\left[z_\alpha \cdot \sqrt{\widehat{var}(\log_e \hat{D})}\right] \tag{3.73}$$

and

$$\widehat{var}(\log_e \hat{D}) = \log_e \left[1 + \frac{\widehat{var}(\hat{D})}{\hat{D}^2}\right] \tag{3.74}$$

This is the method used by Distance, except z_α is replaced by a slightly better constant that reflects the finite and differing degrees of freedom associated with each variance component.

The use of the normal distribution to approximate the sampling distribution of $\log_e(\hat{D})$ is generally good when each component of $\widehat{var}(\hat{D})$ (e.g. $\widehat{var}(n)$ and $\widehat{var}[\hat{f}(0)]$) is based on sufficient degrees of freedom (say 30 or more). However, sometimes the empirical estimate of $var(n)$ in particular is based on less than 10 replicate lines, and hence on few degrees of freedom. When component degrees of freedom are small, it is better to replace z_α by a constant based on a t-distribution approximation. In this case we recommend an approach adapted from Satterthwaite (1946); see also Milliken and Johnson (1984) for a more accessible reference.

Adapting the method of Satterthwaite (1946) to this distance sampling context, z_α in the above log-based confidence interval is replaced by the two-sided alpha-level t-distribution percentile $t_{\text{df}}(\alpha)$ where df is computed as below. The coefficients of variation $cv(\hat{D})$, $cv(n)$, $cv[\hat{f}(0)]$ or $cv[\hat{h}(0)]$, and, where relevant, $cv[\hat{E}(s)]$ are required, together with the associated degrees of freedom. In general, if there are q estimated components in $\hat{D}$, then the computed degrees of freedom are

$$\text{df} = \frac{[cv(\hat{D})]^4}{\sum_{i=1}^{q} [cv_i]^4/\text{df}_i} = \frac{[\sum_{i=1}^{q} [cv_i]^2]^2}{\sum_{i=1}^{q} [cv_i]^4/\text{df}_i} \tag{3.75}$$

This value may be rounded to the nearest integer to allow use of tables of the t-statistic.

For the common case of line transect sampling of single objects using k replicate lines, the above formula for df becomes

$$\text{df} = \frac{[cv(\hat{D})]^4}{[cv(n)]^4/(k-1) + \{cv[\hat{f}(0)]\}^4/(n-p)} \tag{3.76}$$

where p is the number of parameters estimated in $\hat{f}(x)$. This Satterthwaite procedure is used by program Distance instead of the first order z_α approximation. It makes a noticeable difference in confidence intervals for small k, especially if the ratio $cv(n)/cv[\hat{f}(0)]$ is greater than 1; in practice, it is often as high as 2 or 3.

3.6.2 *Replicate lines or points*

Replicate lines or points may be used to estimate the contribution to overall variance of the observed sample size. In line transects, the replicate lines may be defined by the design of the survey; for example if the lines are parallel and either systematically or randomly spaced, then each line is considered a replicate. Surveys of large areas by ship or air frequently do not utilize such a design for practical reasons. In this case, a 'leg' might be defined as a period of search without change of bearing, or all effort for a given day or watch period. The leg will then be treated as a replicate line. When data are collected on an opportunistic basis from, for example, fisheries vessels, an entire fishing trip might be considered to be the sampling unit.

Suppose the number of detections from line or point i is $n_i, i = 1, \ldots, k$, so that $n = \sum n_i$. Then for point transects (or for line transects when the replicate lines are all the same length), the empirical estimate of $var(n)$ is

$$\widehat{var}(n) = \frac{k \sum_{i=1}^{k} (n_i - (n/k))^2}{k-1} \tag{3.77}$$

For line transects, if line i is of length l_i and total line length $= L = \sum_{i=1}^{k} l_i$, then

$$\widehat{var}(n) = \frac{L \sum_{i=1}^{k} l_i (n_i/l_i - n/L)^2}{k-1} \tag{3.78}$$

Encounter rate n/L is often a more useful form of the parameter than n alone; the variance of encounter rate is $\widehat{var}(n)/L^2$. There is a similarity here to ratio estimation in finite population sampling, except that we take all line lengths l_i, and hence L, to be fixed (as distinct from random) values. Consequently, the variance of a ratio estimator does not apply here, and our $\widehat{var}(n/L)$ is a little different from classical finite sampling theory.

If the same line or point is covered more than once, and an analysis of the pooled data is required, then the sampling unit should still be the line or point. That is, the distance data from repeat surveys over a short time period of a given line or point should be pooled prior to analysis. Consider point transects, in which point i is covered t_i times, and in total, n_i objects are detected. Then

$$\widehat{var}(n) = \frac{T \sum_{i=1}^{k} t_i (n_i/t_i - n/T)^2}{k-1} \tag{3.79}$$

where

$$T = \sum_{i=1}^{k} t_i \tag{3.80}$$

The formula for line transects becomes

$$\widehat{var}(n) = \frac{T_L \sum_{i=1}^{k} t_i \cdot l_i \cdot (n_i/(t_i \cdot l_i) - n/T_L)^2}{k-1} \tag{3.81}$$

where

$$T_L = \sum_{i=1}^{k} t_i \cdot l_i \tag{3.82}$$

The calculations may be carried out in Distance by setting sample effort equal to t_i for point i (point transects) or $t_i \cdot l_i$ for line i (line transects). Good design practice seeks to ensure that t_i is the same for every point or line (so that potential bias is avoided), in which case the above formulae simplify. In this case, if $t_i = t$ for all i, then a constant multiplier t can be set in Distance, and sample effort set to one for each point (point transects) or l_i for line i (line transects).

The above provides empirical variance estimates for just one component of eqn (3.3), which may then be substituted in eqn (3.4). A more direct approach is to estimate object density for each replicate line or point. Define

$$\hat{D}_i = \frac{n_i \cdot \hat{E}_i(s)}{a_i \cdot \hat{P}_{a_i}}, \quad i = 1, \ldots, k \tag{3.83}$$

Then for point transects (and line transects when all lines are the same length),

$$\hat{D} = \frac{\sum_{i=1}^{k} \hat{D}_i}{k} \tag{3.84}$$

and

$$\widehat{var}(\hat{D}) = \frac{\sum_{i=1}^{k} (\hat{D}_i - \hat{D})^2}{k(k-1)} \tag{3.85}$$

For line transects with replicate line i of length l_i,

$$\hat{D} = \frac{\sum_{i=1}^{k} l_i \hat{D}_i}{L} \tag{3.86}$$

and

$$\widehat{var}(\hat{D}) = \frac{\sum_{i=1}^{k} \{l_i (\hat{D}_i - \hat{D})^2\}}{L(k-1)} \tag{3.87}$$

In practice, sample size is seldom sufficient to allow this approach, so that resampling methods such as the bootstrap and the jackknife are required.

3.6.3 *The jackknife*

Resampling methods start from the observed data and sample repeatedly from them to make inferences. The jackknife (Gray and Schucany 1972; Miller 1974) is carried out by removing each observation in turn from the data, and analysing the remaining data. It could be implemented for line and point transects by dropping each individual sighting from the data in turn, but it is more useful to define replicate points or lines, as above. The following development is for point transects, or line transects when the replicate lines are all of the same length.

First, delete all data from the first replicate point or line, so that sample size becomes $n - n_1$ and the number of points or lines becomes $k - 1$. Estimate object density using the reduced data set, and denote the estimate by $\hat{D}_{(1)}$. Repeat this step, reinstating the dropped point or line and removing the next, to give estimates $\hat{D}_{(i)}$, $i = 1, \ldots, k$. Now calculate the *pseudovalues*:

$$\hat{D}^{(i)} = k \cdot \hat{D} - (k-1) \cdot \hat{D}_{(i)}, \quad i = 1, \ldots, k \tag{3.88}$$

These pseudovalues are treated as k replicate estimators of density, and eqns (3.84) and (3.85) yield a jackknife estimate of density and variance:

$$\hat{D}_{\mathrm{J}} = \frac{\sum_{i=1}^{k} \hat{D}^{(i)}}{k} \tag{3.89}$$

and

$$\widehat{var}_{\mathrm{J}}(\hat{D}_{\mathrm{J}}) = \frac{\sum_{i=1}^{k} (\hat{D}^{(i)} - \hat{D}_{\mathrm{J}})^2}{k(k-1)} \tag{3.90}$$

For line transects in general, eqn (3.88) is replaced by

$$\hat{D}^{(i)} = \frac{L \cdot \hat{D} - (L - l_i) \cdot \hat{D}_{(i)}}{l_i}, \quad i = 1, \cdots, k \tag{3.91}$$

and the jackknife estimate and variance are found by substitution into eqns (3.86) and (3.87):

$$\hat{D}_{\mathrm{J}} = \frac{\sum_{i=1}^{k} l_i \hat{D}^{(i)}}{L} \tag{3.92}$$

and

$$\widehat{var}_{\mathrm{J}}(\hat{D}_{\mathrm{J}}) = \frac{\sum_{i=1}^{k} \{l_i (\hat{D}^{(i)} - \hat{D}_{\mathrm{J}})^2\}}{L(k-1)} \tag{3.93}$$

An approximate $100(1 - 2\alpha)\%$ confidence interval for density D is given by

$$\hat{D}_{\mathrm{J}} \pm t_{k-1}(\alpha) \cdot \sqrt{\widehat{var}_{\mathrm{J}}(\hat{D}_{\mathrm{J}})} \tag{3.94}$$

where $t_{k-1}(\alpha)$ is from Student's t-distribution with $k-1$ degrees of freedom.

This interval may have poor coverage when the number of replicate lines is small; Buckland (1982) found better coverage using

$$\hat{D} \pm t_{k-1}(\alpha) \cdot \sqrt{\widehat{var}_{\mathrm{J}}(\hat{D})} \tag{3.95}$$

where $\hat{D}$ is the estimated density from the full data set and

$$\widehat{var}_{\mathrm{J}}(\hat{D}) = \frac{\sum_{i=1}^{k} \{l_i (\hat{D}^{(i)} - \hat{D})^2\}}{L(k-1)} \tag{3.96}$$

The jackknife provides a strictly balanced resampling procedure. However, there seems little justification for assuming that the pseudovalues are normally distributed, and the above confidence intervals may be poor when the number of replicate lines or points is small. Further there is little or no control over the number of resamples taken; under the above procedure, it is necessarily equal to the number of replicate lines or points k, and performance may be poor when k is small. Thirdly a resample can never be larger than the original sample, and will always be smaller unless there are no sightings on at least one of the replicate lines or points. The bootstrap does not have these defects, and offers greater flexibility and robustness.

3.6.4 *The bootstrap*

The bootstrap (Efron 1979; Efron and Tibshirani 1993; Davison and Hinkley 1997; Manly 1997) provides a powerful yet simple method for variance and interval estimation. Consider first the nonparametric bootstrap, applied in the most obvious way to a line transect sample. Suppose the data set comprises n perpendicular distances, $y_1, \ldots, y_n$, and the probability density evaluated at zero, $f(0)$, is to be estimated. Then a bootstrap sample may be generated by selecting a sample of size n with replacement from the observed distances. An estimate of $f(0)$ is found from the bootstrap sample using the same model as for the observed sample. A second bootstrap sample is then taken, and the process repeated. Suppose in total B samples are taken. Then the variance of $\hat{f}(0)$ is estimated by the sample variance of bootstrap estimates of $f(0)$, $\hat{f}_i(0)$, $i = 1, \ldots, B$ (Efron 1979). The percentiles of the distribution of bootstrap estimates give approximate confidence limits for $f(0)$ (Buckland 1980; Efron 1981). An approximate $100(1-2\alpha)\%$ central confidence interval is given by $[\hat{f}_{(j)}(0), \hat{f}_{(k)}(0)]$, where $j = (B+1)\alpha$ and $k = (B+1)(1-\alpha)$ and $\hat{f}_{(i)}(0)$ denotes the ith smallest bootstrap estimate (Buckland 1984). To yield reliable confidence intervals, the number of bootstrap samples B should be at least 200, and preferably in the range 400–1000, although around 100 are adequate for estimating standard errors. The value of B may be chosen so that j and k are integer, or j and k may be rounded to the nearest integer values, or interpolation may be used between the ordered values that bracket the required percentile. Generally, $B = 1000$ provides adequate precision. Various modifications to the percentile method have been proposed (Efron and Tibshirani 1993), but the simple method is sufficient for our purposes. The parametric bootstrap is applied in exactly the same manner, except that the bootstrap resamples are generated by taking a random sample of size n from the fitted probability density, $\hat{f}(y)$.

If no polynomial or Fourier series adjustments are made to the fit of a parametric probability density, the above implementation of the bootstrap (whether parametric or nonparametric) yields variance estimates for $\hat{f}(0)$ close to those obtained using the information matrix. Since the bootstrap consumes considerably more computer time (up to B times that required by an analytical method), it would not normally be used in this case. When adjustments are made, precision as measured by the information matrix is conditional on the number of polynomial or Fourier series terms selected by the stopping rule (e.g. AIC). The Fourier series model in particular gives analytical standard errors that are strongly correlated with the number of terms selected (Buckland 1985). The above implementation of the bootstrap avoids this problem by applying the stopping rule independently to each bootstrap data set so that variation arising from estimating the number of terms required is accounted for (Buckland 1982).

In practice the bootstrap is usually more useful when the sampling unit is a replicate line or point, as for the jackknife method. The simplest procedure is to sample with replacement from the replicate lines or points using the nonparametric bootstrap. Unlike the jackknife, the sample need not be balanced, but a degree of balance may be forced by ensuring that each replicate line or point is used exactly B times in the B bootstrap samples (Davison *et al.* 1986). Density D is estimated from each bootstrap sample, and the estimates are ordered, to give $\hat{D}_{(i)}$, $i = 1, \dots, B$. Then

$$\hat{D}_B = \frac{\sum_{i=1}^{B} \hat{D}_{(i)}}{B} \tag{3.97}$$

and

$$\widehat{var}_B(\hat{D}_B) = \frac{\sum_{i=1}^{B} (\hat{D}_{(i)} - \hat{D}_B)^2}{B - 1} \tag{3.98}$$

while a $100(1-2\alpha)\%$ confidence interval for D is given by $[\hat{D}_{(j)}, \hat{D}_{(k)}]$, with $j = (B+1)\alpha$ and $k = (B+1)(1-\alpha)$ as above. (The estimates do not need to be ordered if a confidence interval is not required.) The estimate based on the original data set, $\hat{D}$, is usually used in preference to the bootstrap estimate $\hat{D}_B$, with $var(\hat{D})$ estimated by $\widehat{var}_B(\hat{D}_B)$. An advantage of this implementation of the bootstrap over analytic variance estimates is that we do not assume that the estimates n, $\hat{f}(0)$ and $\hat{E}(s)$ are independent.

If required, an automated model selection procedure such as AIC can be used to select the best approximating model independently for each resample. This is particularly useful when different models give appreciably different density estimates, yet the model selection criteria do not strongly favour any one model. By reselecting the model in each resample, the variability between bootstrap estimates of density reflects uncertainty due to having to estimate which model is appropriate. In other words, the bootstrap variance incorporates a component for model selection uncertainty (Buckland *et al.* 1997). By applying the full estimation procedure to each replicate, components of the variance for estimating the number of adjustment terms and for estimating $E(n)$ and $E(s)$ (and any additional multipliers, if these are estimated independently for each bootstrap resample) are all automatically incorporated. An example of such an analysis is given in Section 5.7.2.

A common misconception is that no model assumptions are made when using the nonparametric bootstrap. However, the sampling units from which resamples are drawn are assumed to be independently and identically distributed. If the sampling units are legs of effort, then each leg should be randomly located and independent of any other leg. In practice, this is seldom the case, but legs should be defined that do not seriously violate the assumption. For example, in marine line transect surveys, the sampling effort might be defined as all effort carried out in a single day. The overnight break in effort will reduce the dependence in the data between one sampling unit and the next, and the total number of sampling units should provide adequate replication except for surveys of short duration.

In line transect sampling, the individual lines should be the sampling units; it is wrong to break long lines into small units and to bootstrap on those units (an example of pseudoreplication). This is because the assumption of independence can be seriously violated, leading to bias in the variance estimate. For example, if long transect lines are designed to be perpendicular to object density contours, subdivision of the lines will lead to overestimation of variance. In the case of point transects, if points are positioned along lines, then each line of points should be considered a sampling unit. If points are randomly distributed or evenly distributed throughout the study area, then individual points may be taken as sampling units. If a single line or point is covered more than once, and an analysis of the pooled data is required, the sampling unit should still be the line or point; it is incorrect to analyse the data as if different lines or points had been covered on each occasion. The methods of Sections 3.6.2–3.6.4 all allow analyses in which the line or point is the sampling unit.

3.6.5 *Estimating change in density*

Frequently, we wish to draw inference on change in density over time, or difference in density between habitats. We consider here simple comparisons between two density estimates. An example involving estimation of trend over time is given in Section 8.4.

Consider two density estimates, $\hat{D}_1$ and $\hat{D}_2$. Suppose first that they are independently estimated. We then estimate the difference in density by $\hat{D}_1 - \hat{D}_2$ with variance

$$\widehat{var}(\hat{D}_1 - \hat{D}_2) = \widehat{var}(\hat{D}_1) + \widehat{var}(\hat{D}_2) \tag{3.99}$$

Distance provides approximate degrees of freedom df_1 for $\hat{D}_1$ and df_2 for $\hat{D}_2$, based on Satterthwaite's approximation. We can use these to obtain an approximate t-statistic:

$$T = \frac{(\hat{D}_1 - \hat{D}_2) - (D_1 - D_2)}{\sqrt{\widehat{var}(\hat{D}_1 - \hat{D}_2)}} \sim t_{\text{df}} \tag{3.100}$$

where

$$\text{df} \simeq \frac{\{\widehat{var}(\hat{D}_1) + \widehat{var}(\hat{D}_2)\}^2}{\{\widehat{var}(\hat{D}_1)\}^2/\text{df}_1 + \{\widehat{var}(\hat{D}_2)\}^2/\text{df}_2} \tag{3.101}$$

Provided df are around 30 or more, the simpler z-statistic provides a good approximation:

$$Z = \frac{(\hat{D}_1 - \hat{D}_2) - (D_1 - D_2)}{\sqrt{\widehat{var}(\hat{D}_1 - \hat{D}_2)}} \sim N(0,1) \tag{3.102}$$

For either statistic, we can test the null hypothesis $H_0 : D_1 = D_2$ by substituting $D_1 - D_2 = 0$ in eqn (3.100) or (3.102), and looking at the resulting value in t-tables or z-tables. Approximate $100(1-2\alpha)\%$ confidence limits for $(D_1 - D_2)$ are given by

$$(\hat{D}_1 - \hat{D}_2) \pm t_{\mathrm{df}}(\alpha)\sqrt{\widehat{var}(\hat{D}_1 - \hat{D}_2)} \tag{3.103}$$

for df < 30, or

$$(\hat{D}_1 - \hat{D}_2) \pm z(\alpha)\sqrt{\widehat{var}(\hat{D}_1 - \hat{D}_2)} \tag{3.104}$$

otherwise.

Often, a single detection function is fitted to pooled data, so that

$$\hat{D}_1 = \frac{n_1 \hat{f}(0)\hat{E}_1(s)}{2L_1} \quad \text{and} \quad \hat{D}_2 = \frac{n_2 \hat{f}(0)\hat{E}_2(s)}{2L_2} \tag{3.105}$$

(where the terms $\hat{E}_i(s)$ are omitted if the objects are not in clusters). Because $\hat{f}(0)$ appears in both equations, we can no longer assume that $\hat{D}_1$ and $\hat{D}_2$ are independent. Instead, we can write

$$\hat{D}_i = \hat{M}_i \hat{f}(0), \quad i = 1, 2 \tag{3.106}$$

where

$$\hat{M}_i = \frac{n_i \hat{E}_i(s)}{2L_i} \tag{3.107}$$

As the M_i are independently estimated, and are assumed to be independent of $\hat{f}(0)$, we can now find the variance of $\hat{D}_1 - \hat{D}_2$ using the delta method:

$$\hat{D}_1 - \hat{D}_2 = (\hat{M}_1 - \hat{M}_2)\hat{f}(0) \tag{3.108}$$

so that

$$\begin{aligned} \widehat{var}(\hat{D}_1 - \hat{D}_2) &= (\hat{D}_1 - \hat{D}_2)^2 \left[\frac{\widehat{var}(\hat{M}_1 - \hat{M}_2)}{(\hat{M}_1 - \hat{M}_2)^2} + \frac{\widehat{var}\{\hat{f}(0)\}}{\{\hat{f}(0)\}^2} \right] \\ &= \{\hat{f}(0)\}^2 \widehat{var}(\hat{M}_1 - \hat{M}_2) + (\hat{M}_1 - \hat{M}_2)^2 \widehat{var}\{\hat{f}(0)\} \end{aligned} \tag{3.109}$$

where

$$\widehat{var}(\hat{M}_1 - \hat{M}_2) = \widehat{var}(\hat{M}_1) + \widehat{var}(\hat{M}_2) \tag{3.110}$$

and

$$\widehat{var}(\hat{M}_i) = \hat{M}_i^2 \left[\frac{\widehat{var}(n_i)}{n_i^2} + \frac{\widehat{var}\{\hat{E}_i(s)\}}{\{\hat{E}_i(s)\}^2} \right], \quad i = 1, 2 \tag{3.111}$$

Note that the second form of eqn (3.109) still applies when $(\hat{M}_1 - \hat{M}_2) = 0$, whereas the first form leads to a ratio of zero over zero.

Inference can now proceed as before, either with additional applications of Satterthwaite's approximation in conjunction with eqn (3.100), if an approximate t-statistic is required, or more usually, by straightforward application of eqn (3.102).

3.6.6 *A finite population correction factor*

We denote the size of the surveyed area within distance w of the line or point by a. If the size of the study area is A, a known proportion a/A is sampled. Moreover, there is a finite population of objects, N, in the area. Thus the question arises of whether a finite population correction (fpc) adjustment should be made to sampling variances. We give here a few thoughts on this matter.

Assume that there is no stratification (or that we are interested in results for a single stratum). Then for strip transect or plot sampling, fpc $= 1 - a/A$. The adjusted variance of $\hat{N}$ is

$$var(\hat{N}) \cdot (1 - a/A) \tag{3.112}$$

where $var(\hat{N})$ is computed from infinite population theory. In distance sampling, not all the objects are detected in the sampled area a, so that the fpc differs from $1 - a/A$. Also, no adjustment is warranted to $var(\hat{P}_a)$ because this estimator is based on the detection distances, which conceptually arise from an infinite population of possible distances, given random placement of lines or points, or different choices of sample period.

Consider first the case where objects do not occur in clusters, and the following simple formula applies:

$$\hat{N} = A \cdot \frac{n}{a \cdot \hat{P}_a} \tag{3.113}$$

for which

$$[cv(\hat{N})]^2 = [cv(n)]^2 + [cv(\hat{P}_a)]^2 \tag{3.114}$$

The fpc is the same whether it is applied to coefficients of variation or variances. Heuristic arguments suggest that the fpc might be estimated by $1 - n/\hat{N}$, or equivalently by $1 - (a \cdot \hat{P}_a)/A$. In the case of a census of sample plots (or strips), $\hat{P}_a \equiv 1$ and the correct fpc is obtained. For the above simple case of distance sampling, $cv(\hat{N})$ corrected for finite population sampling is

$$[cv(\hat{N})]^2 = [cv(n)]^2 \cdot \left[1 - \frac{a \cdot \hat{P}_a}{A}\right] + [cv(\hat{P}_a)]^2 \tag{3.115}$$

This fpc is seldom large enough to make any difference. When it is, then the assumptions on which it is based are likely to be violated. For the correction $1 - (a \cdot \hat{P}_a)/A$ to be valid, the surveyed areas within distance w of each line or point must be non-overlapping. Further, it must be assumed that an object cannot be detected from more than one line or point; if objects are mobile, the fpc $1 - n/\hat{N}$ is inappropriate unless movement from one strip to another is unlikely within the duration of the survey.

If objects occur in clusters, correction is more complicated. First consider when there is no size bias in the detection probability. The above result still applies to $\hat{N}_s$, the estimated number of clusters. However, the number of individuals is estimated as

$$\hat{N} = A \cdot \frac{n}{a \cdot \hat{P}_a} \cdot \bar{s} \tag{3.116}$$

Inference about N is limited to the time when the survey is done, hence to the actual individuals then present. If all individuals were counted ($P_a = 1$), $var(\hat{N})$ should be zero; hence a fpc should be applied to $\bar{s}$ and conceptually, it should be $1 - (n \cdot \bar{s})/\hat{N} = 1 - (a \cdot \hat{P}_a)/A$. Thus for this case we have

$$[cv(\hat{N})]^2 = [\{cv(n)\}^2 + \{cv(\bar{s})\}^2] \cdot \left[1 - \frac{a \cdot \hat{P}_a}{A}\right] + [cv(\hat{P}_a)]^2 \tag{3.117}$$

Considerations are different for inference about $E(s)$. Usually one wants the inference to apply to the population in the (recent) past, present and (near) future, and possibly to populations in other areas as well. If this is the case, $var(\bar{s})$ should not be corrected using the fpc.

Consider now the case of clusters with size-biased detection. The fpc applied to the number of clusters is as above. For inference about $\hat{N}$, the fpc applied to the variance of $\hat{E}(s)$ is still $1 - (n \cdot \bar{s})/\hat{N}$, which is now equal to

$$1 - \frac{\bar{s}}{\hat{E}(s)} \cdot \frac{a \cdot \hat{P}_a}{A} \tag{3.118}$$

Thus the adjusted coefficient of variation of $\hat{N}$ is given by

$$[cv(\hat{N})]^2 = [cv(n)]^2 \cdot \left[1 - \frac{a \cdot \hat{P}_a}{A}\right] + [cv(\hat{P}_a)]^2 + [cv\{\hat{E}(s)\}]^2 \cdot \left[1 - \frac{\bar{s}}{\hat{E}(s)} \cdot \frac{a \cdot \hat{P}_a}{A}\right] \tag{3.119}$$

in the case of size-biased detection of clusters.

We reiterate that these finite population corrections are seldom worth making.

3.7 Stratification and covariates

Two methods of handling heterogeneity in data, and of improving precision and reducing bias of estimates, are stratification and inclusion of covariates in the analysis. Stratification might be carried out by geographic region, environmental conditions, cluster size, time, animal behaviour, detection cue, observer or many other factors. Different stratifications may be selected for different components of the estimation equation. For example, reliable estimation of $f(0)$ or $h(0)$ requires that sample size is quite large. Fortunately, it is often reasonable to assume that these parameters are constant across geographic strata. By contrast, encounter rate or cluster size may vary appreciably across strata, but can be estimated with low bias from small samples. In this case, reliable estimates can be obtained for each geographic stratum by estimating $f(0)$ or $h(0)$ from distance data pooled across strata, and estimating other parameters (usually encounter rate and mean cluster size) individually by stratum, although it may prove necessary to stratify by say cluster size or environmental conditions when estimating $f(0)$ or $h(0)$. In general, different stratifications may be needed for each component of eqn (3.3).

Post-stratification refers to stratification of the data after they have been collected and examined. This practice is often useful, for example to stratify by cluster size or environmental conditions, but care must be taken to avoid bias. For example, if geographic strata are defined to separate areas for which encounter rate was high from those for which it was low, and estimates are given separately for these strata, there will be a tendency to overestimate density in the high encounter rate stratum, and underestimate density in the low encounter rate stratum. Variance will be underestimated in both strata. If prior to the survey, there is knowledge of relative density, geographic strata should be defined when the survey is designed, so that density is relatively homogeneous within each stratum. Survey effort might then be relatively greater in strata for which density is higher (Section 7.2.2.3).

Variables such as environmental conditions, time of day, date or cluster size might enter the analysis as covariates rather than stratification factors.

If the number of potential covariates is large, they might be reduced in some way, for example through stepwise regression or principal components regression. To carry out a covariate analysis, an appropriate model must be defined. For example, the scale parameter of a model for the detection function might be replaced by a linear function of parameters:

$$\beta_0 + \beta_1 \cdot X_{1i} + \beta_2 \cdot X_{2i} + \cdots \tag{3.120}$$

where X_{1i} might be sea state (Beaufort) at the time of detection i, X_{2i} might be cluster size for the detection, and so on.

3.7.1 *Stratification*

The simplest example of a stratified analysis is to estimate abundance independently within each stratum. A more parsimonious approach is to assume that at least one parameter is common across strata, or a subset of strata, an assumption that can be tested. Consider a point transect survey for which points were located in V geographic strata of areas A_v, $v = 1, \ldots, V$. Suppose we assume there is no size bias in detected clusters, and abundance estimates are required by stratum. Suppose further that data are sparse, so that $h(0)$ is estimated by pooling detection distances across strata. From eqn (3.3) with $a \cdot P_a = 2\pi k/h(0)$, we obtain

$$\hat{D}_v = \frac{n_v \cdot \hat{h}(0) \cdot \bar{s}_v}{2\pi k_v} = \frac{\hat{h}(0) \cdot \hat{M}_v}{2\pi} \tag{3.121}$$

where $\hat{M}_v = n_v \bar{s}_v / k_v$ for stratum v. Mean density $\hat{D}$ is then the average of the individual estimates, weighted by the respective stratum areas A_v:

$$\hat{D} = \frac{\sum_v A_v \hat{D}_v}{A} \tag{3.122}$$

with $A = \sum_v A_v$. The variance of any $\hat{D}_v$ may be found from eqn (3.4). However, to estimate $var(\hat{D})$, care must be taken, since one component of the estimation equation is common to all strata (see also Section 3.6.5). The correct equation is

$$\widehat{var}(\hat{D}) = \hat{D}^2 \cdot \left\{ \frac{\widehat{var}(\hat{M})}{\hat{M}^2} + \frac{\widehat{var}[\hat{h}(0)]}{[\hat{h}(0)]^2} \right\} \tag{3.123}$$

where

$$\hat{M} = \frac{\sum_v A_v \hat{M}_v}{A} \tag{3.124}$$

and

$$\widehat{var}(\hat{M}) = \frac{\sum_v A_v^2 \cdot \widehat{var}(\hat{M}_v)}{A^2} \tag{3.125}$$

with

$$\widehat{var}(\hat{M}_v) = \hat{M}_v^2 \cdot \left\{ \frac{\widehat{var}(n_v)}{n_v^2} + \frac{\widehat{var}(\bar{s}_v)}{\bar{s}_v^2} \right\} \tag{3.126}$$

Thus the estimation equation has been separated into two components, one of which ($\hat{M}_v$) is estimated independently in each stratum, and the other of which is common across strata. Population abundance is estimated by $\hat{N} = A \cdot \hat{D} = \sum A_v \hat{D}_v$, with $\widehat{var}(\hat{N}) = A^2 \cdot \widehat{var}(\hat{D})$. Further layers of stratification might be superimposed on a design of this type. In the above example, each stratum might be covered by more than one observer, or several forests might be surveyed, and a set of geographic strata defined in each. Provided the principle of including each independent component of the estimation equation just once in the variance expression is adhered to, the above approach is easily generalized.

If the Satterthwaite method of calculating approximate degrees of freedom is adopted, it can be applied in the above example by first calculating the df associated with each of the $\hat{M}_v$, using eqn (3.75). The df for $\hat{M}$ is then computed using the Satterthwaite approximation for a sum:

$$df = \frac{(\sum \widehat{var}(\hat{M}_v))^2}{\sum [\widehat{var}(\hat{M}_v)]^2 / df_v}$$

where df_v is the degress of freedom for M_v. Finally, df for $\hat{D}$ are found by applying eqn (3.75) again, this time with components corresponding to $\hat{M}$ and $\hat{h}(0)$.

The areas A_v are weights in the above expressions. For many purposes, it may be appropriate to weight by effort rather than area. For example, suppose two observers independently survey the same area in a line transect study. Then density within the study area may be estimated separately from the data of each observer (perhaps with at least one parameter assumed to be common between the observers), and averaged by weighting the respective estimates by length of transect covered by the respective observers. Note that in this case, an average of the two abundance estimates from each stratum is required, rather than a total. If stratification is by factors such as geographic region, cluster size, animal behaviour or detection cue, then the strata correspond to mutually exclusive components of the population and the estimates should be summed, whereas if stratification is by factors such as environmental conditions, observer, time or date (assuming no migration), then each stratum provides an estimate of the whole population, so that an average is appropriate.

Note that the stratification factors for each component of estimation may be completely different provided the components are combined with care. As a general guide, the parameter $f(0)$ or $h(0)$ should only be estimated separately by stratum if there is evidence that the parameter varies between strata, and some assessment should be made of whether the number of strata can be reduced. This policy is recommended since estimation of the parameter is unreliable if sample size is not large. Encounter rate and mean cluster size on the other hand may be reliably estimated from small samples, so if there is doubt, these parameters should be estimated independently within each stratum. Further, if abundance estimates are required by stratum, then both encounter rate and mean cluster size should normally be estimated by stratum. If all parameters can be assumed common across strata, such as observers of equal ability covering a single study area at roughly the same time, stratification is of no benefit. Also, in the special case that strata correspond to different geographic regions, effort per unit area is the same in each region, the parameter $f(0)$ or $h(0)$ can be assumed constant across regions, and estimates are not required by region, stratification is unnecessary. Proration of a total abundance estimate by the area of each region is seldom satisfactory. An example in which stratification was used in a relatively complex way to improve abundance estimation of North Atlantic fin whales is given in Section 8.5.

Further, parsimony may be introduced by noting that $var(n_v)$ is a parameter to be estimated, and $b_v = var(n_v)/E(n_v)$ is often quite stable over strata. Especially if the n_v are small, it is useful to assess whether $b_v = b$ can be assumed for all v. If it can, the number of parameters is reduced. The parameter b can be estimated by

$$\hat{b} = \frac{\sum_v \widehat{var}(n_v)}{n} \tag{3.127}$$

which is equal to the slope of the line obtained from a weighted regression through the origin of $\widehat{var}(n_v)$ on n_v. A more efficient estimate of $var(n_v)$ is then $\widehat{var}_p(n_v) = \hat{b}/n_v$. This approach is illustrated in Section 8.4. The same method might also be applied to improve the efficiency of $\widehat{var}(\bar{s}_v)$.

3.7.2 *Covariates*

Several possibilities exist for incorporating covariates. Ramsey *et al.* (1987) used the effective area, ν, as a scale parameter, and related it to covariates using a log link function:

$$\log_e(\nu) = \beta_0 + \sum_j \beta_j \cdot X_j \tag{3.128}$$

where X_j is the jth covariate. Marques and Buckland (in preparation) extended this approach for use with series models using the half-normal and hazard-rate keys. The methodology is described in detail in Buckland *et al.* (in preparation).

Drummer and McDonald (1987) considered a single covariate X, taken to be cluster size in their example, and incorporated it into detection functions by replacing y by y/X^{γ}, where γ is a parameter to be estimated. Thus the univariate half-normal detection function

$$g(y) = \exp\Big[-\frac{y^2}{2\sigma^2}\Big] \tag{3.129}$$

becomes the 'bivariate' detection function:

$$g(y|X) = \exp\Big[-\frac{y^2}{2(\sigma X^{\gamma})^2}\Big] \tag{3.130}$$

The interpretation is now that $g(y \mid X)$ is the detection probability of a cluster at distance y, given that its size is X. Drummer and McDonald proposed the following detection functions as candidates for this approach: negative exponential, half-normal, generalized exponential, exponential power series and reversed logistic. They implemented the method for the first three, although their software SIZETRAN (Drummer 1991) is prone to convergence problems. This approach is also a special case of the general methodology of Marques and Buckland (in preparation), presented in Buckland *et al.* (in preparation).

Beavers and Ramsey (1998) developed a method of incorporating covariates into detectability that allows standard, non-covariate distance sampling software to be used. They regressed the logarithm of the observed perpendicular distances on the covariates, and then adjusted those distances to the values predicted when the covariates are given their average values. This standardizes the distance data, which may now be analysed conventionally. See also Fancy (1997).

3.8 Efficient simulation of distance data

3.8.1 *The general approach*

To produce simulated distance data requires the Monte Carlo generation of sample size n, detection distances $y = x$ or r, and for the clustered case, cluster sizes s. The efficient way to do this is first to generate sample size according to some discrete distribution, $p(n)$, then generate n distances and cluster sizes based on the bivariate sampling distribution of distance and s. The alternative is first to generate spatially distributed clusters, and independently for each cluster, a cluster size s. Then determine for each cluster whether it gets detected according to some detection function, $g(y \mid s)$. This method is indirect and inefficient. The purpose of this section is twofold: to show how to simulate distance data directly and to outline explicitly how the simulated data of Chapters 4 and 5 were generated.

The following general approach is recommended. First, decide on a detection function; it will be bivariate if cluster size is to vary and there is to be size-biased detection. Otherwise, $g(y)$ depends only on distance. For the clustered case, decide on a probability distribution, $\pi(s)$, of cluster sizes in the yet-to-be-sampled population. Also select a sampling distribution for sample size, $p(n)$, such as Poisson or negative binomial. It is then possible to specify the exact parameterization of $p(n)$. To simulate data for k replicate lines or points, first generate independent sample sizes $n_1, \ldots, n_k$ according to $p(n)$. If objects do not occur in clusters, just generate n_i independent distances, $i = 1, \ldots, k$, according to the probability density function of distances, $f(y)$. This function is determined by, and computed from, $g(y)$, and the context (line or point transects). If cluster size varies but detection is independent of size, then for each generated distance y, produce independently a value of the random variable s according to $\pi(s)$.

The case of size-biased detection requires a two-step process of either generating y from its marginal density function, then s from the size-biased distribution $\pi^*(s \mid y)$, or the reverse (which we recommend): generate an observed cluster size from the marginal size-biased distribution of detections, $\pi^*(s)$, then generate y from $f(y \mid s)$. The general methodology for size-biased detection is covered in Buckland *et al.* (in preparation); here, we quote a few key results.

The distribution of n, $p(n)$, must have, at least implicitly, $E(n)$ as one of its parameters, where

$$E(n) = \frac{2LD_s}{f(0)} \quad \text{(line transects)}$$

$$E(n) = \frac{2\pi k D_s}{h(0)} \quad \text{(point transects)}$$

where $\pi = 3.14159$, and the density D_s denotes density of clusters.

We now summarize quantities that must be specified to simulate distance data. These quantities are interrelated and hence cannot be independently set; in particular, we recommend that either $E(n)$ or sampling effort (L or k) is specified, and the other quantity is computed. Constants to be specified are w and k, and for line transects, line lengths l_i, $i = 1, \ldots, k$, which sum to L. Separately specified parameters are D_s and $E(s)$, and any additional parameters in $p(n)$ other than $E(n)$. Fundamental distributions to specify are $p(n)$ and $\pi(s)$. Finally, there is the detection function, $g(y \mid s)$, which, in conjunction with $\pi(s)$, determines the sampling distributions of y and s. In general we compute $f(0)$, $h(0)$, $\pi^*(s)$, $f(x \mid s)$ and $f(r \mid s)$ numerically. For simulating line transect data, we have

$$f(0 \mid s) = \frac{1}{\int_0^w g(x \mid s)\, dx} \tag{3.131}$$

$$f(0) = \frac{1}{\sum_{s=1}^{\infty} \pi(s)/f(0 \mid s)} \tag{3.132}$$

$$\pi^*(s) = \left[\frac{f(0)}{f(0 \mid s)}\right] \pi(s) \tag{3.133}$$

and

$$f(x \mid s) = f(0 \mid s) \cdot g(x \mid s) \tag{3.134}$$

Formulae necessary in simulation of point transect data are

$$h(0 \mid s) = \frac{1}{\int_0^w r \cdot g(r \mid s)\, dr} \tag{3.135}$$

$$h(0) = \frac{1}{\sum_{s=1}^{\infty} \pi(s)/h(0 \mid s)} \tag{3.136}$$

$$\pi^*(s) = \left[\frac{h(0)}{h(0 \mid s)}\right]\pi(s) \tag{3.137}$$

and

$$f(r \mid s) = h(0 \mid s) \cdot r \cdot g(r \mid s) \tag{3.138}$$

If the distribution of n is assumed to be Poisson, then

$$p(n) = \frac{e^{-E(n)}[E(n)]^n}{n!} \tag{3.139}$$

Alternatively, a useful parameterization of the negative binomial is

$$p(n) = \frac{\Gamma(\theta + n)}{\Gamma(\theta)\Gamma(n+1)}(1-\tau)^n \cdot \tau^\theta, \quad 0 < \theta, \quad 0 < \tau < 1, \quad 0 \le n \tag{3.140}$$

which has $E(n) = \theta \cdot (1-\tau)/\tau$ and $var(n) = E(n)/\tau$.

In point transects all with a fixed observation time, τ and θ can be the same over different points (within a stratum). For line transects, the l_i usually vary, and we recommend keeping τ constant while letting θ vary by line length, because this gives coherent results: $n_1 + \cdots + n_k$ is then distributed as negative binomial with parameters τ and $\theta_1 + \cdots + \theta_k$. Thus, we can arbitrarily vary the line lengths and still have sample sizes (individually and totals) as negative binomial random variables. Under this strategy, $1/\tau$ is the variance inflation factor relative to a Poisson (random) spatial distribution of objects. For the case of $\tau = 1$, use the Poisson distribution for Monte Carlo generation of sample sizes.

Consider the line transect case with one long line (i.e. ignore replicate lines) of length L and objects not clustered. We would first specify $g(x)$, then compute $f(0)$, by numerical integration if need be. It is important to keep straight the units of measurement in a simulation, because with real data, detection distances and line length are often in different units, such as metres and kilometres. Also, the units of $f(0)$ are the reciprocal of the distance units used for x. Say we get $f(0) = 10$, with units on x taken as kilometres. Then effective strip half-width is 0.1 km or 100 m.

In this hypothetical example, next we specify $D = 2$ objects/km^2 and $E(n) = 70$. Now we determine L from $E(n) = 2LD/f(0)$: $70 = 2 \cdot 2 \cdot L/10$, or $L = 175\,\text{km}$. If we want n to be Poisson, then we generate it from a Poisson with mean 70. Given n, generate $x_1, \ldots, x_n$ from the pdf $f(x) = f(0) \cdot g(x)$.

We illustrate the approach in more detail for the simulation of data for the examples of Chapters 4 and 5.

3.8.2 *The simulated line transect data of Chapter 4*

The half-normal bivariate detection function may be expressed as

$$g(x \mid s) = \exp\Big[-\frac{1}{2}\Big(\frac{x}{\sigma(s)}\Big)^2\Big] \tag{3.141}$$

where we model the scale parameter, σ, as a function of cluster size (cf. Quinn 1979; Drummer and McDonald 1987; Ramsey *et al.* 1987; Otto and Pollock 1990). In particular, the form $\sigma(s) = \sigma \cdot s^{\alpha}$ has been much used. This is a reasonable model for data analysis; however, for simplicity of theory, we used a linear form for line transects:

$$\sigma(s) = \sigma_0\Big[1 + b \cdot \frac{s - E(s)}{E(s)}\Big] \tag{3.142}$$

subject to the constraint

$$b \le \frac{E(s)}{E(s) - 1}$$

and assuming $0 \le b$, although to a limited extent, negative values of b are mathematically possible. This form is also suggested because it puts the problem into the framework of generalized linear models (McCullagh and Nelder 1989). The case $b = 0$ corresponds to detection probability independent of cluster size. For $w = \infty$, $f(0 \mid s)$ and $f(x \mid s)$ are closed form:

$$\begin{aligned} f(0 \mid s) &= \sqrt{\frac{2}{\pi}} \cdot \frac{1}{\sigma(s)} \\ f(x \mid s) &= \sqrt{\frac{2}{\pi}} \cdot \frac{1}{\sigma(s)} \cdot \exp\Big[-\frac{1}{2}\Big(\frac{x}{\sigma(s)}\Big)^2\Big] \end{aligned} \tag{3.143}$$

Applying eqn (3.132), we have

$$f(0) = \sqrt{\frac{2}{\pi}} \cdot \frac{1}{\sum_{s=1}^{\infty} \pi(s) \cdot \sigma(s)} = \sqrt{\frac{2}{\pi}} \cdot \frac{1}{E\{\sigma(s)\}} \tag{3.144}$$

The form of $\sigma(s)$ in eqn (3.142) is convenient because we can explicitly evaluate its expectation with respect to $\pi(s)$; in fact, $E\{\sigma(s)\} = \sigma_0$ for any value of the parameter b. Thus, for any extent of size bias under this model,

$$f(0) = \sqrt{\frac{2}{\pi}} \cdot \frac{1}{\sigma_0} \tag{3.145}$$

From eqn (3.133),

$$\begin{aligned} \pi^*(s) &= \left[\frac{f(0)}{f(0 \mid s)}\right] \pi(s) = \frac{\sigma(s)}{\sigma_0} \pi(s) \\ &= \left[1 + b \cdot \frac{s - E(s)}{E(s)}\right] \pi(s), \quad s = 1, 2, \ldots \end{aligned} \tag{3.146}$$

The expected value of s in the sample of detected clusters is

$$E^*(s) = \sum s \cdot \pi^*(s) = E(s) + b \cdot \frac{var(s)}{E(s)} \tag{3.147}$$

A simple choice for $\pi(s)$ is to let $s - 1$ have a Poisson distribution with mean $E(s) - 1$:

$$\pi(s) = \frac{e^{-[E(s)-1]} \cdot [E(s) - 1]^{s-1}}{(s-1)!}, \quad s = 1, 2, \ldots \tag{3.148}$$

We used this $\pi(s)$ to generate the example data of Chapter 4. Using eqn (3.146), we created a table of values of $\pi^*(s)$ for specified $E(s)$, b and σ_0. The detected cluster size was then generated by standard Monte Carlo methods. Given that value of s, x was generated according to the distribution of eqn (3.143). This was done by generating a standard normal deviate z, and calculating $x = \sigma(s) \cdot z$.

The distribution of sample detections chosen for the Chapter 4 examples was negative binomial, with $\tau = 0.4$, so that the variance inflation factor is 2.5. The choice of 12 replicate lines was arbitrary. Other choices made: $\sigma_0 = 10\,\text{m}$ and $E(n) = 96$. It was then convenient to use $L = 48\,\text{km}$, and to keep the encounter rate constant over replicate lines (whose lengths varied). These choices and decisions produce, by design, the result

$$E(n_i) = \theta_i \frac{1-\tau}{\tau} = 2l_i, \quad i = 1, \ldots, 12 \tag{3.149}$$

which implies $\theta_i = 64l_i/48$.

Equation (3.144) yields $f(0) = 0.0798\,\text{m}^{-1}$, or $79.8\,\text{km}^{-1}$. Hence density of clusters is

$$D = \frac{96 \times 79.8}{2 \times 48} = 79.8\,\text{clusters/km}^2 \tag{3.150}$$

The simulation was set up in such a way that density of individuals is $79.8 \cdot E(s) = 239.4$ regardless of the value of b; b only determines the degree of size bias. Values of b used were 0 (no size bias) and 1, both with the same set of $n_1, \ldots, n_{12}$.

3.8.3 *The simulated size-biased point transect data of Chapter 5*

The generation of the simulated data for point transects with size bias used the same half-normal bivariate detection function as for the line transect case:

$$g(r \mid s) = \exp\left[-\frac{1}{2}\left(\frac{r}{\sigma(s)}\right)^2\right] \tag{3.151}$$

However, the relevant formulae are now eqns (3.135)–(3.138). In particular, we have for $w = \infty$,

$$h(0 \mid s) = \frac{1}{[\sigma(s)]^2} \tag{3.152}$$

and

$$f(r \mid s) = \frac{r}{[\sigma(s)]^2} \cdot \exp\left[-\frac{1}{2}\left(\frac{r}{\sigma(s)}\right)^2\right] \tag{3.153}$$

Using eqn (3.136) with the above expression for $h(0 \mid s)$, we have

$$h(0) = \frac{1}{\sum_{s=1}^{\infty} \pi(s) \cdot \{\sigma(s)\}^2} = \frac{1}{E[\{\sigma(s)\}^2]} \tag{3.154}$$

Thus we choose to parameterize $\sigma(s)$ as

$$\{\sigma(s)\}^2 = \sigma_0^2 \cdot \left[1 + b \cdot \frac{s - E(s)}{E(s)}\right] \tag{3.155}$$

subject to the constraint

$$b \leq \frac{E(s)}{E(s) - 1} \tag{3.156}$$

This model gives

$$h(0) = \frac{1}{\sigma_0^2} \tag{3.157}$$

and

$$h(0 \mid s) = \frac{1}{\sigma_0^2 \cdot [1 + b \cdot (s - E(s))/E(s)]} \tag{3.158}$$

Hence, from eqn (3.137),

$$\begin{aligned} \pi^*(s) &= \left[\frac{h(0)}{h(0 \mid s)}\right] \pi(s) \\ &= \frac{\sigma(s)}{\sigma_0} \cdot \pi(s) \\ &= \left[1 + b \cdot \frac{s - E(s)}{E(s)}\right] \pi(s), \quad s = 1, 2, \ldots \end{aligned} \tag{3.159}$$

which is identical to $\pi^*(s)$ for the line transect case in eqn (3.146).

For the size-biased example of Chapter 5, we choose $\pi(s)$ to be the geometric distribution,

$$\pi(s) = (1-\beta)^{s-1} \cdot \beta, \quad 0 < \beta < 1, \ \ s = 1, 2, \ldots \tag{3.160}$$

For this distribution, $E(s) = 1/\beta$. From the above expression for $\pi^*(s)$, we have for this example

$$\pi^*(s) = \left[1 + b \cdot \frac{s - E(s)}{E(s)}\right] \cdot (1-\beta)^{s-1} \cdot \beta, \quad s = 1, 2, \ldots \tag{3.161}$$

Note that $b = 1$ corresponds to considerable size bias and gives the simple form

$$\pi^*(s) = s \cdot (1-\beta)^{s-1} \cdot \beta^2, \quad s = 1, 2, \ldots \tag{3.162}$$

For the example in Section 5.8, $E(s) = 1.85$, for which $\beta = 0.54054$, and $b = 0.75$. These values serve to specify $\pi^*(s)$ completely. Also, we set $\sigma_0 = 30\,\text{m}$, which, together with $b = 0.75$, serves to specify $g(r \mid s)$, $h(0 \mid s)$ and $f(r \mid s)$. The latter is given by eqn (3.153); that density function has cumulative distribution function

$$F(r \mid s) = 1 - \exp\left[-\frac{1}{2}\left(\frac{r}{\sigma(s)}\right)^2\right] \tag{3.163}$$

Consequently, for this example, a random r, given an s drawn from eqn (3.161), was produced as

$$r = \sigma(s) \cdot \sqrt{-2 \cdot \log_e(1-u)} \tag{3.164}$$

where u is a uniform pseudo-random variable on the interval 0 to 1.

Counts were generated from the negative binomial distribution with variance inflation (dispersion) factor set at 2.65, so that $\tau = 0.37736$. The encounter rate per point was set at 1.6, and k was set at 60 points, giving $E(n) = 96$. In terms of eqn (3.140) and associated results, this means that the count at a given point was generated by specifying

$$E(n_i) = \frac{\theta(1-\tau)}{\tau} = 1.6 \tag{3.165}$$

and

$$var(n_i) = \frac{E(n_i)}{\tau} = 2.65\, E(n_i), \quad i = 1, \ldots, k \tag{3.166}$$

Hence, given $\tau = 0.37736$, we obtain $\theta = 0.52416$. As a consequence of the choice of model and parameters in this example, the density of clusters is

$$\frac{96 \times [1/30]^2}{2\pi \times 60} \times 1\,000\,000 = 283\,\text{clusters/km}^2 \tag{3.167}$$

and density of individuals is $1.85 \times 2.83 = 523\,\text{objects/km}^2$.

3.8.4 *Discussion*

We have recommended simulating the pair (y, s) by generating s from its marginal distribution $\pi^*(s)$, and then y from the conditional distribution, $f(y \mid s)$. The algebra for this was straightforward in the above two examples. Consider, however, the reverse process for the point transect example above: generate r from $f(r)$ then s from $\pi^*(s \mid r)$. The relevant formulae are

$$f(r) = \frac{1}{\sigma_0^2}\left[\sum_{s=1}^{\infty}(1-\beta)^{s-1} \cdot \beta \cdot r \cdot \exp\left\{-\frac{1}{2}\left(\frac{r}{\sigma(s)}\right)^2\right\}\right] \tag{3.168}$$

and

$$\pi^*(s \mid r) = \frac{(1-\beta)^{s-1} \cdot \exp[-\frac{1}{2}(r/\sigma(s))^2]}{\sum_{s=1}^{\infty}(1-\beta)^{s-1} \cdot \exp[-\frac{1}{2}(r/\sigma(s))^2]} \tag{3.169}$$

Use of these formulae would entail much more computing than the use of $\pi^*(s)$ and $f(r \mid s)$. Heuristically, this is because there is only a finite (and small, usually) number of possible values for s, whereas infinitely many values of r can occur. Therefore, with each new r, one must recompute the entire distribution $\pi^*(s \mid r)$ before s can be generated.

In some real applications, cluster sizes potentially range from one to thousands, for example dolphin surveys on some species. To simulate the essence of such applications, it is not necessary for s to vary over this set of values. A set of a hundred (or fewer) values should suffice (e.g. s taking values 1–10, 15, 20, 30, 50, 75, 100–900 by 100, 1000–5000 by 500). Keeping the range set of s small will greatly speed up simulations.

Closed-form expressions underlying simulations will be the exception. Even in the above line transect examples, if we take w as finite, numerical integration must be used to find the necessary quantities given by eqns (3.131)–(3.134). Expect to use numerical integration; fortunately for purposes of simulation, the numerical methods need not be highly sophisticated.

Sometimes we only want to explore statistical properties of estimators of $f(0)$, $h(0)$, $g(x \mid s)$, $g(r \mid s)$ and $E(s)$, and not properties of $\hat{D}_s$ and $\hat{D}$. In this event, it is not necessary to generate a random sample size n for each replicate. In fact, it is better to fix n and do say 1000 replicates at that n, and repeat the process over a set of values of n.

3.9 Exercises

1. If all the assumptions of a line transect experiment are met except that $g(0) < 1$, and an analysis is conducted assuming that $g(0) = 1$, then what is the approximate expected value of $\hat{D}$ as a function of $g(0)$ and D?

2. A line transect survey of objects that occur singly is conducted, yielding $cv(n) = \sqrt{3}/10$ ($= 0.1732$) and $cv\{\hat{f}(0)\} = 0.1$.

(a) What is the value of $cv(\hat{D})$?

(b) What is $\widehat{cv}$ $(\hat{D})$?

(c) What percentage of the variance in $\hat{D}$ is due to variation in n?

3. Suppose perpendicular distances $x_1, \ldots, x_n$ are recorded during a line transect survey. The biologist wishes to fit the following negative exponential detection function to the data.

$$g(x) = \exp(-\lambda x), \quad 0 \leq x < \infty$$

(a) Derive the maximum likelihood estimator $\hat{f}(0)$ of $f(0)$, the density function of perpendicular distances evaluated at distance zero.

(b) Derive an estimator for the variance of $\hat{f}(0)$.

(c) Assume that n follows a Poisson distribution. Using the result $\hat{D} = n\,\hat{f}(0)/2L$ and the delta method, or otherwise, obtain expressions for the estimators $\hat{D}$ and $\hat{var}(\hat{D})$.

4. The negative exponential model was derived in the early days of development of line transect methodology:

$$g(x) = \exp(-x/\mu), \quad 0 \leq x < \infty$$

The maximum likelihood estimator of μ is $\hat{\mu} = \bar{x}$. Under this model, it may be shown that density is given by $D = n/(2L\mu)$ and the effective strip half-width is μ.

Some time ago, researchers used this model to estimate rabbit densities. They found that $\hat{D}$ and $\hat{\mu}$ showed a strong negative correlation, and concluded that probability of detection was strongly influenced by rabbit density. This conclusion is a fallacy. Explain the flaw in the researchers' argument.

5. (a) Derive the result from point transect sampling:

$$D = \frac{E(n)\,h(0)}{2\pi k}$$

where $D =$ animal density, $n =$ number of animals detected, $h(0) =$ the slope of the density function of sighting distances, evaluated at distance zero and $k =$ number of points surveyed.

(b) Suppose sighting distances are squared prior to analysis. Show that the transformation $s = r^2$ yields:

$$D = \frac{E(n) f_s(0)}{\pi k}$$

where $f_s(s)$ is the density function of s.

(c) Why does the result of part (b) allow line transect software to be used to analyse point transect data? What is the disadvantage of this approach?

6. Describe, with the aid of diagrams, the techniques of line and point transect sampling, and list the assumptions required by each method. Outline circumstances in which (a) line transect sampling would be preferable to point transect sampling, and (b) point transect sampling would be preferable to line transect sampling. Show that the number of animals detected in a line transect survey, denoted by n, may be converted to an estimate of animal density D using the equation

$$\hat{D} = \frac{n\hat{f}(0)}{2L}$$

where L is the total length of transects and $\hat{f}(0)$ is the estimated density function of perpendicular distances x evaluated at $x = 0$. Explain, with the aid of diagrams or otherwise, why the equivalent derivation for point transects is less simple.

7. In point transect sampling, the half-normal detection curve is given by

$$g(r) = \exp\left(\frac{-r^2}{2\,\sigma^2}\right), \quad 0 \le r < \infty$$

(a) Explain intuitively, with the aid of a diagram, why the probability density function of detection distances, $f(r)$, is proportional to $r\,g(r)$. (A rigorous proof is not required, but can be used if preferred.)

(b) Hence or otherwise show that for the half-normal model,

$$f(r) = \frac{r\,g(r)}{\sigma^2}$$

(c) Find the maximum likelihood estimator of σ^2.

(d) Hence show that animal density D is estimated by

$$\hat{D} = \frac{n^2}{\pi\,k\,\sum_{i=1}^{n} r_i^2}$$

where r_i is the ith detection distance, $i = 1, \ldots, n$, and k is the number of points sampled.

(e) Give two reasons why the above model is not suitable as a general model for the analysis of point transect data sets.

4
Line transects

4.1 Introduction

The purpose of this chapter is to illustrate the application of the theory of Chapter 3 to line transect data, and to present the strategies for analysis outlined in Section 2.5. In general, the parameter $f(0)$ in a line transect analysis does not have a closed form estimator. That is, for most models $\hat{f}(0)$ cannot be expressed as a formula that can be evaluated by simple substitution of the data. Instead, numerical methods are required. We therefore rely on specialized computer software to analyse distance sampling data. With such software, many more analysis options are possible and more reliable estimates can be obtained than if analysis is attempted without the aid of a computer.

This chapter uses a series of examples in which complexities are progressively introduced. The examples come from simulated data for which the parameters are known; this makes comparisons between estimates and true parameters possible. However, in every other respect, the data are treated as any real data set undergoing analysis, where the parameters of interest are unknown. A simple data set, for which each object represents an individual animal, plant, nest or whatever, is first introduced. Truncation of the distance data, modelling the spatial variation of objects to estimate variance in sample size, grouping of data, and model selection philosophy and methods are then addressed. Once an adequate model has been selected, we focus on statistical inference given that model, to illustrate estimation of density and measures of precision. Finally, the objects are allowed to be clusters (coveys, schools and flocks). Cluster size is first assumed to be independent of distance and then allowed to depend on distance.

The example data are chosen to be 'realistic' from a biological standpoint. The data (sample size, distances and cluster sizes) are generated stochastically and simulate the case where the assumptions of line transect sampling are true. Thus, no objects went undetected on the line ($g(0) = 1$), no movement occurred prior to detection, and data were free of measurement error (e.g. heaping at zero distance). In addition, the sample size n was adequate. The assumptions, and survey design to ensure that they are met, are discussed in Chapter 7. Examples illustrating analysis of real data where some of these assumptions fail are provided in Chapter 8. The example data of this chapter are analysed using various options of program Distance.

4.2 Example data

This simulated example comprises an area of size A, whose boundary is well defined, and traversed by a sample of 12 parallel lines, randomly spaced, or systematically spaced with a random start. The area is irregularly shaped, so that the lines, which run from boundary to boundary, are of unequal lengths, $(l_1, l_2, \ldots, l_{12})$. We assume that no stratification is required and that the population was sampled once by a single observer using an exacting field protocol; hence the key assumptions have been met. The distance data, recorded in metres, were taken without a fixed transect width (i.e. $w = \infty$), and ungrouped, to allow analysis of the data as either grouped or ungrouped. The detection function $g(x)$ was a simple half-normal with $\sigma = 10\,\text{m}$, giving $f(0) = \sqrt{2/(\pi\sigma^2)} = 0.079788\,\text{m}^{-1}$. The n_i were drawn from a negative binomial distribution such that the spatial distribution of objects was somewhat clumped (i.e. $var(n) > n$). Specifically, $var(n_i \mid l_i) = 2E(n_i \mid l_i)$.

The total length ($L = \sum l_i$) of the 12 transects was 48 000 m and $n = 105$ objects were detected. $E(n) = 96$, so that rather more were observed than expected. The true density is

$$D = \frac{N}{A} = \frac{E(n) \cdot f(0)}{2L} \text{ objects/m}^2 \tag{4.1}$$

where all measurements are in metres. To convert density from numbers per m^2 into numbers per km^2, multiply D by 1 000 000. The true density is known in this simulated example to be 79.788 objects per km^2, or roughly 80 objects/km^2.

The first step is to examine the distance data by plotting histograms using a range of choices of distance categories. It is often informative to plot a histogram with many fine intervals (Fig. 4.1). Here one can see the presence of a broad shoulder, no evidence of heaping, and no indication of evasive movement prior to detection; the data appear to be 'good'.

4.3 Truncation

4.3.1 *Right-truncation*

Inspection of the histogram in Fig. 4.1 shows the existence of an extreme observation or 'outlier' at around 35 m. If data are untruncated in the field, a useful rule of thumb at the analysis stage is to truncate around 5% of the data from the right-hand tail of the detection function. Here, the six most extreme distances are 19.27, 19.42, 19.44, 19.46, 21.21 and 35.82 m.

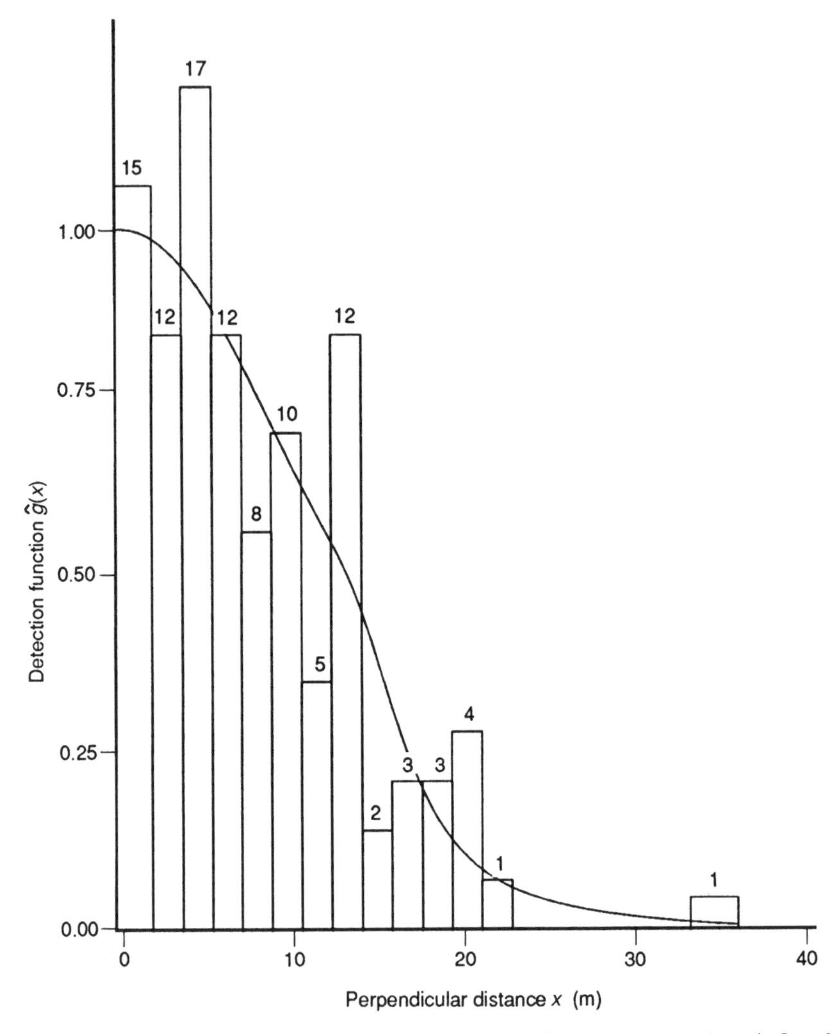

Fig. 4.1. Histogram of the example data using 20 distance categories. A fitted hazard-rate key with one cosine adjustment term is shown as a candidate model for the detection function, $g(x)$.

Thus, a reasonable choice for w might be 19 m. In fact, for these data, there is likely to be little benefit in truncating the distances between 19 and 20 m, as these are clearly not extreme, so that we might choose to take $w = 20$ m. An alternative is to fit a reasonable preliminary model to the data, compute $\hat{g}(x)$ to find the value of x such that $\hat{g}(x) = 0.15$, and use this value of x as the truncation point for further analysis.

As an illustration, the half-normal key function was fitted to the ungrouped, untruncated data and found to fit well. This approach suggests that $w = 19\,\text{m}$ is a reasonable choice, as the half-normal model yields $\hat{g}(19) = 0.13$, slightly below the suggested value of 0.15. The deletion of outliers is useful because these extreme observations provide little information for estimating $f(0)$, the density function at $x = 0$, but can be difficult to model. The series expansions often require additional adjustment terms to fit the few data in the tail of the distance distribution, which may unnecessarily increase the sampling variance of the density estimate. In this example, both truncation rules suggest a value around $w = 19\,\text{m}$, leaving $n = 99$ observations. For the rest of this chapter we will concentrate on results with $w = 19\,\text{m}$, but estimates corresponding to no truncation will also be given and compared. The choice of truncation point is not a critical decision for these example data where all the assumptions are met and the true detection function is simple.

For the true model, the quantity $E(n) \cdot f(0)$ remains unchanged as the truncation point is varied. Consequently, for good data (i.e. data satisfying the assumptions) and a reasonable model for $g(x)$, the product $n \cdot \hat{f}(0)$ is quite stable over a range of truncation points. With increased truncation, n decreases, but $\hat{f}(0)$ increases to compensate. The estimate $n \cdot \hat{f}(0)$ under the half-normal model is 8.477 if data are truncated at 19 m, and 8.417 without truncation.

Truncation of the data at $w = 19\,\text{m}$ removed only six detections. If a series expansion model is used, up to three fewer parameters are required to model the truncated data than the untruncated data (Table 4.1). (Note that the truncation distance w supplied to Distance must be finite; by 'untruncated data', we mean that w was at least as large as the largest recorded distance.) Outliers in the right tail of the distance distribution required additional adjustment parameters and the inclusion of such terms tended to increase the sampling variance of $\hat{f}(0)$ and hence $\hat{D}$ when a robust but poor model was used (Table 4.2).

Truncation of the distance data for analysis deletes outliers and facilitates modelling of the data. However, as some data are discarded, one might ask if the uncertainty in $\hat{D}$ increases. First, this issue is examined when the true model is known and used (i.e. the half-normal in this case). The coefficient of variation increased about 1% when the data were truncated at 19 m relative to untruncated (Table 4.2). Thus, little precision was lost due to truncation if the data were analysed under the model used to generate the data. Of course, one never knows the true detection function except for computer simulation examples.

When series expansion models are used for the analysis of the example data, the uniform key function with either cosine or polynomial adjustments gives a smaller coefficient of variation when the data are truncated (Table 4.2). This small increase in precision is because only one parameter was required for a good model fit when $w = 19\,\text{m}$, whereas two to four parameters were required to fit the untruncated data

Table 4.1. Summary of AIC values for two truncation values (w) for the example data analysed as ungrouped and at three different groupings (five groups of equal width, 20 groups of equal width and five unequal groups such that the number detected was nearly equal in each group). For each analysis, the model with the smallest AIC is indicated by an asterisk (AIC values cannot be compared across analyses with different grouping or different truncation distances)

		$w = 19\,\text{m}$			w = largest observation		
		No. of parameters			No. of parameters		
Data type	Model (key + adjustment)	Key	Adjustments	AIC	Key	Adjustments	AIC
Ungrouped	Uniform + cosine	0	1	562.98	0	2	636.48*
	Uniform + polynomial	0	1	563.28	0	4	638.18
	Half-normal + Hermite	1	0	562.60*	1	0	636.98
	Hazard-rate + cosine	2	0	565.22	2	0	639.16
Grouped (5 equal)	Uniform + cosine	0	1	300.91	0	3	224.77
	Uniform + polynomial	0	1	301.09	0	4	226.75
	Half-normal + Hermite	1	0	300.63*	1	0	222.13*
	Hazard-rate + cosine	2	0	303.18	2	0	224.21
Grouped (20 equal)	Uniform + cosine	0	1	563.58	0	2	520.88
	Uniform + polynomial	0	1	563.40	0	3	524.78
	Half-normal + Hermite	1	0	563.03*	1	0	520.31*
	Hazard-rate + cosine	2	0	565.80	2	1	523.53
Grouped (5 unequal)	Uniform + cosine	0	1	323.05*	0	2	345.13
	Uniform + polynomial	0	1	324.45	0	4	348.56
	Half-normal + Hermite	1	0	323.35	1	0	342.54*
	Hazard-rate + cosine	2	0	324.32	2	0	344.95

Table 4.2. Summary of estimated density ($\hat{D}$) and coefficient of variation (cv) for two truncation values (w) for the example data. Estimates are derived for four robust models of the detection function. The data analysis was based on ungrouped data and three different groupings (five groups of equal width, 20 groups of equal width, and five unequal groups such that the number detected was nearly equal in each group)

		Truncation			
		$w = 19\,\text{m}$		w = largest observation	
Data type	Model (key + adjustment)	$\hat{D}$	cv (%)	$\hat{D}$	cv (%)
Ungrouped	Uniform + cosine	90.38	15.9	80.52	16.8
	Uniform + polynomial	78.95	14.8	84.53	20.0
	Half-normal + Hermite	88.31	16.7	87.68	15.3
	Hazard-rate + cosine	84.23	18.4	72.75	15.6
Grouped (5 equal)	Uniform + cosine	88.69	15.9	94.09	16.7
	Uniform + polynomial	79.37	15.2	88.39	19.1
	Half-normal + Hermite	86.94	16.8	92.16	15.8
	Hazard-rate + cosine	84.49	19.6	80.80	16.7
Grouped (20 equal)	Uniform + cosine	89.95	15.8	80.06	15.1
	Uniform + polynomial	79.10	14.9	74.43	15.3
	Half-normal + Hermite	87.98	16.6	86.87	15.7
	Hazard-rate + cosine	85.81	19.2	84.06	18.1
Grouped (5 unequal)	Uniform + cosine	86.14	16.3	81.40	19.0
	Uniform + polynomial	78.60	15.8	86.91	17.8
	Half-normal + Hermite	85.12	17.0	88.84	16.3
	Hazard-rate + cosine	86.83	20.3	82.54	17.7

(Table 4.2). Precision was better for the untruncated data for the hazard-rate model.

The effect of truncation on the point estimates was relatively small, and estimates were not consistently smaller or larger than when data were untruncated (Table 4.2). The various density estimates ranged from 72.75 to 94.09, and their coefficients of variation ranged from 14.8% to 20.3%. The true parameter value was $D = 80$ objects/km^2.

In general, some right-truncation is recommended, especially for obvious outliers. Although some precision might be lost due to truncation, it is usually slight. Often, precision is increased because fewer parameters are required to model the detection function. Most importantly, truncation will often reduce bias in $\hat{D}$ or improve precision, or both, by making the data easier to model. Extreme observations in the right tail of the distribution may arise from a different detection process (e.g. a deer seen at some distance from the observer along a forest trail, or a whale breaching near the horizon), and are generally not informative, in addition to being difficult to model. Truncation is an important tool in the analysis of distance sampling data.

4.3.2 *Left-truncation*

In some surveys, such as aerial surveys with an inadequate view of the line, probability of detection on the line may be uncertain. In this instance, a histogram of the distance data will tend to show too few detections near the line. A simple solution is to offset the line, to a distance at which detection is believed to be certain. Any observations closer to the observer are then truncated. This is termed 'left-truncation' because it is observations from the left end of a conventional plot of the detection function that are truncated. Another method of left-truncation is to retain the line at its original location, but to truncate data within a given distance of the line. The detection function is fitted to the remaining data, and extrapolated back to zero distance (Alldredge and Gates 1985). An example of both approaches is given in Section 8.4.3.

Left-truncation is sometimes used to alleviate other problems with data on or near the line. If there are either too many or too few detections close to the line, left-truncation might be considered. However, the reasons for the problem should be understood before a decision on whether to left-truncate is made. If there are too few detections near the line because objects move away from the observer before they are detected, then there will be too many observations further away, and left-truncation would lead to overestimation of density. Similarly, if there are too many observations close to the line because of movement towards the observer, or because perpendicular distances have been rounded to zero, then left-truncation would cause us to underestimate abundance. On the other hand, if transects are conducted along tracks (Section 7.8.5), a surfeit or deficit of detections near the line may simply reflect a higher or lower density of animals along the tracks, in which case left-truncation may yield estimates of density that are more representative of the survey region.

4.4 Estimating the variance in sample size

Before the precision of an estimate of density can be assessed, attention must be given to the spatial distribution of the objects of interest. If the n detected objects came from a sample of objects that were randomly (i.e. Poisson) distributed in space, then $var(n) = E(n)$ and $\widehat{var}(n) = n$. Because most biological populations exhibit some degree of clumping or loose aggregation, one expects $var(n) > E(n)$. Thus, empirical estimation of the sampling variance of n is recommended. This makes it nearly essential to sample using a number of lines, preferably $k > 20$. Variation in n_i, the number of detections found by line, provides a valid estimate of $var(n)$ without having to resort to the unrealistic Poisson assumption and risk what may be a substantial underestimate of the sampling variance of the estimator of density. The 12 lines used in the example is too few for precise estimation of $var(n)$, but serves to illustrate the approach.

After truncating at 19 m, the line lengths in km and numbers of detections (l_i, n_i) for the $k = 12$ lines were: $(5, 14)$, $(2, 2)$, $(6, 8)$, $(4, 8)$, $(3, 3)$, $(1, 4)$, $(4, 10)$, $(4, 8)$, $(5, 17)$, $(7, 20)$, $(3, 0)$ and $(4, 5)$. The estimator for the empirical variance of n is (from Section 3.6.2)

$$\widehat{var}(n) = \frac{L \sum_{i=1}^{k} l_i (n_i/l_i - n/L)^2}{k-1} = 195.8 \tag{4.2}$$

This estimate is based on $k - 1 = 11$ degrees of freedom. The corresponding standard error of n is estimated as $\widehat{se}(n) = \sqrt{\widehat{var}(n)} = 14.0$. The ratio of the empirical variance to the estimated Poisson variance is $195.8/99 = 1.98$, indicating some spatial aggregation of objects.

Equivalently, if we take the units of encounter rate to be objects/km, the sampling variance of the encounter rate (n/L) is

$$\widehat{var}(n/L) = \frac{\sum_{i=1}^{k} (l_i/L)(n_i/l_i - n/L)^2}{k-1} = 0.0850 \tag{4.3}$$

The standard error of encounter rate is given by $\widehat{se}(n/L) = \sqrt{\widehat{var}(n/L)} = 0.292$ objects/km. Note that $\widehat{se}(n) = L \cdot \widehat{se}(n/L) = 48 \times 0.292 = 14.0$ objects as above.

4.5 Analysis of grouped or ungrouped data

Analysis of the ungrouped data is recommended for the example because it is known that the assumptions of line transect sampling hold. Little efficiency is lost by grouping data, even with as few as five or six well-chosen distance intervals. Grouping of the data can be used to improve robustness in the estimator of density in cases of heaping and movement prior to detection (Chapter 7).

For the example, changes in the estimates of density under a given model were in most cases slight (much smaller than the standard error) whether the analysis was based on the ungrouped data or one of the three sets of grouped data. This is a general result if the assumptions of distance sampling are met. If heaping, errors in measurement, or evasive movement prior to detection are present, then appropriate grouping will often lead to improved estimates of density and better model fit. Grouping the data is a tool for the analysis of real data to gain estimator robustness. When heaping occurs, cutpoints should be selected to avoid favoured rounding distances as far as possible. Thus, if values tend to be recorded to the nearest 10 m, cutpoints might be defined at 5, 15, 25 m, ... The first cutpoint is the most critical. If assumptions hold, the first interval should be relatively narrow, so that the first cutpoint is on the shoulder of the detection function. However, it is not unusual for 10% or more of detections to be recorded as on the centreline, especially when perpendicular distances are calculated from sighting distances and angles. In this circumstance, a relatively wide first interval should be chosen, to ensure that few detections are erroneously allocated to the first interval through measurement error, and in particular, through rounding a small sighting angle to zero.

4.6 Model selection

4.6.1 *The models*

Results for fitting the detection function are illustrated using the uniform, half-normal and hazard-rate models as key functions. Cosine and simple polynomial expansions are used with the uniform key, Hermite polynomials are used with the half-normal key, and a cosine expansion is used with the hazard-rate key. Thus, four models for $g(x)$ are considered for the analysis of these data. Modelling in this example can be expected to be relatively easy as the data are well-behaved, exhibit a shoulder, and the sample size is adequate ($n = 105$ before truncation). With such ideal data, the choice of model is unlikely to affect the abundance estimate much, whereas if survey design or data collection is poor, different models might yield substantially different estimates.

From an inspection of the data in Fig. 4.1, it is clear that the uniform key function will require at least one cosine or polynomial adjustment term. The data here were generated under a half-normal detection function so we might expect the half-normal key to be sufficient without any adjustment terms. However, the data were stochastically generated; the addition of a Hermite polynomial term is quite possible, although it would just fit 'noise'. The hazard-rate key has two parameters and seldom requires adjustment terms when data are good. In general, a histogram of the untruncated data using 15–20 intervals will reveal the characteristics of the data. Such a histogram will help identify outliers, heaping, measurement errors, and evasive movement prior to detection.

4.6.2 *Akaike's Information Criterion*

Akaike's Information Criterion (AIC) provides an objective, quantitative method for model selection. Burnham and Anderson (1998) address the use of AIC in detail and Akaike (1985) gives the underlying theory. The criterion is

$$\text{AIC} = -2 \cdot \log_{\text{e}}(\mathcal{L}) + 2q \tag{4.4}$$

where $\mathcal{L}$ is the maximized likelihood and q the number of estimated parameters.

The adjustment of Hurvich and Tsai (1989, 1995) can be useful for small samples:

$$\mathrm{AICc} = \mathrm{AIC} + \frac{2q(q+1)}{n-q-1} \tag{4.5}$$

where n is the number of objects or object clusters detected. The adjustment is small provided $n \gg q$, say $n > 20q$. AIC (or AICc) is computed for each candidate model, and that with the lowest AIC is selected for analysis and inference. Having selected a model, one should check that it fits as judged by the usual χ^2 goodness of fit statistic. Visual inspection of the estimated detection function plotted on the histogram is also informative because one can better judge the model fit near the line, and perhaps discount some lack of fit in the right tail of the data.

AIC was computed for the four models noted above for both grouped and ungrouped data with and without truncation (Table 4.1). For computational reasons, w was set equal to the largest observation in the case of no truncation. Three sets of cutpoints were considered for each model for illustration. Set 1 had five groups of equal width, set 2 had 20 groups of equal width, and set 3 had five groups whose width increased with distance, such that the number detected in each distance category was nearly equal. Note that AIC can only be used to compare models fitted to exactly the same data. It cannot therefore not be used to select a truncation distance w, or to compare models if different truncation distances have been used, or to assess whether, or how, to group distance data.

AIC values in Table 4.1 indicate that the half-normal model is the best of the four models considered. Here, we know that the data were generated from this model. All four models have generally similar AIC values within any set of analyses of Table 4.1. Still, AIC selects the half-normal model in three of the four instances, both with truncation and without. Thus, the main analysis will focus on the ungrouped data, truncated at $w = 19\,\mathrm{m}$, under the assumption that $g(x)$ is well modelled by the half-normal key function with no adjustment parameters. The fit of this model is shown in Fig. 4.2. One might suspect that all four models would provide valid inference because of the similarity of the AIC values. Often, AIC will identify a subset of models that are clearly inferior and these should be discarded from further consideration.

AIC is a useful tool for surveys that use a stratified design, where for example the different strata might correspond to different survey blocks, species, observers, habitats or environmental conditions. One analysis can be carried out by fitting the detection function to distance data pooled across strata, and separate analyses can be done by stratum. If the sum of the AIC values across strata is less than the AIC value from the pooled analysis, this suggests that the detection function differs between strata, and that it should be fitted separately in each stratum. An example of such an

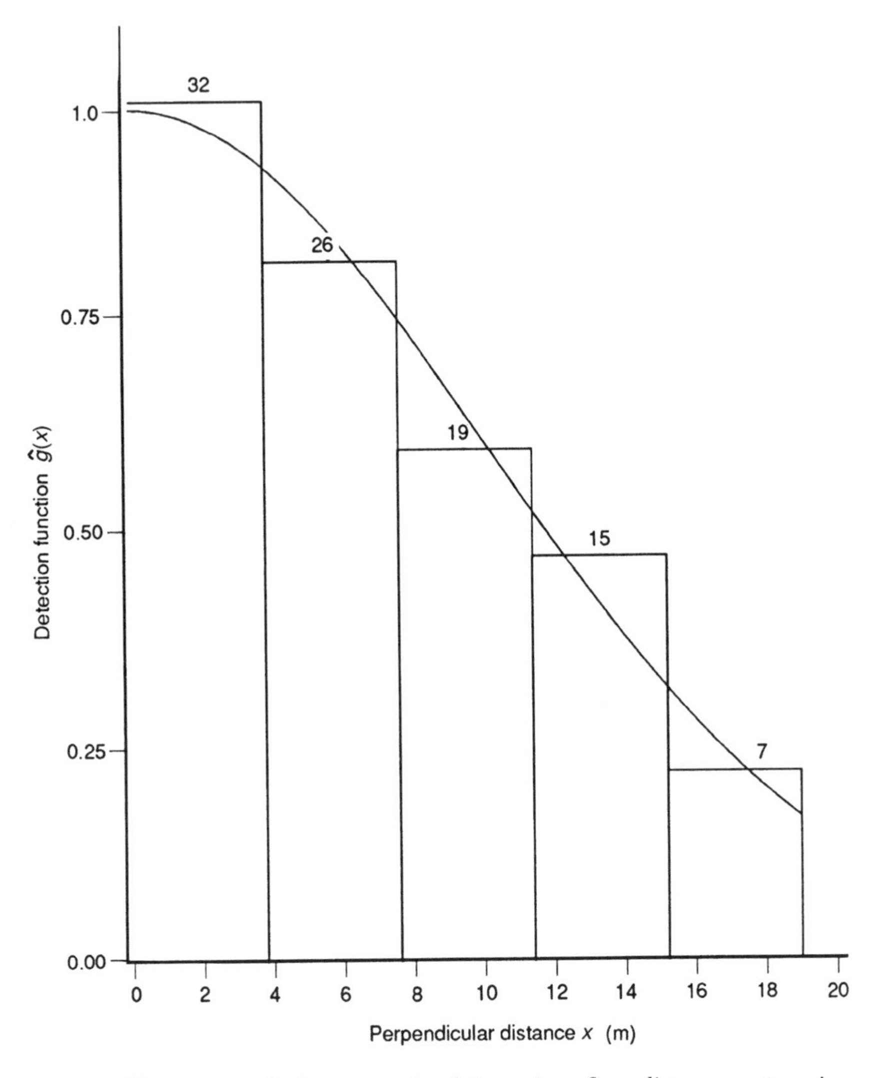

Fig. 4.2. Histogram of the example data using five distance categories. A half-normal detection function, fitted to the ungrouped data with $w = 19\,\text{m}$, is shown and was used as a basis for final inference from these data.

analysis, where the strata correspond to different cluster sizes, is given in Section 4.8.2.3.

4.6.3 *Likelihood ratio tests*

The addition of adjustment terms to a given key function can be judged using likelihood ratio tests. We recommend AIC or AICc in preference to the likelihood ratio test, for the pragmatic reason that it allows comparison of models with different keys. By contrast, the likelihood ratio test only allows testing between models when the parameters of one model are a subset of those for the other. We also prefer the philosophical basis underlying AIC to that underlying hypothesis tests.

We illustrate the likelihood ratio test on the ungrouped, untruncated data from our example (Fig. 4.1). Assume the key function is the two-parameter hazard-rate model. This model is fitted to the distance data to provide the maximum likelihood estimates of the hazard-rate parameters. Does an adjustment term significantly improve the fit of the model of $g(x)$ to the data? Let $\mathcal{L}_0$ be the maximized value of the likelihood for the two-parameter hazard-rate model and $\mathcal{L}_1$ be its value for the three-parameter model (hazard-rate model + one cosine adjustment term). Then, the test statistic for this likelihood ratio test is

$$\chi^2 = -2 \log_e(\mathcal{L}_0/\mathcal{L}_1) \tag{4.6}$$

If the two-parameter model is the true model, this statistic is distributed asymptotically as χ^2, with degrees of freedom given by the difference in number of parameters between the two models ($3 - 2 = 1$ df). If the additional term makes a significant improvement in the fit, then the test statistic will be 'large'. We obtain $\log_e(\mathcal{L}_0) = -259.898$ and $\log_e(\mathcal{L}_1) = -258.763$. Which is the better model?

$$\chi^2 = -2[\log_e(\mathcal{L}_0) - \log_e(\mathcal{L}_1)] = -2[-259.898 - (-258.763)] = 2.27$$

Comparing this with χ_1^2, we find $p = 0.132$. If as proposed in Section 3.4.3, we set the size of the test at $\alpha = 0.15$, then the two-parameter model is rejected in favour of the three-parameter model. The procedure is repeated to examine the adequacy of the three-parameter model against a four-parameter model with two cosine terms. Note that this illustration used the untruncated data; additional terms are frequently needed to model the right tail of the distance data if proper truncation has not been done.

4.6.4 *Goodness of fit*

Goodness of fit is described briefly in Section 3.4.4. Some goodness of fit statistics for the example with $w = 19$ m and 20 groups are given in Table 4.3. These data were taken when all the assumptions were met; all four models fit the data well and yield similar estimates of density.

Real data are often heaped, so that no parsimonious model seems to fit the data well, as judged by the χ^2 test. The goodness of fit statistic is very sensitive to heaping in the data, but this sensitivity can be reduced substantially by appropriate choice of cutpoints for the χ^2 test. If possible, there should be one favoured rounding distance per interval, and it should

Table 4.3. Goodness of fit statistics for models fitted to the example data with $w = 19\,\mathrm{m}$ and 20 groups

Model	χ^2	df	p
Uniform + cosine	14.58	17	0.62
Uniform + polynomial	13.11	17	0.73
Half-normal + Hermite	13.63	17	0.69
Hazard-rate + cosine	14.91	16	0.53

occur at the midpoint of the interval. The grouped nature of the (rounded) data is then correctly recognized in the analysis. If cutpoints are badly chosen, heaping will generate spurious significant χ^2 values.

For data that meet all the assumptions, goodness of fit provides a useful assessment of the adequacy of a fitted model. Often though, a good model will give a significant goodness of fit statistic due to one or more assumption failures. Thus if no model appears to fit well, this is usually an indication of problems with the data, and the goodness of fit test, along with histograms of the data, allow exploration of these problems.

4.7 Estimation of density and measures of precision

4.7.1 *The standard analysis*

Preliminary analysis leads us to conclude that the half-normal model is an adequate model of the detection function, with truncation of the distance data at $w = 19\,\mathrm{m}$, fitted to ungrouped data, and using the empirical variance of n.

Replacing the parameters of eqn (3.7) by their estimators and simplifying, estimated density becomes

$$\hat{D} = \frac{n \cdot \hat{f}(0)}{2L} \tag{4.7}$$

where n is the number of objects detected, L is the total length of transect line, and $\hat{f}(0)$ is the estimated probability density evaluated at zero perpendicular distance. For the example data, and adopting the preferred analysis, program Distance yields $\hat{f}(0) = 0.08563$ with $\widehat{se}\{\hat{f}(0)\} = 0.007601$. The units of measure for $f(0)$ are m^{-1}. Often, estimates of the effective strip half-width, $\mu = 1/f(0)$, are given in preference to $\hat{f}(0)$, since it has an intuitive interpretation. It is the perpendicular distance from the line for which the number of objects closer to the line that are missed is equal to the number of objects farther from the line (but within the truncation distance w) that are detected.

For the half-normal model, if data are neither grouped nor truncated, a closed-form expression for $\hat{f}(0)$ exists (Section 3.3.4.1):

$$\hat{f}(0) = \sqrt{\frac{2}{\pi\hat{\sigma}^2}} \tag{4.8}$$

where $\hat{\sigma}^2 = \sum x_i^2/n$.

Similarly, closed-form (though not maximum likelihood) expressions exist for $f(0)$ under the Fourier series model (uniform key + cosine adjustments) of Burnham *et al.* (1980: 56–61). However, generally the maximum likelihood estimator of $f(0)$ must be computed numerically because no closed-form equation exists.

The estimate of density for the example data truncated at 19 m, and using $\hat{f}(0)$ from Distance, is

$$\hat{D} = \frac{n \cdot \hat{f}(0)}{2L} = \frac{99 \times 0.08563}{2 \times 48} = 0.0883 \,\text{objects/m} \cdot \text{km}$$

Since the units of $\hat{f}(0)$ are m^{-1} and those for L are km, multiplying by 1000 gives $\hat{D} = 88.3$ objects/km^2.

The estimator of the sampling variance of this estimate is

$$\widehat{var}(\hat{D}) = \hat{D}^2 \cdot \{[cv(n)]^2 + [cv\{\hat{f}(0)\}]^2\} \tag{4.9}$$

where

$$[cv(n)]^2 = \frac{\widehat{var}(n)}{n^2} = \frac{195.8}{99^2} = 0.01998$$

and

$$[cv\{\hat{f}(0)\}]^2 = \frac{\widehat{var}\{\hat{f}(0)\}}{\{\hat{f}(0)\}^2} = \frac{0.007601^2}{0.08563^2} = 0.007879$$

Then

$$\widehat{var}(\hat{D}) = (88.3)^2[0.01998 + 0.007879] = 217.2$$

and

$$\widehat{se}(\hat{D}) = \sqrt{\widehat{var}(\hat{D})} = 14.74$$

The coefficient of variation of estimated density is $cv(\hat{D}) = \widehat{se}(\hat{D})/\hat{D} = 0.167$, or 16.7%, which might be adequate for many purposes. An approximate 95% confidence interval could be set in the usual way as $\hat{D} \pm 1.96 \cdot \widehat{se}(\hat{D})$, resulting in the interval [59.4, 117.2]. Log-based confidence intervals (Burnham *et al.* 1987: 211–3) offer improved coverage by allowing for the asymmetric shape of the sampling distribution of $\hat{D}$

for small n. The procedure allows lower and upper 95% bounds to be computed as

$$\hat{D}_L = \frac{\hat{D}}{C} \quad \text{and} \quad \hat{D}_U = \hat{D} \cdot C \tag{4.10}$$

where

$$C = \exp\left\{1.96 \cdot \sqrt{\log_e(1 + [cv(\hat{D})]^2)}\right\}$$

This method gives the interval [63.8, 122.2], which is wider than the symmetric interval, but is a better measure of the precision of the estimate $\hat{D} = 88.3$. The imprecision in $\hat{D}$ is primarily due to the variation associated with n; approximately 72% (i.e. 0.01998/(0.01998 + 0.007879)) of $\widehat{var}(\hat{D})$ is due here to the sampling variation in n.

The use of 1.96 in constructing the above confidence intervals is only justified if the degrees of freedom of all variance components in $\widehat{var}(\hat{D})$ are large, say greater than 30. In this example, the degrees of freedom for the component $\widehat{var}(n)$ are only 11. If there were only one variance component, it would be standard procedure to use the t-distribution on the relevant degrees of freedom, rather than the standard normal distribution, as the basis for a confidence interval. When the relevant variance estimator is a linear combination of variances, there is a procedure using an approximating t-distribution as the basis for the confidence interval. This more complicated procedure is explained in Section 4.7.4 below, and is used automatically by program Distance.

The effective strip half-width is estimated by $\hat{\mu} = 1/\hat{f}(0) = 11.68\,\text{m}$. The unconditional probability of detecting an object in the surveyed area, $a = 2wL$, is estimated by $\hat{P}_a = \hat{\mu}/w = 0.61$, which is simply the ratio of the effective strip half-width to the truncation distance, $w = 19\,\text{m}$. These are maximum likelihood estimates as they are one-to-one transformations of the maximum likelihood estimate of $f(0)$.

In summary, we obtain $\hat{D} = 88.3$, $\widehat{se}(\hat{D}) = 14.7$, $cv = 16.7\%$, with a 95% confidence interval of [63.8, 122.2]. Recalling that the true parameter $D = 80$, this particular estimate is a little high, largely because the sample size ($n = 105$, untruncated) happened to be above that expected ($E(n) = 96$). This is not unusual, given the large variability in n due to spatial aggregation of the objects, and the confidence interval easily covers the parameter. Some alternative analyses and issues and their consequences will now be explored.

4.7.2 *Ignoring information from replicate lines*

If the Poisson assumption ($\widehat{var}(n) = n$) had been used with $w = 19\,\text{m}$ and $L = 48\,\text{km}$, then the estimate of density would not change, but $\widehat{se}(\hat{D})$ would be underestimated at 11.84, with 95% confidence interval for D

of [67.98, 114.70]. While this interval happens to cover D, the method underestimates the uncertainty of the estimator $\hat{D}$; if many data sets were generated, the true coverage of the interval would be well below 95%. This procedure cannot be recommended; one should estimate the variance associated with sample size empirically from the counts on the individual replicate lines, including those lines with zero counts. For example, line 11 had no observations ($n_{11} = 0$), which must be included in the analysis as a zero count.

4.7.3 *Bootstrap variances and confidence intervals*

The selected model for $g(x)$ for the example data was the half-normal, with $w = 19$ m, fitted to ungrouped distance data. The maximum likelihood estimate of $f(0)$ was 0.08563 with $\widehat{se}\{\hat{f}(0)\} = 0.007601$. The bootstrap procedure (Section 3.6.4) can be used to obtain a more robust estimate of this standard error. Suppose a resample is generated by selecting n distances *with replacement* from the original n. The required number of series expansion terms can be estimated in each resample, and variance due to this estimation, ignored in the analytical method, is then a component of the bootstrap variance. As an illustration, 1000 bootstrap replications were performed, yielding an average $\hat{f}(0) = 0.08587$ with $\widehat{se}\{\hat{f}(0)\} = 0.00748$. In this simple example where the true model is the fitted model without any adjustment terms, the two procedures yield nearly identical results. We can go a step further and select between each of several possible models when analysing each resample, so that a component of variance corresponding to model selection uncertainty is incorporated.

A better use of the bootstrap in line transect sampling is to sample with replacement from the replicate lines, until either the number of lines in the resample equals the number in the original data set, or the total effort in the resample approximately equals the total effort in the real data set. These two methods only differ if lines vary in length. If the model selection procedure is automated, it can be applied to each resample, so that model selection uncertainty can be incorporated in the estimate of precision. Further, the density D may be estimated for each resample, so that robust standard errors and confidence intervals may be set that automatically incorporate variance in sample size (or equivalently, encounter rate) and cluster size if relevant, as well as in the estimate of $f(0)$. The method is described in Section 3.6.4, and an example of its application to point transect data is given in Section 5.7.2.

A possible analysis strategy is to carry out model selection and choice of truncation distance first, and then to evaluate bootstrap standard errors only after a particular model has been identified. Although model selection uncertainty is then ignored, the bootstrap is computationally intensive, and its use at every step in the analysis may sometimes be prohibitive.

4.7.4 *Satterthwaite degrees of freedom for confidence intervals*

For the log-based confidence interval approach, there is a method to allow for the finite degrees of freedom of each estimated variance component in $\widehat{var}(\hat{D})$. This procedure dates from Satterthwaite (1946); a more accessible reference is Milliken and Johnson (1984). Assuming the log-based approach, $[\log_e(\hat{D}) - \log_e(D)]/cv(\hat{D})$ is well approximated by a t-distribution with degrees of freedom computed in the case of two variance components by the formula

$$\text{df} = \frac{[cv(\hat{D})]^4}{[cv(n)]^4/(k-1) + [cv\{\hat{f}(0)\}]^4/(n-p)} \tag{4.11}$$

where

$$[cv(\hat{D})]^2 = \frac{\widehat{var}(\hat{D})}{\hat{D}^2}$$

$$[cv(n)]^2 = \frac{\widehat{var}(n)}{n^2}$$

$$[cv\{\hat{f}(0)\}]^2 = \frac{\widehat{var}\{\hat{f}(0)\}}{\{\hat{f}(0)\}^2}$$

Given the computed degrees of freedom, one finds the two-sided 100(1–2α)% percentile of the t-distribution with these degrees of freedom; df is in general non-integer, but may be rounded to the nearest integer. Usually $\alpha = 0.025$, giving a 95% confidence interval. Then one uses the value of $t_{\text{df}}(0.025)$ in place of 1.96 in the confidence interval calculations, so that

$$\hat{D}_L = \frac{\hat{D}}{C} \quad \text{and} \quad \hat{D}_U = \hat{D} \cdot C \tag{4.12}$$

where

$$C = \exp\left\{t_{\text{df}}(0.025) \cdot \sqrt{\log_e\{1 + [cv(\hat{D})]^2\}}\right\}$$

This lengthens the confidence interval noticeably when the number of replicate lines is small.

We illustrate this procedure with the current example for which $[cv(n)]^2 = 0.01998$ on 11 degrees of freedom, and $[cv\{\hat{f}(0)\}]^2 = 0.007879$ on 98 degrees of freedom. Thus $[cv(\hat{D})]^2 = 0.027859$. The above formula for df gives

$$\text{df} = \frac{0.0007761}{0.0003992/11 + 0.00006208/98} = 21.02$$

which we round to 21 for looking up $t_{21}(0.025) = 2.08$ in tables. Using 2.08 rather than 1.96, we find that $C = 1.4117$, and the improved 95% confidence interval is [62.6, 124.7], compared with [63.8, 122.2] using $z = 1.96$. Confidence intervals in Distance use the Satterthwaite procedure.

This procedure for computing the degrees of freedom for an approximating t-distribution generalizes to the case of more than two components, for example when there are three parameter estimates, n, $\hat{f}(0)$ and $\hat{E}(s)$. Section 3.6.1 gives the general formula.

4.8 Estimation when the objects are in clusters

Often the objects are detected in clusters (flocks, coveys or schools) and further considerations are necessary in this case. The density of clusters (D_s), the density of individual objects (D), and average cluster size $E(s)$ are the biological parameters of interest in surveys of clustered populations, and several intermediate parameters are of statistical interest (e.g. $f(0)$). Here, we will assume that the clusters are reasonably well defined; populations that form loose aggregations of objects are more problematic.

It is assumed that the distance measurement is taken from the line to the geometric centre of the cluster. If a truncation distance w is adopted in the field (as distinct from in the analysis), then a cluster is recorded if its centre lies within distance w and a count made of all individuals within the cluster, including those individuals that are at distances greater than w. If the geometric centre of the cluster lies at a distance greater than w, then no measurement should be recorded, even if some individuals in the cluster are within distance w of the line. The sample size of detected objects n is the number of clusters, not the total number of individuals detected.

4.8.1 *Observed cluster size independent of distance*

If the size of detected clusters is independent of distance from the line (i.e. $g(x)$ does not depend on s), then estimation of D_s, D and $E(s)$ is relatively simple. The sample mean $\bar{s}$ is taken as an unbiased estimator of the average cluster size. Then, $\hat{E}(s) = \bar{s} = \sum s_i/n$, where s_i is the size of the ith cluster. In general the density of clusters D_s and measures of precision are estimated exactly as given in Sections 4.3–4.7. Then, $\hat{D} = \hat{D}_s \cdot \bar{s}$; the estimator of the density of individuals is merely the product of the density of clusters times the average cluster size. Alternatively, the expression can be written

$$\hat{D} = \frac{n \cdot \hat{f}(0) \cdot \bar{s}}{2L} \tag{4.13}$$

The example data set used throughout this chapter is now reconsidered in view of the (now revealed) clustered nature of the population. The distribution of true cluster size in the population was simulated from a Poisson distribution and the size of detected clusters was independent of distance from the line. The value of s was simulated as 1+ a Poisson variable with a mean of two or, equivalently, $(s-1) \sim \text{Poisson}(2)$, so that $E(s) = 3$. Theoretically, $var(s) = var(s-1) = E(s-1) = E(s) - 1 = 2$. Under the independence assumption, the sample mean $\bar{s}$ is an unbiased estimate of $E(s)$. The true density of individuals was 240. Estimates of D_s (called D in previous sections), $f(0)$, effective strip half-width and the various measures of precision are exactly those derived in Section 4.7.

The estimated average cluster size, $\bar{s}$, for the example data with $w = 19\,\text{m}$ is 2.859 ($\bar{s} = 283/99$) and the empirical sampling variance on $n - 1 = 98$ degrees of freedom is

$$\widehat{var}(\bar{s}) = \frac{\sum_{i=1}^{n}(s_i - \bar{s})^2}{n(n-1)} = 0.02062$$

so that

$$\widehat{se}(\bar{s}) = \sqrt{0.02062} = 0.1436$$

These empirical estimates compare quite well with the true parameters; $var(\bar{s}) = 2/n = 2/99 = 0.0202$, $se(s) = 0.142$. If one uses the knowledge that cluster sizes were based on a Poisson process, one could estimate this true standard error as $\sqrt{(\bar{s}-1)/n} = \sqrt{1.859/99} = 0.137$, which is also close to the true value. The advantage of the empirical estimate is that it remains valid when the Poisson assumption is false.

A plot of cluster size s_i versus distance x_i (Fig. 4.3) provides only weak evidence of dependence ($r = 0.16$, $p = 0.10$). In this case, we take $\hat{E}(s) = \bar{s}$, the sample mean. Thus, the density of individuals is estimated as

$$\hat{D} = \hat{D}_s \cdot \bar{s} = 88.3 \times 2.859 = 252.4 \text{ individuals/km}^2$$

Then, for large samples,

$$\begin{aligned}\widehat{var}(\hat{D}) &= \hat{D}^2 \cdot \{[cv(n)]^2 + [cv\{\hat{f}(0)\}]^2 + [cv(\bar{s})]^2\} \\ &= 252.4^2 \cdot [0.01998 + 0.007879 + 0.002523] \\ &= 1936\end{aligned}$$

so that

$$\widehat{se}(\hat{D}) = \sqrt{1936} = 44.0$$

This gives $\hat{D} = 252.4$ individuals/km^2 with $cv = 17.4\%$. The log-based 95% confidence interval, using the convenient multiplier $z = 1.96$, is [179.8, 354.3]. This interval is somewhat wide, due primarily to the spatial variation in n; $\widehat{var}(n)$ makes up 66% of $\widehat{var}(\hat{D})$, while $\widehat{var}\{\hat{f}(0)\}$ contributes 26% and $\widehat{var}(\bar{s})$ contributes only 8% (e.g. $66\% = \{0.01998/(0.01998 + 0.007879 + 0.002523)\} \times 100$).

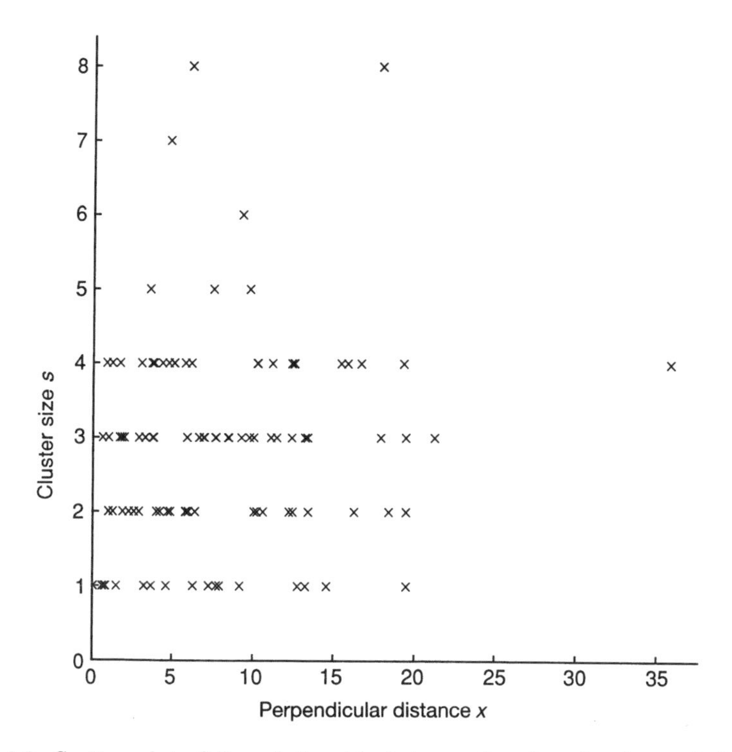

Fig. 4.3. Scatter plot of the relationship between the size of a detected cluster and the distance from the line to the geometric centre of the cluster for the example in which size and detection distance are independent. The correlation coefficient is 0.16.

A theoretically better confidence interval is one based not on the standard normal distribution (i.e. on $z = 1.96$) but rather on a t-distribution with degrees of freedom computed here as

$$\begin{aligned} \text{df} &= \frac{[cv(\hat{D})]^4}{[cv(n)]^4/(k-1) + [cv\{\hat{f}(0)\}]^4/(n-p) + [cv(\bar{s})]^4/(n-1)} \\ &= \frac{0.0009235}{0.0003992/11 + 0.00006208/98 + 0.000006366/98} \\ &= 24.58 \end{aligned}$$

which we round to 25. Given this computed value for the degrees of freedom, we find the two-sided $100(1{-}2\alpha)\%$ percentile $t_{\text{df}}(\alpha)$ of the t-distribution. In this example, $t_{25}(0.025) = 2.06$. Using 2.06 rather than 1.96 in the log-based confidence interval procedure, $C = 1.4282$, giving an improved 95% confidence interval of [176.7, 360.5]. It is this latter interval which Distance computes, applying the procedure of Satterthwaite (1946) to $\log_e(\hat{D})$.

Plots and correlations should always be examined prior to proceeding as if cluster size and detection distance were independent. In particular, some truncation of the data will often have the added benefit of weakening the dependence between s_i and x_i. If truncation is appropriate, then $\hat{E}(s)$ should be based on only those clusters within $[0, w]$. Although our experience suggests that data from surveys of many clustered populations can be treated as if s_i and x_i are independent, using Distance, there is little to be lost from using the default of a regression correction for size-bias (see below), and potentially much to gain.

4.8.2 *Observed cluster size dependent on distance*

The analysis of survey data where the cluster size is dependent on the detection distance is more complicated because of difficulties in obtaining an unbiased estimate of $E(s)$ (Drummer and McDonald 1987; Drummer *et al.* 1990; Otto and Pollock 1990). The dependence arises because large clusters might still be seen at some distance from the line (near w), while small clusters might remain undetected. This phenomenon causes an overestimation of $E(s)$ because too few small clusters are detected (i.e. they are under-represented in the sample). Thus, $\hat{D} = \hat{D}_s \cdot \bar{s}$ is also an overestimate. Another complication is that large clusters might be more easily detected near w than small clusters, but their size might be underestimated due to reduced detectability of individuals at long distances. This phenomenon has a counter-balancing effect on the estimates of $E(s)$ and D. The dependence of x_i on s_i is a case of size-biased sampling (Cox 1969; Patil and Ord 1976; Patil and Rao 1978; Rao and Portier 1985).

The data used in this section are sampled from the same population as in earlier sections of this chapter (i.e. $L = 48$, $D_s = 80$, $f(0) = 0.079$, $E(s) = 3$ and $E(n) = 96$, so that $D = 240 = 3 \times 80$). The half-normal detection function was used, as before, but σ was allowed to be a function of cluster size:

$$\sigma(s) = \sigma_0\left(1 + b \cdot \frac{s - E(s)}{E(s)}\right) \tag{4.14}$$

where $b = 1$ and $E(s) = 3$ in the population. Selecting $b = 1$ represents a strong size bias and corresponds to Drummer and McDonald's (1987) form with $\alpha = 1$. Cluster size in the entire population (detected or not) was distributed as $s - 1 \sim$ Poisson. Given a cluster size s, the detection distance was generated from the half-normal detection function $g(x_i \mid s_i)$. Because of the dependence between cluster size and detection distance, the distance data differ from those in the earlier parts of this chapter. In particular, some large clusters were detected at greater distances (e.g. one detection of 5 objects at 50.9 m). Because of the dependence on cluster size, the bivariate detection function $g(x, s)$ is not half-normal. This detection function is monotone non-increasing in x and monotone non-decreasing in s. In addition, the detected cluster sizes do not represent a random sample from the population of cluster sizes, as small clusters tend to remain undetected except at short distances. Thus, the size of detected clusters is not any simple function of a Poisson variate. Generation of these data is treated in Section 3.8.2.

A histogram of the distance data indicates little heaping and a somewhat long right tail (Fig. 4.4). Truncation at 20 m seemed reasonable and eliminated only 16 observations, leaving $n = 89$ (15% truncation). Truncation makes modelling of the detection function easier and reduces the correlation between detection distance and cluster size. Three robust models were chosen as candidates for modelling $g(x)$: uniform + cosine, half-normal + Hermite polynomial and hazard-rate + cosine. All three models fit the truncated data well. AIC suggested the use of the uniform + cosine model by a small margin (506.26 versus 506.96 for the half-normal), and both models gave very similar estimates of cluster density. The hazard-rate model (AIC = 509.21) provided rather high estimates of cluster density with less precision, although confidence intervals easily covered the true parameter. The uniform + cosine model and the half-normal model both required only a single parameter to be estimated from the data, while the hazard-rate required a cosine adjustment, so used three parameters in all. This may account for some of the increased standard error of the hazard-rate model, but the main reason for the high estimate and standard error is that the hazard-rate model attempts to fit the spike in the histogram of Fig. 4.4 in the first distance category. Because we know the model used to generate these data, we know the spike is spurious, and arises because for this data set, more simulated values occurred within 2.5 m than would be expected. Generally, if such a spike is real, the hazard-rate model yields lower bias (but also higher variance) than most series expansion models, whereas its performance is poor if the spike is spurious. Since AIC selected the uniform + cosine model, we use it below to illustrate methods of analysis of the example data.

The uniform + cosine model for the untruncated data required five cosine terms to fit the right tail of the data adequately (Fig. 4.4). Failure to truncate the data here would have resulted in lower precision, the model would have required five cosine terms instead of just one (Fig. 4.5), and the mean cluster size would have been less reliably estimated (below).

The uniform key together with one cosine adjustment fit the truncated data well (Fig. 4.5, $0.38 \leq p \leq 0.71$, depending on the grouping used for the χ^2 test). The estimated density of clusters was 81.56 ($\widehat{se} = 12.43$), close to the true value (80). The mean cluster size from the sample data was 3.258 ($\widehat{se} = 0.134$) which is surely too high in view of the size-biased sampling caused by the correlation between cluster size and detection distance. However, truncation at $w = 20$ reduced this correlation from

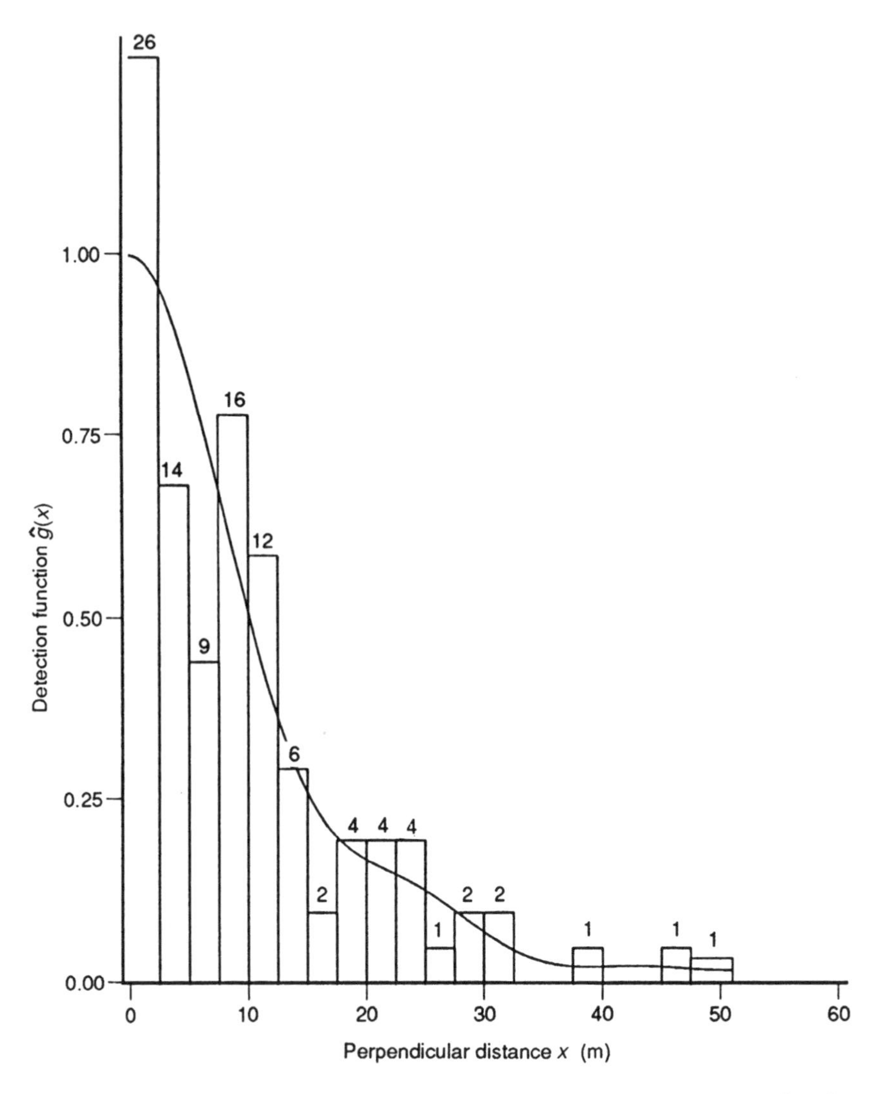

Fig. 4.4. Histogram of the example data using 20 distance categories for the case where cluster size and detection distance are dependent. The fit of a uniform + five-term cosine model is shown.

0.485 to 0.224 so this uncorrected estimate of $E(s)$ may not be heavily biased. Multiplying the density of clusters by the uncorrected estimate of mean cluster size (3.258), the density of individuals is estimated as 265.7 ($\widehat{se}$ = 41.9; 95% ci = [195.4, 361.4]), which is a little high, but still a quite acceptable estimate for these data, for which actual density was 240 individuals/km^2.

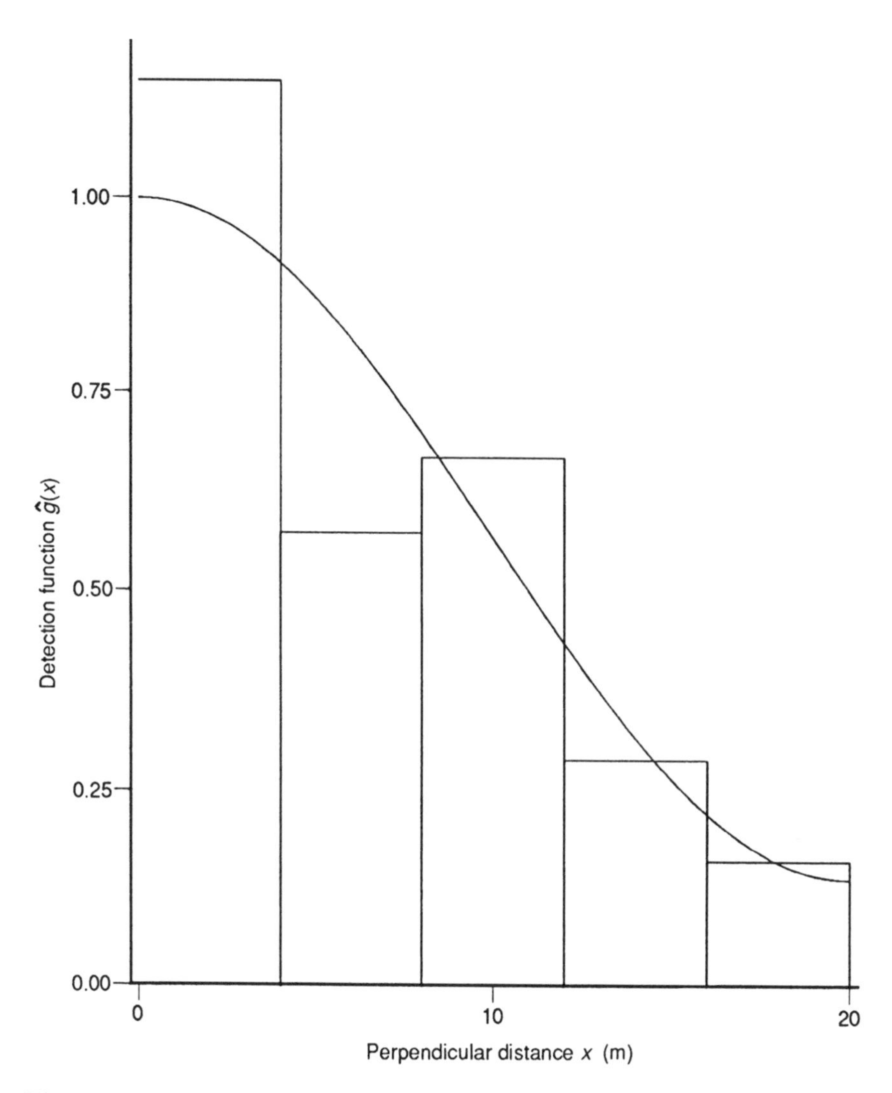

Fig. 4.5. Histogram of the example data using five distance categories and truncation at $w = 20\,\text{m}$ for the case where cluster size and detection distance are dependent. The fit of a uniform + one-term cosine model is shown.

We now consider five approaches for correcting for size bias in the detections.

4.8.2.1 *Truncation*

Observed mean cluster sizes and standard errors corresponding to several levels of truncation of the distance data are shown in Table 4.4. The detection function $g(x)$ was fitted to distance data truncated at w and mean

Table 4.4. Observed mean cluster sizes and standard errors for various truncation distances v. Probability of detection at the truncation distance for cluster size estimation, $\hat{g}(v)$, was estimated from a uniform + one-term cosine model with $w = 20$ m (Fig. 4.5) for $v \leq 20$ m, and from a uniform + five-term cosine model with $w = 51$ m for $v = 51$ m

Truncation distance v (m)	n	$\bar{s}$	$\widehat{se}(\bar{s})$	$\hat{g}(v)$
51.0	105	3.581	0.150	0.02
20.0	89	3.258	0.134	0.14
13.6	80	3.188	0.144	0.30
10.5	69	3.116	0.152	0.50
9.1	61	3.098	0.166	0.60
7.6	49	3.061	0.192	0.70
6.0	44	3.114	0.206	0.80
4.1	37	3.081	0.214	0.90

cluster size was estimated after truncating distance data at distance v, $v \leq w$. The gain in precision by reducing the truncation distance from 51 to 20 m arises because predominantly large clusters are truncated, and variability in the size of remaining clusters is reduced. It is clear from Table 4.4 that 20 m is too large a truncation distance for unbiased estimation of mean cluster size, since mean cluster size continues to fall when the truncation distance is reduced further. The choice of truncation distance is a compromise between reducing bias and retaining adequate precision. Here, mean cluster size appears to stabilize at a truncation distance of 10.5 m, for which $\hat{g}(10.5) = 0.5$. Thus, mean cluster size is estimated to be 3.116 with $\widehat{se} = 0.152$. Replacing the estimates $\bar{s} = 3.258$ and $\widehat{se} = 0.134$ by these values, density is estimated as 254.1 individuals/km^2, with $\widehat{se} = 40.7$ and approximate 95% confidence interval [186.1, 347.0] (based on $z = 1.96$ rather than Satterthwaite's correction). This estimate is closer to the true value of 240 individuals/km^2, and precision is almost unaffected, because the contribution to the overall variance due to variation in cluster size is slight.

Hence if sample size is large, we may select a truncation point v for the estimation of $E(s)$ that is smaller than the truncation point w for the estimation of $f(0)$ to reduce the size bias in detected clusters. For example, $E(s)$ might be estimated by the mean size of clusters detected within a distance x of the line, where $\hat{g}(x) = 0.6$. Often bias reduction is more important than optimizing precision in estimating mean cluster size because the relative contribution of $var(\bar{s})$ to $var(\hat{D})$ is usually small, as in this example.

4.8.2.2 *Replacement of clusters by individuals*

If a cluster of size s_i is replaced by s_i objects at the same distance, the assumption that detections are independent is violated. This compromises analytic variance estimates and model selection procedures. The first difficulty may be overcome by using robust methods for variance estimation (e.g. estimate variance in encounter rate from replicate lines, or apply the nonparametric bootstrap by resampling replicate lines), but the standard model selection tools are also compromised by lack of independence. A solution is to select a model taking clusters as the sampling unit, then refit the model (with the same series terms, if any) to the data with object as the sampling unit. For our example, this leads us to fit a uniform + one-term cosine model to the distance data truncated at 20 m, giving the following estimates. Number of objects detected, $n = 290$. Estimated density, $\hat{D} = 255.6$ objects/km^2, with analytic $\widehat{se} = 38.3$ and 95% confidence interval [184.7, 353.8]. These estimates are very close to those obtained assuming cluster size is independent of distance, although the point estimate is rather closer to the true density of 240 objects/km^2. Average cluster size can be estimated by the ratio of estimated object density (255.6) to estimated cluster density (81.56), giving 3.134. The precision of this estimate could be quantified using the bootstrap. In each bootstrap resample, both densities, and hence their ratio, would be estimated, and a variance and confidence interval obtained as described in Section 4.7.3.

This procedure cannot generally be recommended. However, it may be useful if the population being sampled occurs in loose aggregations, rather than tight, easily defined clusters. The distance to each individual object should ideally be measured in this case, although it may be sufficient to record positions and sizes of smaller groups within a cluster. The method will often perform poorly unless sample size is fairly large.

4.8.2.3 *Stratification*

Choice of number of strata is determined largely by sample size. The more strata, the greater the reduction in size bias, but an adequate sample size for estimating $f(0)$ is required in each stratum (perhaps at least 20–30 per stratum). Defining two strata, corresponding to cluster sizes 1–3 and ≥ 4, sample sizes before truncation are 52 and 53, respectively. If four strata are defined, for cluster sizes 1–2, 3, 4 and ≥ 5, sample sizes before truncation are 29, 23, 26 and 27. The data were analysed for both choices of stratification.

Results are summarized in Table 4.5. In this case, no precision is lost by stratification, despite the small samples from which $f(0)$ was estimated, and the estimated densities were closer to the true value of 240 objects/km^2 than for the case without stratification. In our experience, loss of precision arising from stratification by cluster size is seldom large, provided sample size in each stratum does not fall below 20, and the method is a simple way of reducing the effects of size-biased sampling. Mean cluster size may

Table 4.5. Summary of results for different stratification options. The model was uniform with cosine adjustments; distance data were truncated at $w = 20$ m. True $D = 240$ objects/km^2

Cluster sizes	Sample size after truncation	Effective strip half-width (m)	AICc	$\hat{D}$	$\widehat{se}(\hat{D})$	95% confidence interval for D
All	89	11.4	506.3	265.8	41.9	(191.1, 369.5)
1–3	51	11.0	286.1	112.4	19.4	
4–9	38	12.1	221.8	147.4	35.0	
All			508.0	259.8	40.0	(190.1, 355.0)
1–2	29	6.6	148.6	84.0	23.4	
3	22	13.7	131.6	50.1	16.3	
4	22	11.7	127.4	78.2	21.0	
5–9	16	20.0	95.9	43.2	11.2	
All			503.4	255.5	37.1	(191.6, 340.8)

be estimated by a weighted average of the mean size per stratum, with weights equal to the estimated density of clusters by stratum. Alternatively, $E(s)$ may be estimated as overall $\hat{D}$ from the stratified analysis divided by $\hat{D}_s$ from the unstratified analysis. For two strata, this yields $\hat{E}(s) = 259.8/81.56 = 3.185$, and for four strata, $\hat{E}(s) = 255.5/81.56 = 3.133$. Both estimates are close to the true mean cluster size of 3.0. The bootstrap could be used to estimate the precision of these estimates.

Inspection of AICc values in Table 4.5 suggests that the analysis based on four strata is to be preferred. By fitting separate detection functions in each stratum, this statistic suggests that we have been able to fit the data appreciably better than for analyses without stratification or with just two strata. Thus we estimate density by $\hat{D} = 255.5$ objects/km^2, with 95% confidence interval [191.6, 340.8].

The reader is referred to Drummer (1985) and Quinn (1985) for further information on stratification.

4.8.2.4 *Regression estimator*

The procedure we recommend in most cases is a regression of s_i or $\log_e(s_i)$ on $\hat{g}(x_i)$ (Section 3.5.4). This allows an estimate of $E(s)$ at locations where $\hat{g}(x_i) = 1$; that is, locations at which detectability is certain, where size bias should not occur. Proper truncation of the distance data should be considered prior to the regression analysis (e.g. $g(x) \doteq 0.15$). Applying this method to the example, with dependent variable $\log_e(s_i)$, yields $\hat{E}(s) = 2.930$ and $\widehat{se}\{\hat{E}(s)\} = \sqrt{\widehat{var}\{\hat{E}(s)\}} = 0.139$, which is close to the true mean cluster size of 3.0 with good precision. The corresponding density estimate

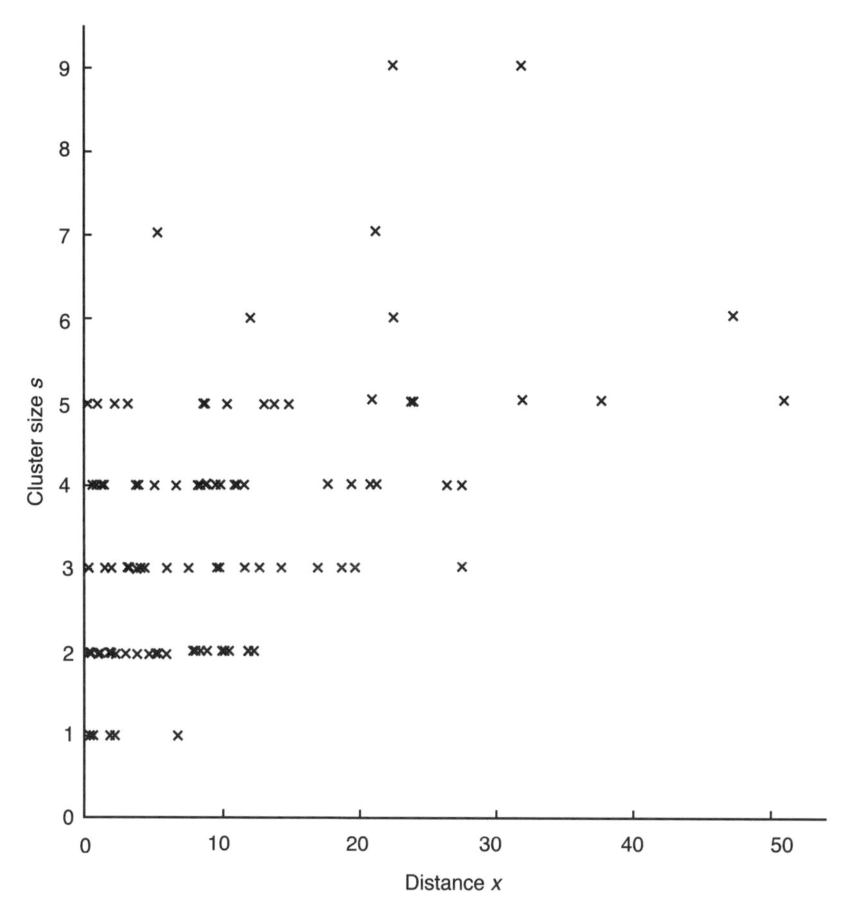

Fig. 4.6. Scatter plot of the relationship between the size of a detected cluster and the distance from the line to the geometric centre of the cluster for the example in which probability of detection is a function of cluster size. The correlation coefficient is 0.485 ($w = \infty$).

is 239.0 individuals/km^2, with $\widehat{se} = 38.1$, $cv = 16.0\%$ and 95% confidence interval [175.1, 326.1]. The regression approach allows $f(0)$ to be estimated using all the robust theory available and then treats the estimation of mean cluster size as a separate problem. A scatter plot of cluster size against distance or estimated detection probability can be used to investigate the form of the relationship, although the scatter can be wide (Fig. 4.6).

The regression method of estimating $E(s)$ generally reduces bias. By modelling cluster size as a function of detection probability, precision often improves relative to estimating $E(s)$ by the sample mean, $\bar{s}$. Even when variance increases, $var\{\hat{E}(s)\}$ is typically a small component of $var(\hat{D})$, so that little precision is lost by applying an adjustment for size bias. Use of $\bar{s}$ as an estimate of $E(s)$ is not recommended if dependence between cluster size and detection distance is suspected.

4.8.2.5 *Bivariate models for the detection function*

Drummer and McDonald (1987) proposed three bivariate parametric models for the detection function. Independently, Ramsey *et al.* (1987) developed a covariate approach for distance sampling, in which one of the covariates might be cluster size. These methods are discussed in Section 3.7.2. Other authors to consider this issue include Otto and Pollock (1990) and Quang (1991). Marques and Buckland (in preparation) generalized the methods of Ramsey *et al.* and these methods now supersede those of Drummer and McDonald, whose software SIZETRAN (Drummer 1991) was restricted to ungrouped, untruncated data, and often failed to converge. Bivariate and general covariate models are covered in Buckland *et al.* (in preparation).

4.9 Assumptions

Assumptions of line transect sampling are covered in detail in Section 2.1, and further discussion is given in Chapter 7. We outline a few issues here, to counter some of the more common misconceptions about what is assumed in deriving valid density or abundance estimates.

4.9.1 *Independence*

Most model selection procedures and some variance estimation procedures assume that objects are randomly and independently distributed throughout the study area. Provided lines are randomly located, or a systematic grid of lines is randomly positioned in the study area, we do not require this assumption to estimate object density. If variance is estimated from replicate lines, or by resampling lines, we again do not need the assumption that objects are distributed independently of each other.

Whether or not objects are distributed independently, we assume that they are detected independently. Usually, departures from this assumption are not too problematic. Robust variance methods should be adopted, with care taken to ensure that the correct sampling unit is selected; whether a detection associated with one sampling unit is made should be largely independent of detections made in other sampling units. Often, all effort carried out by a single observer in a single session comprises a suitable sampling unit. If random lines are used, the appropriate sampling unit is the line, and all data associated with it. If the design comprises a systematic grid of lines, use of lines as sampling units should again prove satisfactory. These issues are discussed further in Section 3.6. The most extreme departure from independent detections is when objects occur in clusters. Strategies for this case are outlined in Section 4.8.

Intuitively, it seems that dependence between detections might generate bias in estimated object density. Suppose density varies substantially. Detected objects in areas of high density will tend to have more near-neighbours than those in areas of low density. If the detection of one object makes it more likely that nearby objects will also be detected, then additional detections will tend to be triggered, and hence the effective width of the strip increased, in the higher density areas. Provided detection on the line or at the point is certain, this does not feed through to bias in estimated object density; the pooling robustness property (defined in Chapter 2) can be shown mathematically to apply. Examples where this property is especially useful include songbird surveys if the presence of one singing bird triggers another to sing, or surveys in which the observer physically measures the distance to detected objects (e.g. dung piles), as more objects might become visible when the observer leaves the line to measure the distance to a detected object. In each case, there will be a greater tendency to record so-called secondary detections in areas of higher object density.

4.9.2 *Detection on the line*

Estimation of $g(0)$ for line transect surveys is considered in Buckland *et al.* (in preparation). Whenever possible, surveys should be designed to ensure that $g(0) = 1$. The solution of guarding the centreline can be counterproductive if this gives rise to two detection processes. If one process generates detections at large distances, but is such that $g(0)$ is appreciably less than one, and the other generates detections only at small distances, then the composite detection function will be impossible to model adequately. If such a field strategy is adopted, data from the two processes should be recorded and analysed separately, although problems are likely to remain. If it is suspected that $g(0)$ is less than one, methods that might increase it include using more observers to cover the line, travelling more slowly along the line, using only experienced observers, improving the training of observers, and upgrading optical aids. For terrestrial surveys in which animals are flushed, trained dogs can be an effective aid, allowing a wider area to be efficiently searched.

4.9.3 *Movement prior to detection*

Random movement of objects before detection generates positive bias in estimates of object density. Hiby (1986) showed that bias is small provided that object movement is slow relative to that of the observer (less than one half of the observer's speed). A strategy for line transect analysis of fast-moving objects is outlined in Section 6.5. Movement in response to the observer is problematic, and is discussed in Section 2.1. From a practical viewpoint, field procedures should be developed that ensure that most detections occur at distances for which responsive movement is unlikely to have occurred. In other words, the observer should strive to detect the object before the object is able to move far from its initial position in response to the observer's presence. If this is not possible, the methods of Turnock and Quinn (1991) or of Borchers *et al.* (1998a) (which are also designed to allow estimation of $g(0)$) might be attempted.

If the distance data appear to have a distinct mode away from the origin, the analysis is problematic. This might happen by chance, as a result of heaping, or through the presence of evasive movement prior to detection. Some robust models will attempt to fit the data near the origin, so that the mode of the density function is to the right of the origin. In these cases, it is often prudent to constrain the estimated detection function to be monotone non-increasing in an attempt to minimize bias. A weak constraint is to impose the condition $\hat{g}(x) \leq g(0) = 1$ for all $x > 0$. This condition is often sufficient to achieve a monotone non-increasing function and satisfactory estimates of $f(0)$. The alternative is a constraint forcing a strict non-increasing function such that $\hat{g}(x_i) \geq \hat{g}(x_{i+1})$, where $x_i < x_{i+1}$. The default option of Distance imposes the strong constraint. To reduce the computational cost of applying the constraint, the estimated density is evaluated at a relatively small number of points, with more points close to zero. The fit is modified if the constraint is not satisfied for all successive pairs of these points. The user may instead select the weak constraint, or override both constraints. Distance warns the user if a constraint has caused the model fit to be modified. In this case, the bootstrap estimate of $var(\hat{f}(0))$ is recommended, since the assumptions on which the analytic variance is based are violated.

4.9.4 *Inaccuracy in distance measurements*

Bias in distance estimation should be avoided. If distances are consistently overestimated by 10%, densities are underestimated by 9%; if they are underestimated by 10%, densities are overestimated by 11%. Chen (1998) notes that, even if distance estimation is unbiased on average, measurement errors lead to downward bias in $\hat{D}$ that does not reduce as sample size increases. For shipboard surveys, reticles or graticules (Section 7.4.1.1) are almost essential for accurate distance estimation. In terrestrial surveys, short distances can often be measured, and survey lasers are accurate to within a metre for distances from about 20 m to several hundred metres. Distance categories can be accurately determined in aerial surveys by lining up markers on the windows with markers or streamers on the wing struts, although the height of the aircraft must be accurately measured and constant. In hilly terrain, perpendicular distances from the aircraft must be determined by other means.

Heaping is a common feature of distance data for which measurement is inaccurate. A disproportionate number of distances are recorded as round values such as 10 or 100 m. Unless heaping is severe, reliable estimation is generally possible except when a high proportion (say $> 10\%$) of distances is recorded as zero, which generates a 'spiked' histogram. This commonly occurs in shipboard surveys when perpendicular distances are calculated as $x = r \sin(\theta)$, where r is the sighting distance and θ the sighting angle. If the sighting angle is not measured accurately, rounding of small angles to zero is common, so that x is calculated to be zero. Surveys of dung piles also frequently show a high proportion of zero distances; for such surveys, detection distances are typically very small, and especially if the line isn't accurately located, many are recorded as on the line. It is not possible to estimate object density reliably from such data, so every effort should be made to avoid rounding distances to zero.

4.10 Summary

Data analysis is relatively easy if the survey is well designed, sample size adequate, and the data are collected so as to satisfy the main assumptions. The analysis of small samples, especially where some assumptions have been violated, is more problematic. Adequate analysis cannot be carried out without specialist software.

An objective strategy must be followed, such as that outlined in this chapter and Section 2.5. The data must be checked for recording or data-entry errors. Plotting the distance data as histograms will often reveal anomalies that must be further considered. Truncation of some observations in the right tail of the distance data should always receive consideration. Several candidate, robust models should be considered. The use of AIC and other criteria are helpful in selecting the best model, or a small subset of good models, for final analysis and inference. Once a model is selected, maximum likelihood estimation is used to obtain parameter estimates and measures of their precision. With good data (adequate sample size and validity of the key assumptions), inference using two or three good, robust models is likely to yield similar estimates. Where this is not the case, the analyst might wish to consider model averaging, which provides a mechanism for drawing inference that is not conditional on a single selected model (Buckland *et al.* 1997; Burnham and Anderson 1998).

If objects on the centreline are missed, $E(\hat{D})$ will be too low. If 20% of objects on the centreline are missed, the density estimate can be expected to show a negative bias of around 20%. Movement prior to detection is also problematic. Measurement errors should be reduced to a minimum. If measurements are subject to rounding but roughly unbiased, provided there is not substantial rounding at zero, estimation is unlikely to be seriously compromised. Bias in measurements generates a similar degree of bias in the estimate of object density, and should be avoided. Valid inference is only possible with good field design and careful attention to field methods, to ensure that the assumptions are met. While analysis procedures are robust to some types of assumption failure, there is no substitute for quality data that meet all the key assumptions. Searching should be conducted such that the distance data have a broad shoulder. The presence of a shoulder makes model selection less important and improves the quality of inference. The reader is urged to study the material in Chapter 7 prior to the conduct of a survey involving distance sampling.

The extension of these strategies for analysis carry over to more complicated surveys involving stratification, surveys repeated in time using the same lines, multiple observers, aerial or underwater platforms, or samples of very large areas. Some of these issues are illustrated in Chapter 8 (and by Burnham *et al.* 1980: 41–55), and specialized theory is extended in Buckland *et al.* (in preparation).

Surveys of clustered populations require additional care in counting the number of individuals in each cluster detected and addressing the possible size-biased aspects of such sampling. Plotting the cluster sizes s_i against the x_i distances is always recommended. Our experience suggests that size bias is often a minor issue if cluster size is not too variable; proper truncation of perpendicular distance data can often allow simple models to provide valid inference concerning the density of clusters and individuals. However, if the largest cluster is say more than five times the size of the smallest, correction for possible size bias should be investigated. When cluster size is highly variable (e.g. from one or two individuals to many thousands, as in some species of marine mammals), then very careful modelling and analysis of the data is required.

Populations in large, loose aggregations, scattered around the sample area, are problematic. Theory and software are readily available for the analysis of sample data from populations of individuals randomly distributed in space, and the same is true of populations distributed under some regular stochastic process that generates some degree of spatial aggregation, by computing $var(n)$ empirically. Good theory and software also exist for the analysis of populations that occur in definable clusters where the cluster size is not highly variable. Difficulties arise when populations are spatially distributed in loose clusters whose boundaries, and therefore size, must be determined subjectively. If at all possible, the location of each individual object should be recorded in this circumstance, so that the method of Section 4.8.2.2 can be applied, but the cluster to which each individual belongs should also be noted, to allow comparative analyses of clusters. Populations in large or highly variable groups require great care in estimating $E(s)$ in ways that minimize or avoid bias. Estimation of average cluster size must receive special emphasis in the design of the survey and the pilot study (e.g. temporarily leaving the planned centreline in aerial surveys of cetaceans to count individuals more accurately).

The following is intended as a crude checklist of the stages required to carry out a full analysis of line transect data. Not all steps are necessarily required in any given analysis, especially if similar data sets have been analysed previously.

1. Key in and validate the data. The data should not be aggregated in any way prior to entry. Thus if distances are ungrouped, they should not be entered as grouped data, even if they are subsequently to be grouped for analysis. Distances should be entered by line, so that individual lines can be defined as the sampling units. For stratified designs, these lines should be allocated to their strata. Distance guides the user through the data entry process.
2. Plot histograms of the perpendicular distance data, using different choices for the cutpoints, and fit a preliminary model to the data. Examine the histograms for evidence of failure of assumptions. Review the required assumptions, and consider which might be dubious, given the biology and behaviour of the species and the field protocol. Interpret the histograms in the light of this review. If data are ungrouped, assess whether they should be grouped before analysis, selecting group cutpoints to reduce the effect of heaping, or to alleviate the effects of a spurious spike in the data at zero distance (Section 4.5). Identify a suitable truncation point w for perpendicular distances, preferably such that $\hat{g}(w) \simeq 0.15$, although truncation of roughly 5% of observations is often satisfactory (Section 4.3). Assess from the histograms whether this truncation distance is reasonable; if not, select one or more alternatives.
3. Fit several models that satisfy the model robustness, shape and estimator efficiency criteria. We recommend suitable combinations of a half-normal, uniform or hazard-rate key with simple or Hermite polynomial or cosine adjustments. Select a single model, for example using AIC, and assess its adequacy using goodness of fit (Section 4.6). If the fit is poor, investigate the reasons, and evaluate possible solutions. Assess the sensitivity of estimation to the model selected; if sensitivity is high (e.g. the detection curve is excessively spiked under one or more models), examine whether the estimates from the selected model should be replaced or supplemented by those from other models that yield adequate fits. If necessary, consider using model averaging techniques.
4. If the detections are of clusters of objects, assess whether there is evidence of size bias, and if necessary, try one or more of the methods of Section 4.8 to correct for it.

5. Having identified a model for the perpendicular distance data, review the options for variance estimation, for stratifying some or all components of estimation, and for including covariates. Select options that are likely to reduce bias; of the options remaining, select those that yield the most efficient estimation. Fit the data using the preferred model(s) and options.

4.11 Exercises

1. (a) In a line transect survey, perpendicular distances to detected objects are measured accurately, and truncated at distance w, such that $g(w) \simeq 0.15$. All main assumptions are met. One analyst chooses to analyse the exact distances, whereas a second chooses to group the data into ten equal intervals, with cutpoints at $0, 0.1w, 0.2w, \ldots, 0.9w, w$. They argue over the relative merits of the analyses. Do you think this decision is likely to have much effect on estimated density $\hat{D}$ or its variance?

(b) If perpendicular distance data were grouped in the field into eight intervals, does this necessarily mean that the assumption that distances are correctly recorded is violated? Explain.

(c) Which of the following schemes for grouping the data into eight intervals is best? Which is worst? Explain your answer.

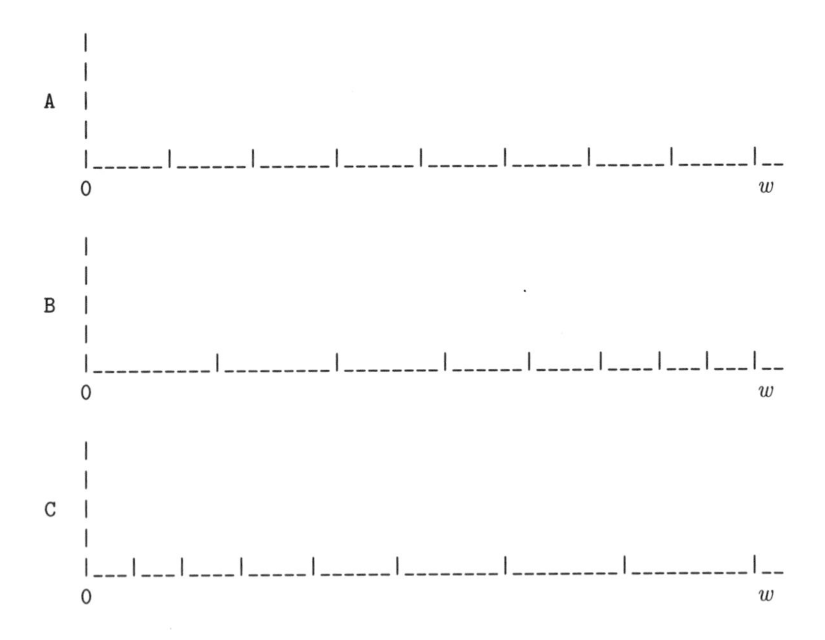

2. If a line transect survey is to be conducted on each of the following species, with the mode of travel as stated, which assumption (if any) is most likely to be violated?

(a) Blue whales (aerial)

(b) Deer in scrub habitat (walking)

(c) Deer in scrub habitat (aerial)

(d) Rabbits (horseback)

(e) Partridges (walking, with dog)

(f) Saguaro cacti (walking)

(g) Monkeys in rain forest (walking)

(h) Bottom fish (SCUBA diving)

3. In a line transect survey, numbers of ruffed grouse detected by 5 m perpendicular distance band were:

Distance (m)	0-5	5-10	10-15	15-20	20-25	25-30
Frequency	103	62	28	12	9	4

The negative exponential and hazard-rate models were fitted to these data using Distance. The negative exponential model gave a χ^2 goodness of fit statistic of 1.70 (4 df) and an AIC value of 587.95. The equivalent values for the hazard-rate model were 0.62 (3 df) and 588.86, respectively. The following was extracted from the output files.

```
   Negative Exponential key, k(y) = exp(-y/A(1))


           Point      Standard   Percent Coef.      95 Percent
Parameter  Estimate    Error     of Variation    Confidence Interval
---------  --------   --------   ------------   -------------------
  f(0)      .13331     .01033       7.75         .11455     .15513
  p         .25005     .01937       7.75         .21487     .29099
  ESW       7.5015     .58121       7.75         6.4461     8.7297
  n/L       2.1800     .14765       6.77         1.9093     2.4891
  D         1.4530     .14953      10.29         1.1883     1.7768
---------  --------   --------   ------------   -------------------

  Hazard Rate key, k(y) = 1 - Exp(-(y/A(1))**-A(2))

           Point      Standard   Percent Coef.      95 Percent
Parameter  Estimate    Error     of Variation    Confidence Interval
---------  --------   --------   ------------   -------------------
  f(0)      .09552     .00808       8.46         .08094     .11272
  p         .34897     .02954       8.46         .29571     .41183
  ESW       10.469     .88617       8.46         8.8713     12.355
  n/L       2.1800     .14765       6.77         1.9093     2.4891
  D         1.0412     .11287      10.84         .84238     1.2868
---------  --------   --------   ------------   -------------------

 Measurement Units
 ---------------------------------
 Density: Numbers/Hectare
     ESW: metres
```

(a) Why do the estimates of the three parameters '`f(0)`', '`p`' and '`ESW`' all have the same coefficient of variation?

(b) Define AIC as implemented within Distance, and explain how it is used. Which of the two models gave the larger likelihood in the above analyses?

(c) Summarize the relative merits of the two models for analysing these data. Which model would you choose? Why?

(d) If the analyst believed that there was no rational basis for preferring one model over the other, explain how the bootstrap may be used so that model selection uncertainty is reflected in variance estimates and confidence intervals. The above confidence intervals for $f(0)$ do not overlap. Assuming that you use the percentile method to set 95% confidence limits from your bootstrap estimates, indicate roughly the range of values you would expect your interval to span, and why.

4. (a) Tables 1–3 contain estimates from Distance of a line transect survey of a desert tortoise population in two geographic strata ('North' and 'South'). Estimates labelled 'All' are from data pooled over both strata. The animals occur in clusters. Observers walked along marked lines and used laser range-finders to measure distances to tortoises.

Table 1. Estimates of $f(0)$ (in km^{-1}). cv is the coefficient of variation, n is sample size, AIC is Akaike's Information Criterion. Estimation was performed using interval data (five intervals in all cases). pr($>\chi^2$) is the reported probability of obtaining a greater χ^2 goodness-of-fit test statistic than that observed, with degrees of freedom indicated by df

Stratum	$f(0)$	cv	n	AIC	pr($>\chi^2$)	df
All	0.319	0.10	125	396.41	0.91	2
North	0.296	0.11	78	251.11	0.86	2
South	0.415	0.30	47	147.91	0.50	2

Table 2. Expected cluster size estimates. $\hat{E}(s)$ is the expected cluster size from a regression of $\log(s)$ on $\hat{g}(x)$. pr($T < t$) and df show the significance of the regression, and the corresponding degrees of freedom

Stratum	Param.	Estimate	Std. error	df	pr($T < t$)
All	$\hat{E}(s)$	3.4	0.19	123	0.84
	$\bar{s}$	3.2	0.15		
North	$\hat{E}(s)$	3.3	0.26	76	0.73
	$\bar{s}$	3.1	0.20		
South	$\hat{E}(s)$	3.4	0.28	45	0.45
	$\bar{s}$	3.3	0.23		

Table 3. Encounter rates (animals per km of transect)

Stratum	n/L	cv	Surface area (km^2)
All	0.60	0.12	1680
North	0.59	0.13	1040
South	0.61	0.24	640

(i) Decide whether to estimate $f(0)$ separately in each stratum or from the pooled data. Give reasons.

(ii) Decide whether to estimate mean cluster size separately in each stratum or from the pooled data. Give reasons.

(iii) Decide whether to estimate mean cluster size from the regression, or using the observed mean. Give reasons.

(iv) Irrespective of what you decided above, estimate individual abundance and its cv using stratified encounter rates, the $f(0)$ estimate from the pooled data, and the associated mean cluster size estimate from regressing $\log(s)$ on $\hat{g}(x)$.

(b) Desert tortoises are a protected species. They move very slowly and never very far, live a long time, reproduce very slowly, and spend a large proportion of their time underground. Older animals are noticeably larger than young animals; sex can only be determined by close examination. How suitable is the line transect method for estimating the abundance of this species? Consider its main assumptions, and comment on how likely they are to be met. Discuss what further information would be required to yield a reliable estimate of abundance. How might this information be obtained?

5. (a) Shown below are selected parts of analyses, carried out in Distance, corresponding to three repeat counts of line transect surveys for red foxes. Interpret the analyses. What evidence is there that an assumption has failed for these data?

(b) The transects in this study were carried out by driving along minor roads after dark, using spotlights to search for foxes. Discuss the possible reasons for the assumption failure noted in part (a). For each reason you suggest, discuss whether the analyses as carried out are likely to be reliable, or whether a simple adjustment can be carried out to the analyses to make them more reliable.

```
Effort         :   144.0000
 # samples     :     3
 Width         :    165.0000
 # observations:   132
Model
  Half-normal key, k(y) = Exp(-y**2/(2*A(1)**2))
  Hermite polynomial adjustments of order(s) :  4
```

```
************************************
*  Probability Function Estimation *
*   Detection Probability Plot     *
************************************
```

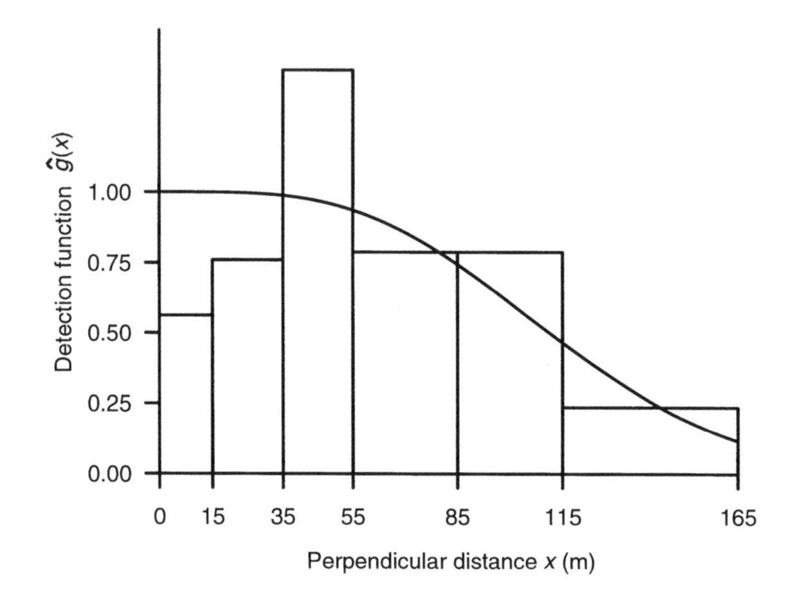


```
************************************
*  Probability Function Estimation *
* Chi-Square Goodness of Fit Test  *
************************************
```

Cell i	Cut Points		Observed Values	Expected Values	Chi-square Values
1	0.00	15.0	10.0	17.77	3.395
2	15.0	35.0	18.0	23.59	1.325
3	35.0	55.0	34.0	22.90	5.377
4	55.0	85.0	28.0	30.27	0.170
5	85.0	115.	28.0	21.58	1.912
6	115.	165.	14.0	15.89	0.226

```
Total Chi-square value =    12.4041  Degrees of Freedom =   3

Probability of a greater chi-square value, P = 0.00612

**************************************************************
  Model
    Hazard Rate key, k(y) = 1 - Exp(-(y/A(1))**-A(2))
```

```
************************************
*  Probability Function Estimation *
*   Detection Probability Plot     *
************************************
```

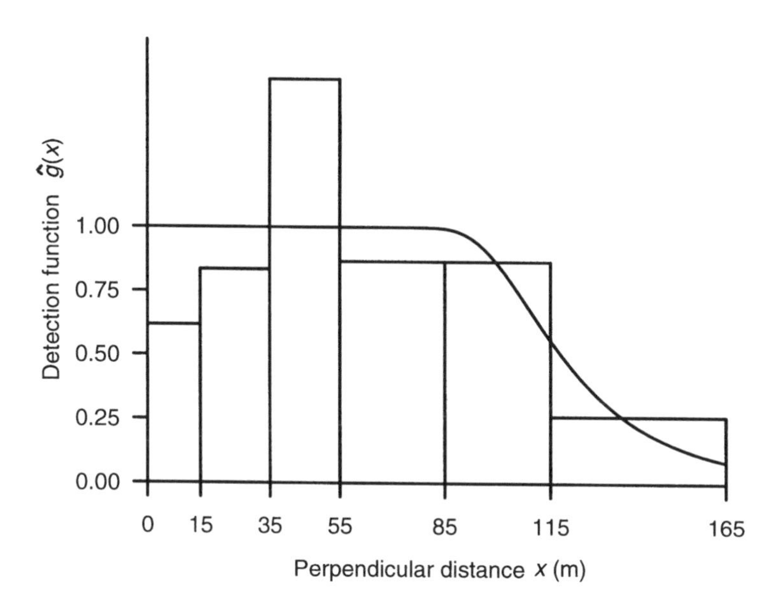


```
************************************
*  Probability Function Estimation *
* Chi-Square Goodness of Fit Test  *
************************************
```

Cell i	Cut Points		Observed Values	Expected Values	Chi-square Values
1	0.00	15.0	10.0	16.18	2.360
2	15.0	35.0	18.0	21.57	0.592
3	35.0	55.0	34.0	21.57	7.159
4	55.0	85.0	28.0	32.35	0.584
5	85.0	115.	28.0	26.82	0.052
6	115.	165.	14.0	13.51	0.018

```
Total Chi-square value =   10.7652  Degrees of Freedom =   3
```

```
Probability of a greater chi-square value, P = 0.01307
```

```
                           ******************************
                           *     Estimation Summary     *
                           *   Detection probability    *
                           ******************************


Combined Estimates:
                           Estimate     %CV     df   95% Confidence  Interval
                          ------------------------------------------------------
Half-normal/Hermite
               m          2.0000
               AIC        1319.4
               Chi-p    0.61196E-02
               f(0)     0.89737E-02   15.36    130  0.66344E-02  0.12138E-01
               p        0.67538       15.36    130  0.49932      0.91351
               ESW       111.44       15.36    130   82.388      150.73
Hazard/Cosine
               m          2.0000
               AIC        1316.2
               Chi-p    0.13066E-01
               f(0)     0.81713E-02    5.22    130  0.73701E-02  0.90596E-02
               p        0.74170        5.22    130  0.66897      0.82233
               ESW       122.38        5.22    130   110.38       135.68



                           ******************************
                           *     Estimation Summary     *
                           *     Density/Abundance      *
                           ******************************


                           Estimate     %CV     df   95% Confidence Interval
                          ------------------------------------------------------
 Stratum: count 1
 Half-normal/Hermite
                DS         4.3934      21.18   130   2.9028          6.6493
                D          4.5450      21.29   135   2.9973          6.8919
 Hazard/Cosine
                DS         4.0005      15.49   130   2.9498          5.4255
                D          4.1787      15.64   164   3.0744          5.6796
 Stratum: count 2
 Half-normal/Hermite
                DS         3.7390      22.04   130   2.4301          5.7529
                D          3.9226      22.23   139   2.5411          6.0551
 Hazard/Cosine
                DS         3.4047      16.65   130   2.4547          4.7224
                D          3.5815      16.89   168   2.5720          4.9874
 Stratum: count 3
 Half-normal/Hermite
                DS         4.2064      21.40   130   2.7675          6.3934
                D          4.5315      21.63   140   2.9695          6.9152
 Hazard/Cosine
                DS         3.8303      15.79   130   2.8078          5.2252
                D          4.1926      16.09   173   3.0580          5.7481
```

```
******************************
*     Estimation Summary     *
*      Density/Abundance     *
******************************
```

Pooled Estimates:

		Estimate	%CV	df	95% Confidence Interval	
Half-normal/Hermite						
	DS	4.1129	16.07	98	2.9961	5.6461
	D	4.3330	16.07	98	3.1563	5.9485
Hazard/Cosine						
	DS	3.7452	7.04	10	3.2017	4.3808
	D	3.9843	7.27	8	3.3703	4.7100

6. (a) In line transect sampling, the half-normal detection function is given by $g(x) = \exp(-x^2/2\sigma^2), 0 \leq x < \infty$, with $\mu = \int_0^\infty g(x)\,dx = \sqrt{\pi\sigma^2/2}$.

(i) Given that the estimated density of objects may be expressed as $\hat{D} = n\hat{f}(0)/2L$, show that

$$\hat{D} = \left[2\pi L^2 \sum_{i=1}^{n} \frac{x_i^2}{n^3}\right]^{-0.5}$$

where n is the number of objects detected, x_i is the perpendicular distance to the ith detected object, $i = 1, \ldots, n$, $f(x)$ is the pdf of perpendicular distances x, and L is the total length of transects.

(ii) How does the above formula change if objects occur in clusters, n clusters are detected, and mean cluster size in the population is estimated by mean cluster size in the sample?

(b) A data set together with the table of estimates from Distance are given below.

(i) Give a brief description of the parameters estimated by Distance.

(ii) Why are the confidence intervals asymmetric?

(iii) Use the result of part (a)(i) to evaluate the density of clusters and of objects.

(iv) Why do your values from part (iii) differ slightly from those given by Distance?

Summary of data:

```
perpendicular distances in metres (cluster sizes in parentheses)
Line 1:  78.6 (1); 102.4 (4); 124.4 (2);  37.6 (3);  47.8 (7);  84.5 (3);
         13.4 (2);  58.0 (2);  74.5 (5); 114.5 (3);   8.6 (4);  92.3 (6);
        125.1 (4);  60.8 (8)
Line 2:  91.5 (1);  63.8 (2); 212.1 (3)
Line 3:  38.5 (4); 125.5 (4);  47.0 (2); 178.6 (3); 145.5 (1);  51.1 (4);
         42.1 (2);  35.8 (5)
Line 4: 111.6 (4); 122.3 (2);  18.5 (3); 358.2 (4);  26.1 (2);  61.5 (4);
         97.4 (3);  40.1 (2);  97.3 (5)
```

```
Truncation distance w=240m.
Model = half-normal
```

Table of estimates from output file:

Parameter	Point Estimate	Standard Error	Percent Coef. of Variation	95 Percent Confidence Interval	
f(0)	.8720E-02	.1257E-02	14.4	.6512E-02	.1168E-01
p	.4778	.6886E-01	14.4	.3568	.6399
ESW	114.7	16.53	14.4	85.63	153.6
n/L	.9848E-01	.1932E-01	19.6	.5307E-01	.1827
DS	.4294	.1045	24.3	.2387	.7723
SBAR	3.333	.2876	8.6	2.797	3.972
D	1.431	.3696	25.8	.7967	2.571
N	1514.	391.0	25.8	843.0	2721.

7. In a field test of line transect sampling, a known number of objects, N_a, was distributed uniformly through a surveyed strip of half-width w and length L, so that the surveyed area was $a = 2wL$. Two teams independently surveyed the strip. Both detected exactly half of the objects present, and recorded the distances from the line of the objects detected. These distances were analysed using conventional line transect methods, and P_a, the proportion of objects detected, was estimated for each team. Team A obtained $\hat{P}_a = 0.48$, close to the true value. However, team B obtained $\hat{P}_a = 1.0$. Explain how this is possible, and sketch a detection function for each team that is consistent with these results.

8. In a line transect survey, perpendicular distances were estimated by eye. Up to 10.5 m, these estimates were rounded to the nearest metre. Above 10.5 m, they were recorded by two-metre interval. Analyses of these data, carried out in Distance, are shown below. The χ^2 statistic is given for the half-normal model, but results are very similar for other plausible models.

(a) Why is the goodness of fit statistic highly significant? What assumption has failed? Will this adversely affect $\hat{D}$?

(b) Why does the histogram show a spike corresponding to zero distance when the frequency for 0 m is less than that for 1 m?

(c) Should we try to fit models with more adjustment terms to improve the fit?

(d) Would the problem be resolved by collecting more data?

(e) Is there evidence that the detection function has a shoulder?

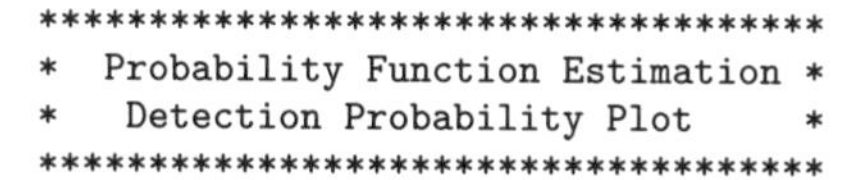

```
************************************
*  Probability Function Estimation *
*   Detection Probability Plot     *
************************************
```

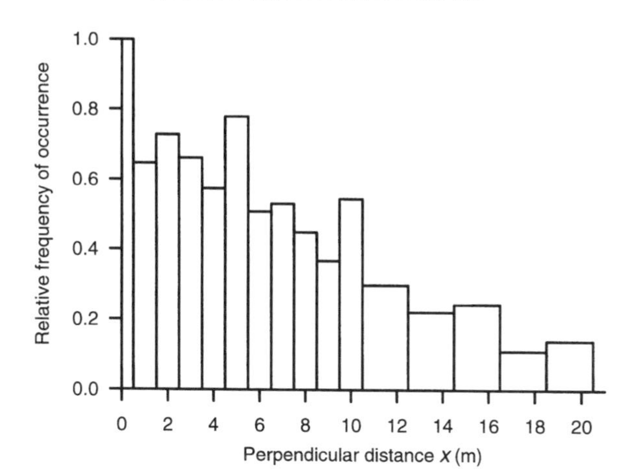


```
************************************
*  Probability Function Estimation *
* Chi-Square Goodness of Fit Test  *
************************************
```

Cell i	Cut Points		Observed Values	Expected Values	Chi-square Values
1	.000	.500	68.0	47.74	8.599
2	.500	1.50	88.0	94.98	.513
3	1.50	2.50	99.0	93.51	.322
4	2.50	3.50	90.0	91.11	.013
5	3.50	4.50	78.0	87.85	1.104
6	4.50	5.50	106.0	83.82	5.866
7	5.50	6.50	69.0	79.16	1.304
8	6.50	7.50	72.0	73.98	.053
9	7.50	8.50	61.0	68.42	.805
10	8.50	9.50	50.0	62.62	2.545
11	9.50	10.5	74.0	56.73	5.261
12	10.5	12.5	81.0	95.96	2.332
13	12.5	14.5	60.0	74.02	2.654
14	14.5	16.5	66.0	54.77	2.305
15	16.5	18.5	30.0	38.87	2.025
16	18.5	20.5	38.0	26.47	5.023

Total Chi-square value = 40.7237 Degrees of Freedom = 14

Probability of a greater chi-square value = .00020

9. If detections and distances are recorded for one side of the line only, how do formulae and analyses change (if at all)?

5

Point transects

5.1 Introduction

Songbird surveys often utilize point transects rather than line transects for several reasons. Once the observer is at the point, he or she can concentrate solely on detecting, locating and identifying birds, without the need to traverse what may be difficult terrain; he or she can take the easiest route into and away from the point, whereas good line transect practice dictates that the observer follows routes determined in advance and according to a randomized design. Further, patchy habitats can be sampled more easily by point transects. Frequently, density estimates are required for each habitat type, or estimation is stratified by habitat to improve precision. Even with the advent of geographic information systems, because the geometry of point sampling is simpler than that for line sampling, designing a point transect survey so that each habitat type is represented in the desired proportions is easier than for line transects. It is also easier to model density as a function of habitat if point samplers are used rather than line samplers. Other advantages of point transects are that known distances from the (stationary) observer may be flagged, to aid distance estimation, and only the observer-to-object distance is required, which is easier to estimate than the perpendicular distance required in line transect sampling if the observer is far from that part of the line closest to the object.

Point transect sampling is used mostly for bird surveys, although it is occasionally used for surveys of plants and of terrestrial mammals. Essentially the same theory is used for the cue count (Section 6.4) and trapping web (Section 6.8) methods. The disadvantages of point transect sampling that make it unsuitable for many purposes include the following. Objects may be disturbed or flushed by an observer approaching the point. It is difficult to determine which of these would have been detected from the point, but if they are ignored, density will be underestimated. The observer may detect many objects and waste much time while travelling between points; for line transects, a higher proportion of time in the field is spent surveying, and a higher proportion of detections is made while surveying. Thus, point transects may be inefficient for objects that occur at low densities. Point transect sampling, as usually implemented, is also less robust to object movement, whether responsive or not, than is line transect sampling. Conceptually, the objects around a point should be located at an instant in time. In practice, a count is carried out over several minutes, to allow the observer to detect and identify objects. If the objects are mobile, this leads to overestimation of density, both because the objects tend to be more detectable when they are closer to the observer, and because new objects enter the area of detectability while the count is being conducted. In some surveys, this has led to density estimates that are an order of magnitude too high. To avoid the problem, a snapshot moment should be defined, and only objects whose position is known (at least approximately) at that moment should be recorded. This leads to a lower sample size, but also to lower bias. The reduction in sample size can be minimized by allowing the observer perhaps two or three minutes before the snapshot moment, to locate and identify objects around the point, and a few minutes more after the snapshot, to verify whether objects detected before that moment are still present, and to identify any remaining unidentified objects.

This chapter illustrates point transect analysis through simulated data sets for which the parameters are known. As in Chapter 4, a simple data set is first introduced. Truncation of the distance data, modelling the spatial variation of objects to estimate $var(n)$, grouping of data, and model selection philosophy and methods are then addressed. Having selected an appropriate model, estimation of density and measures of precision are discussed. In a final example, the objects are assumed to occur in clusters (e.g. family parties or flocks).

5.2 Example data

The example data were generated from a half-normal detection function, $g(r) = \exp(-r^2/2\sigma^2)$, $0 \leq r < \infty$, with $\sigma = 10$ m. There were $k = 30$ points, and the number of detections per point followed a Poisson distribution with parameter $E(n_i) = 5$, $i = 1, \ldots, k$. Each detection is of a single object. Thus $E(n) = 5 \times 30 = 150$, $h(0) = 1/\sigma^2 = 0.01$ and true density is

$$D = \frac{E(n) \cdot h(0)}{2\pi k} = 0.00796 \text{ objects/m}^2 = 79.6 \text{ objects/ha}$$

Untruncated data generated from this model, together with the fitted half-normal model, are shown in Fig. 5.1. Data truncated at 20 m and the corresponding fit of the half-normal are shown in Fig. 5.2. For comparison, the fit of the uniform detection function with one simple polynomial adjustment is shown in Fig. 5.3.

The histograms of Figs 5.1 and 5.2 illustrate two methods of presenting point transect data. In Figs 5.1b and 5.2b, the frequency of distances is shown by distance interval, as for a conventional histogram. The curve is the fitted probability density function (pdf) of recorded distances, with

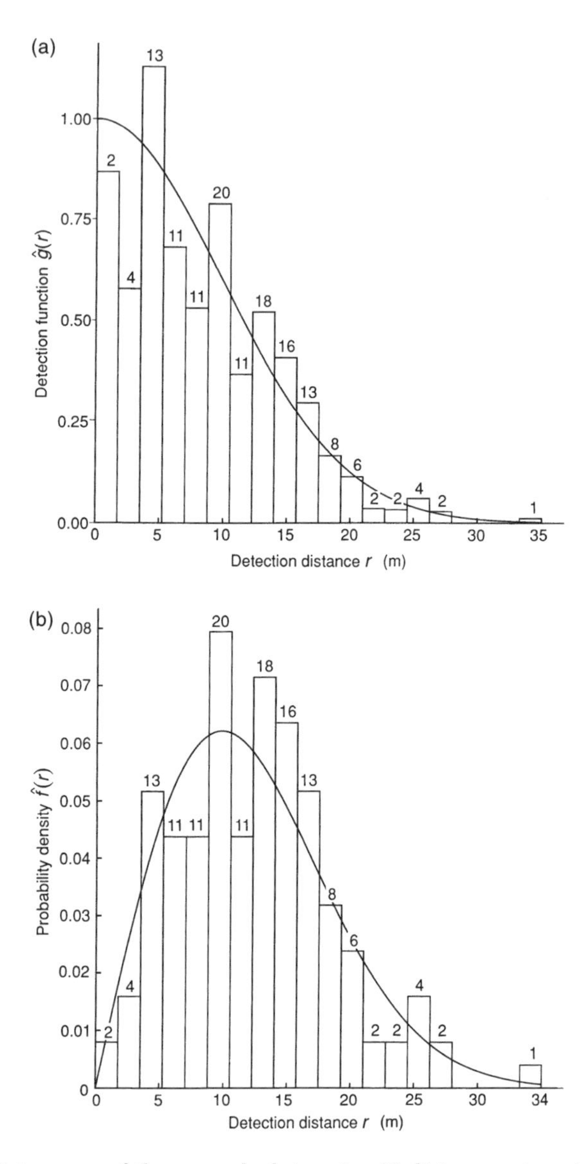

Fig. 5.1. Histograms of the example data using 20 distance categories. The fit of the half-normal detection function to untruncated data is shown in (a), in which frequencies are divided by detection distance, and the corresponding density function is shown in (b).

scale chosen to match that of the frequency data. The parameter $h(0)$ is estimated by the slope of this curve at distance $r = 0$. At small distances, the function increases because area surveyed at a given distance increases with distance from the point. For example, the area surveyed between r and

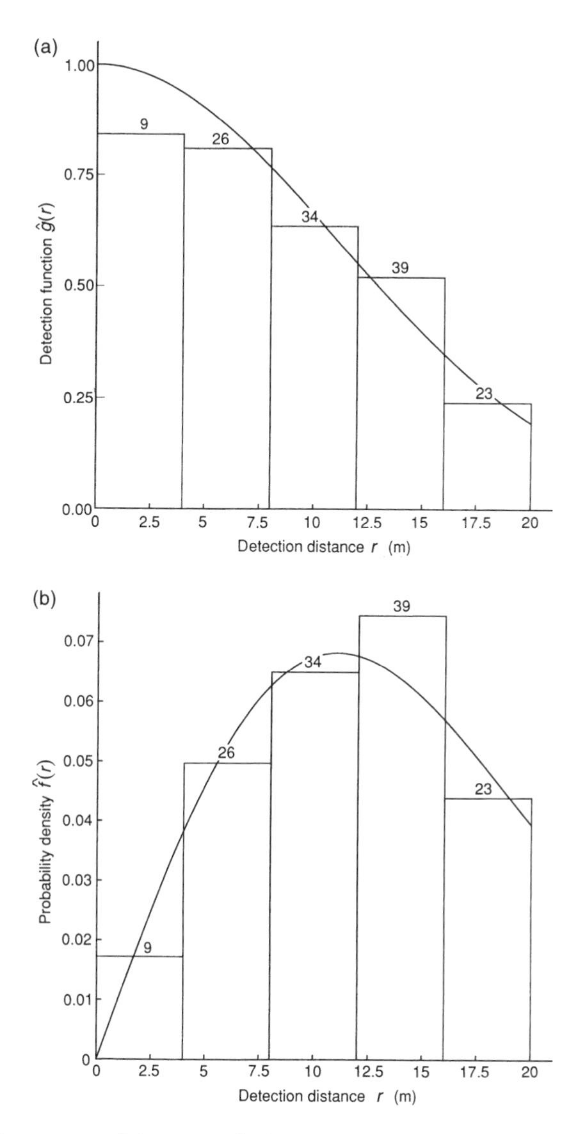

Fig. 5.2. Histograms of the example data, truncated at 20 m, using five distance categories. The half-normal model, fitted to the ungrouped data, is shown and was used for final analysis of these data. The fitted detection function is shown in (a), in which frequencies are divided by detection distance, and the corresponding density function is shown in (b).

$r+\delta$, where δ is small, is approximately $2\pi r\delta$, whereas the area between $2r$ and $2r+\delta$ is roughly $4\pi r\delta$. To examine how probability of detection falls off with distance, we must 'correct' for this increase in area, by assigning the ith detection distance r_i a weight $1/r_i$. Distance shows the fit of the

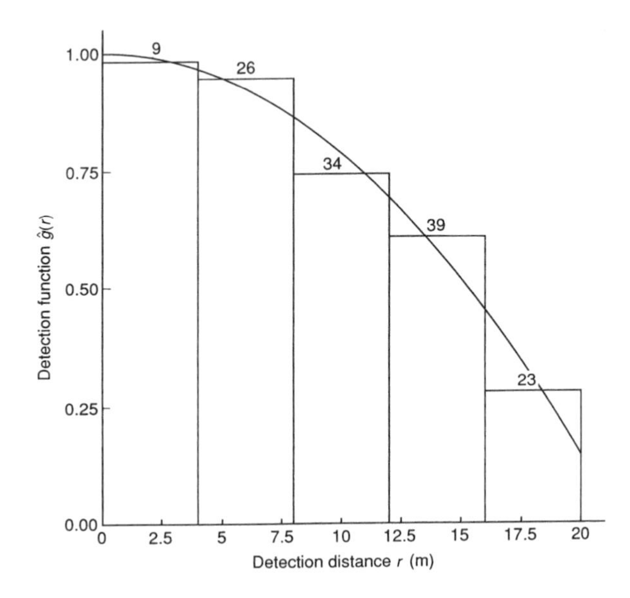

Fig. 5.3. The fit of the detection function using a uniform key with a single simple polynomial adjustment term to the example data, ungrouped and truncated at $w = 20$ m.

estimated detection function on a histogram of scaled frequencies (Figs 5.1a and 5.1b). For each distance interval of the histogram, the height of the histogram bar is calculated as follows. First, if the interval extends from r_B to r_E, then the area that falls between these distances from a single point is $\pi r_E^2 - \pi r_B^2$. For k points, the required area is thus $k\pi(r_E^2 - r_B^2)$. The number of objects detected in the interval divided by the above area is therefore the density of *detected* objects in that interval. Distance divides this by the estimated density of all objects (79.6 objects/ha for our example), and uses this as the height of the histogram bar. Note that the detection function may sometimes appear to fit badly at small distances, as in Fig. 5.2; this is not a programming error, but arises because of the deceptive nature of point transect data. Relatively few distances are recorded close to the point, where area surveyed is small, so the fit of the model is not heavily influenced by distances close to zero. Scaling frequencies by $1/r_i$ causes the first histogram bar, corresponding to the few short distances, to increase in size the most, making these distances appear more influential than they are. The corresponding probability density function is plotted with untransformed frequencies, and should appear to fit the data well, as in Fig. 5.2b, provided heaping is not severe, adequate truncation is carried out and an appropriate model is selected.

5.3 Truncation

5.3.1 *Right-truncation*

The largest detection distance in the example data was 34.16 m, considerably greater than the second largest of 26.87 m (Fig. 5.1). Unless the true detection function is somehow known (as it is for this simulated example), large distances can prove difficult to model, and the extra terms required increase the variance in $\hat{h}(0)$. If the uniform + simple polynomial model is fitted to the untruncated data of Fig. 5.1, four polynomial terms are required, and a less plausible shape for the detection function is obtained than for the single term fit of Fig. 5.3. In Chapter 4, we suggested as rules of thumb that either roughly 5% of observations be truncated or truncation distance w be chosen such that $\hat{g}(w) \doteq 0.15$. These rules do not carry across to point transects, for which a higher proportion of detections occurs in the tail of the detection function. This can be seen in Fig. 5.1; Fig. 5.1a shows that probability of detection is as small as 0.12 at a distance of 20 m, yet 13 of 144 observations (9%) lie beyond 20 m (Fig. 5.1b). We suggest that roughly 10% of observations should be truncated for point transects, or alternatively, w should be chosen such that $g(w) \doteq 0.1$, where $g(w)$ is estimated from a preliminary fit of a plausible model to the data. In this example, a truncation distance of 20 m roughly satisfies both criteria, and is used subsequently for the example data.

Truncation of the data at $w = 20$ m removed 13 detections. If a series expansion model is used, up to three fewer parameters are required to model the truncated data than the untruncated data (Table 5.1). Outliers in the right tail of the distance distribution required additional adjustment parameters. Except for the hazard-rate model, which performed relatively poorly on these data, density estimates varied more when data were untruncated (Table 5.2). The poor performance of the uniform + polynomial model when fitted to untruncated data divided into 20 groups was because Distance failed to converge when attempting to fit a better model; convergence problems are encountered more commonly when appropriate truncation is not carried out. If the generating model is known and used, then truncation is not necessary provided the measurements are exact and no evasive movement prior to detection occurs. However, our models are at best robust approximations to the underlying detection process in real field surveys.

Truncation of the distance data deletes outliers and facilitates model fitting. However, as some data are discarded, the uncertainty in $\hat{D}$ may increase. For the example data, when the generating model was used (i.e. the half-normal), the coefficient of variation was around three percentage points higher for analyses of truncated data in three of the four analyses,

Table 5.1. Summary of AIC values for two truncation values (w) for the example data analysed as ungrouped and at three different groupings (five groups of equal width, 20 groups of equal width and five unequal groups such that the number detected was nearly equal in each group). For each analysis, the model with the smallest AIC is indicated by an asterisk

		$w = 20$ m			w = largest observation		
		No. of parameters			No. of parameters		
Data type	Model (key + adjustment)	Key	Adjustments	AIC	Key	Adjustments	AIC
Ungrouped	Uniform + cosine	0	1	765.51	0	2	918.46*
	Uniform + polynomial	0	1	764.48	0	4	922.32
	Half-normal + Hermite	1	0	764.31*	1	0	919.16
	Hazard-rate + cosine	2	0	767.22	2	1	919.79
Grouped (5 equal)	Uniform + cosine	0	1	403.36	0	2	374.06*
	Uniform + polynomial	0	1	400.83*	0	4	377.78
	Half-normal + Hermite	1	0	401.97	1	1	374.22
	Hazard-rate + cosine	2	0	403.52	2	1	375.91
Grouped (20 equal)	Uniform + cosine	0	1	768.16	0	2	764.20*
	Uniform + polynomial	0	1	766.90	0	2	830.04
	Half-normal + Hermite	1	0	766.85*	1	0	764.34
	Hazard-rate + cosine	2	1	769.94	2	1	765.25
Grouped (5 unequal)	Uniform + cosine	0	1	426.28	0	2	468.50
	Uniform + polynomial	0	1	426.69	0	4	476.91
	Half-normal + Hermite	1	0	425.93*	1	0	467.17*
	Hazard-rate + cosine	2	0	428.04	2	0	472.39

Table 5.2. Summary of estimated density ($\hat{D}$) and coefficient of variation (cv) for two truncation values (w) for the example data. Estimates are derived for four robust models of the detection function. The data analysis was based on ungrouped data and three different groupings (five groups of equal width, 20 groups of equal width, and five unequal groups such that the number detected was nearly equal in each group)

		Truncation			
		$w = 20\,\text{m}$		w = largest observation	
Data type	Model (key + adjustment)	$\hat{D}$	cv (%)	$\hat{D}$	cv (%)
Ungrouped	Uniform + cosine	75.05	14.4	74.13	10.6
	Uniform + polynomial	60.76	12.1	70.88	18.0
	Half-normal + Hermite	70.82	15.7	79.62	12.6
	Hazard-rate + cosine	62.36	18.7	71.02	18.1
Grouped (5 equal)	Uniform + cosine	73.74	14.5	73.77	11.5
	Uniform + polynomial	62.01	12.7	70.14	25.7
	Half-normal + Hermite	69.06	16.0	64.53	26.3
	Hazard-rate + cosine	52.14	14.5	79.13	17.2
Grouped (20 equal)	Uniform + cosine	75.54	14.0	74.24	10.7
	Uniform + polynomial	61.09	12.2	42.26	9.0
	Half-normal + Hermite	71.30	15.6	80.25	12.3
	Hazard-rate + cosine	82.98	26.9	70.85	18.6
Grouped (5 unequal)	Uniform + cosine	74.91	14.3	74.36	15.8
	Uniform + polynomial	61.93	12.8	57.45	37.0
	Half-normal + Hermite	71.57	15.8	80.73	13.0
	Hazard-rate + cosine	84.76	37.2	57.50	13.9

but ten percentage points lower (16.0%, compared with 26.3%) in one of the analyses of grouped data. When the other models were fitted, the coefficient of variation increased after truncation in eight analyses and decreased in four.

The true density in this example was 79.6 objects/ha. In exactly one-half of the 16 analyses of Table 5.2, the estimate was closer to the true density after truncation than before. The case for truncation is therefore not compelling for these simulated data. However, real data tend to be less well behaved, and if no truncation is imposed in the field, truncation at the analysis stage is advisable.

5.3.2 *Left-truncation*

In line transect sampling, if detection on the line is not certain, observations near the line can be 'left-truncated' either by relocating the line to a distance at which detection is believed to be certain or by fitting the detection function only to data beyond that distance, and extrapolating back to distance zero. In standard point transect sampling, we have only the second option (but see Quang 1993 for an alternative approach). In point transect sampling, we may need to left-truncate data if some objects (usually larger birds) flush as the observer walks into the point. Field protocol usually leads to exclusion of these detections, and hence too few detections near the point. (A better protocol might be to include these objects if it is believed that they would have stayed in place had the observer not caused them to flee.) A similar reason for left-truncation is if some objects (usually smaller birds) near the point freeze or hide while the observer is present, thus avoiding detection. Left-truncation is readily carried out in Distance. An example of left-truncation in the closely related method of cue counting is given in Section 6.4.

5.4 Estimating the variance in sample size

If objects were known to be distributed at random, the distribution of sample size n would be Poisson with $var(n) = E(n)$, so that $\widehat{var}(n) = n$. Most biological populations exhibit some degree of clumping, so that $var(n) > E(n)$. If the survey is well designed so that points are spread either systematically or randomly throughout the study area, or within each stratum if the study area is divided into strata, then point transect methods are ideally suited to estimating $var(n)$ empirically, from the variability in sample size between individual points. For the example data, there were $k = 30$ points, and sample sizes n_i within the truncation distance of $w = 20\,\text{m}$ were 1, 1, 5, 6, 3, 8, 7, 5, 3, 4, 1, 8, 3, 1, 2, 7, 4, 4, 6, 7, 6, 8, 4, 3, 5, 2, 4, 2, 9 and 2.

From Section 3.6.2,

$$\widehat{var}(n) = \frac{k \sum_{i=1}^{k} (n_i - \bar{n})^2}{k - 1} \tag{5.1}$$

with $k-1$ degrees of freedom (df). All counts here are positive; had any been zero, they would be retained when calculating the variance. That is, any $n_i = 0$ must be included in the calculation of eqn (5.1). For the example, the above formula yields $\widehat{var}(n) = 172.7$, or $\widehat{se}(n) = 13.14$. Equivalently, the variance of the mean number of objects per point, $\bar{n} = n/k = 4.367$, may be estimated as

$$\widehat{var}(\bar{n}) = \frac{\sum_{i=1}^{k} (n_i - \bar{n})^2}{k \cdot (k - 1)} \tag{5.2}$$

so that $\widehat{var}(\bar{n}) = 0.1919 = \widehat{var}(n)/k^2$. Since $n = 131$ after truncation, $\widehat{var}(n) > n$, indicating possible clumping of objects. In this simulated example, we know that the distribution of n is Poisson, for which

$var(n) = E(n)$, so that $\widehat{var}(n) > n$ only by chance. Indeed, the variance-mean ratio does not differ significantly from one:

$$(k-1) \cdot \frac{\widehat{var}(n)}{n} = 38.2 \tag{5.3}$$

which is a value from $\chi^2_{k-1} = \chi^2_{29}$ if the true distribution of n is Poisson, giving $p = 0.12$. Real populations typically show greater clumping, so that $var(n) > E(n)$.

5.5 Analysis of grouped or ungrouped data

Because the assumptions of point transect sampling are known to hold for the example, analysis of ungrouped data is preferred. Generally, little efficiency is lost by grouping data prior to analysis, even with as few as five or six well-chosen intervals. If recorded distances tend to be rounded to favoured values (heaping), or if there is evidence of movement of objects in response to the observer before detection, appropriate grouping of data can lead to more robust estimation of density (Chapter 7). Often, there are sound practical reasons for recording data by distance interval, instead of measuring each individual detection distance, in which case the field methods determine the analysis option.

For the example, estimated densities tended to be rather more variable between models when analysis was based on grouped data, although coefficients of variation were not consistently higher (Table 5.2). Provided distances can be measured accurately, and movement in response to the observer before detection is not a problem, we recommend that analysis should be of ungrouped data. Otherwise, analysis of grouped data should be considered. If heaping occurs, group cutpoints should be selected so that favoured distances for rounding tend to occur midway between cutpoints. Choice of group interval is often more critical than for line transect sampling, since a smaller proportion of detections occurs near zero distance, yet it is the value of a function at zero distance, $h(0)$, that must be estimated. It is this difficulty that gives rise to the relatively large variability in density estimates in Table 5.2. On the other hand, although poor practice, it is not uncommon for 10% or more of perpendicular distances to be recorded as on the line in line transect sampling. It is rare for an object to be recorded as at the point ($r = 0$) in point transect sampling, so that spurious spikes in the detection function at small distances are uncommon.

5.6 Model selection

5.6.1 *The models*

The same four models for the detection function are considered as in Chapter 4. Thus, the uniform, half-normal and hazard-rate models are used as key functions. Cosine and simple polynomial expansions are used with the uniform key, Hermite polynomials are used with the half-normal key and a cosine expansion is used with the hazard-rate key. The data were generated under a half-normal detection function so we might expect the half-normal key to be sufficient without any adjustment terms. However, the data were stochastically generated, so that the addition of a Hermite polynomial term in one analysis of Table 5.1 is not particularly surprising. A histogram of the data using 15–20 intervals, as in Fig. 5.1, tends to reveal the characteristics of the data, such as outliers, heaping, measurement errors and evasive movement prior to detection.

5.6.2 *Akaike's Information Criterion*

Akaike's Information Criterion (AIC) provides a quantitative method for model selection, whether models are hierarchical or not (Section 3.4.2). The adequacy of the selected model should still be assessed, for example using the usual χ^2 goodness of fit statistics and visual inspection of both the estimated detection function and the corresponding density plotted on histograms of the data, as shown in Figs 5.1 and 5.2. The plots allow the fit of the model near the point to be assessed; some lack of fit in the right tail of the data can be tolerated.

AIC is defined as

$$\text{AIC} = -2 \cdot \log_{\text{e}}(\mathcal{L}) + 2q \tag{5.4}$$

where $\mathcal{L}$ is the maximized likelihood and q the number of parameters in the model. For small samples, if observations are approximately normally distributed, the adjustment of Hurvich and Tsai (1989, 1995) is preferred:

$$\text{AICc} = \text{AIC} + \frac{2q(q+1)}{n-q-1} \tag{5.5}$$

where n is the number of objects or object clusters detected. AIC (or AICc) is computed for each candidate model, and that with the lowest AIC is selected for analysis and inference.

AIC was computed for the four models for both grouped and ungrouped data, with truncation distance w set first to 20 m (13 observations truncated) and then to the largest observation, selected so that no observations were truncated (Table 5.1). Three sets of cutpoints were considered for grouped analyses under each model. Set 1 had five equal groups, set 2 had 20 equal groups and set 3 had five groups whose width varied, such that the number detected in each distance category was nearly equal. AIC cannot be used to select between models if the truncation distances w differ, or, in the case of an analysis of grouped data, if the cutpoints differ. It also cannot be used to decide whether or not to group recorded distances prior to analysis. Thus AIC values can only be compared within each of the eight sets of results in Table 5.1.

The AIC values in Table 5.1 select the half-normal (generating) model in four of the eight sets of results. The uniform key with cosine adjustments is selected three times, and the uniform key with simple polynomial adjustments once. Since the half-normal model is selected for the preferred analysis of ungrouped data, truncated at 20 m, the main analysis will be based upon it. However, the AIC value for the uniform + polynomial model is almost the same as for the half-normal + Hermite model, and might equally well be adopted on this basis. We examine the consequences of selecting this model later. The only model that might reasonably be excluded from further consideration on the basis of its AIC value is the hazard-rate + cosine model.

An example of the use of AIC in a stratified design to assess whether it is better to estimate a pooled detection function across strata is given in Section 5.8.3.

5.6.3 *Likelihood ratio tests*

Consider the example data, ungrouped and with $w = 20$ m, analysed using the uniform key with a single polynomial adjustment (Table 5.1). How was it determined that a single adjustment was required for this model? We recommend use of AIC for selecting the number of adjustment terms. However, because adjustment terms can be added sequentially within a hierarchical modelling structure, we can also use likelihood ratio tests. Let $\mathcal{L}_0$ be the value of the log-likelihood for fitting a uniform key alone, let $\mathcal{L}_1$ be the maximum value of the likelihood when a single polynomial term is added, and $\mathcal{L}_2$ be the value after fitting two polynomial terms. Program Distance gives $\log_e(\mathcal{L}_0) = -394.986$, $\log_e(\mathcal{L}_1) = -381.239$ and $\log_e(\mathcal{L}_2) = -381.061$. The likelihood ratio test of the hypothesis that the uniform key provides an adequate description of the data against the alternative that a single polynomial adjustment to the key provides a better fit is carried out by calculating

$$\begin{aligned}\chi^2 &= -2 \log_e \left(\frac{\mathcal{L}_0}{\mathcal{L}_1}\right) = -2 \left[\log_e(\mathcal{L}_0) - \log_e(\mathcal{L}_1)\right] \\ &= -2 \left[-394.986 + 381.239\right] = 27.49\end{aligned}$$

If the generating model is the uniform key without adjustment, this statistic is distributed asymptotically as χ^2_1. In general, the degrees of freedom for this test statistic are the difference in the number of parameters between the two models being tested. A value of 27.49 is much larger than would be expected if the distribution really was χ^2_1 ($p < 0.001$), suggesting that a uniform detection function is not an adequate description of the data, a conclusion that is obvious from Fig. 5.3. Less obvious is whether an additional polynomial term should be fitted. The above test is now carried out, but with $\mathcal{L}_1$ replacing $\mathcal{L}_0$ and $\mathcal{L}_2$ replacing $\mathcal{L}_1$:

$$\chi^2 = -2\,[\log_e(\mathcal{L}_1) - \log_e(\mathcal{L}_2)] = 0.36$$

Again comparing with χ_1^2, this test statistic is not significant ($p = 0.55$), so a further term does not improve the fit of the model significantly. Our experience suggests that a larger value than the conventional $\alpha = 0.05$ is often preferable for the size of the test, and we suggest $\alpha = 0.15$ (Section 3.4.3).

If the likelihood ratio test indicates that a further term is not required but goodness of fit (below) indicates that the fit is poor, the addition of two terms (using Distance option 'look-ahead' set to two) rather than just one may provide a significantly better fit. Another option is to change the adjustment term selection method in Distance from the default of 'sequential' to the more computer-intensive options of 'forward' or 'all'.

5.6.4 *Goodness of fit*

Goodness of fit is another useful tool for model assessment (Section 3.4.4). Goodness of fit statistics for the example data without grouping, with $w = 20$ m, and using 20 groups of equal width to evaluate the χ^2 statistic, are given in Table 5.3. These data were taken when all the assumptions were met, and all four models fit the data well. If a model was to be selected from these results, there might be a marginal preference for the half-normal + Hermite polynomial model, which we know to be the correct choice in this case. Heaping in real data sets generally means that fewer than 20 groups should be used, with perhaps six to eight usually being reasonable. If heaping is severe, fewer groups might be required, ideally with each preferred rounding distance falling near the middle of each group. The grouped nature of the (rounded) data is then correctly recognized in the analysis. If cutpoints are badly chosen, heaping will lead to spurious significant χ^2 values. If data are collected as grouped, the group cutpoints are determined before analysis, although consecutive groups may be merged.

Table 5.3. Goodness of fit statistics for models fitted to the example data with $w = 20$ m and 20 groups

Model	χ^2	df	p
Uniform + cosine	20.34	17	0.26
Uniform + polynomial	19.84	17	0.28
Half-normal + Hermite	19.20	17	0.32
Hazard-rate + cosine	20.32	16	0.21

5.7 Estimation of density and measures of precision

5.7.1 *The standard analysis*

The preferred analysis from the above considerations comprises the fit of the half-normal key without adjustments to ungrouped data, truncated at $w = 20$ m. The variance of n is estimated empirically.

Replacing the parameters of eqn (3.14) by their estimators and simplifying under the assumption that objects do not occur in clusters, estimated density becomes

$$\hat{D} = \frac{n \cdot \hat{h}(0)}{2\pi k} \tag{5.6}$$

where n is the number of objects detected, k is the number of point transects sampled, and $\hat{h}(0)$ is the slope of the estimated density $\hat{f}(r)$ of observed detection distances evaluated at $r = 0$; $\hat{h}(0) = 2\pi/\hat{\nu}$, where $\hat{\nu}$ is the effective area of detection.

For the example data, and adopting the preferred analysis, program Distance yields $\hat{h}(0) = 0.01019$, with $\widehat{se}\{\hat{h}(0)\} = 0.001233$ (based on approximately $n = 131$ df). The units of $\hat{h}(0)$ are m^{-2}. Thus

$$\hat{D} = \frac{131 \times 0.01019}{2\pi \times 30} = 0.00708 \text{ objects/m}^2, \text{ or } 70.8 \text{ objects/ha}$$

The estimator of the sampling variance of this estimate is

$$\widehat{var}(\hat{D}) = \hat{D}^2 \cdot \{[cv(n)]^2 + [cv\{\hat{h}(0)\}]^2\} \tag{5.7}$$

where

$$[cv(n)]^2 = \frac{\widehat{var}(n)}{n^2} = \frac{172.7}{131^2} = 0.010065 \tag{5.8}$$

and

$$[cv\{\hat{h}(0)\}]^2 = \frac{\widehat{var}\{\hat{h}(0)\}}{\{\hat{h}(0)\}^2} = \frac{0.001233^2}{0.01019^2} = 0.01464 \tag{5.9}$$

Then

$$\widehat{var}(\hat{D}) = (70.8)^2[0.010065 + 0.01464] = 123.84$$

and

$$\widehat{se}(\hat{D}) = \sqrt{\widehat{var}(\hat{D})} = 11.13$$

The coefficient of variation of estimated density is $cv(\hat{D}) = \widehat{se}(\hat{D})/\hat{D} = 15.7\%$, which is likely to be adequate for some purposes. Note that even with a sample size of $n = 131$ after truncation, the coefficient of variation is over 15%. A 95% confidence interval could be calculated as $\hat{D} \pm 1.96(\widehat{se}(\hat{D}))$, giving the interval [49.0, 92.6]. Log-based confidence intervals offer improved coverage by allowing for the asymmetric shape of the sampling distribution of $\hat{D}$ for small n. Applying the procedure of Section 3.6.1, the interval [52.1, 96.2] is obtained, which is slightly wider than the symmetric interval, but is a better measure of the uncertainty in the estimate $\hat{D} = 70.8$. In line transect sampling, the variance of $\hat{D}$ is usually primarily due to the variance in n, but this is less often the case in point transect sampling, where precision in $\hat{h}(0)$ can be poor; here, variance in n accounts for 41% of the total variance estimate.

If the uniform model with polynomial adjustments is adopted, estimated density is 60.8 objects/ha, with 95% log-based confidence interval [48.2, 77.0]. The true parameter value, $D = 79.6$ objects/ha lies above the upper limit of this interval. We return to this example later, to show how the bootstrap may be used to estimate variances and to determine confidence limits that incorporate model selection uncertainty.

For some purposes it is convenient to have a measure of detectability. For example, it may be useful to assess whether the detectability for a species is a function of habitat, which may have implications for survey design. The effective radius of detection $\rho = \sqrt{\nu/\pi}$, estimated by $\hat{\rho} = \sqrt{2/\hat{h}(0)}$, may be used for this purpose. For long-tailed detection functions, ρ may be considerably larger than intuition would suggest, because large numbers of objects are detected at far distances, where the area surveyed is great, relative to close distances, where the surveyed area is small. A parameter that is unaffected either by this phenomenon or (unlike ρ) by choice of truncation distance w is $r_{1/2}$, the distance at which the probability of detecting an object is one-half. For any fitted detection function $\hat{g}(r)$, it may be estimated by solving $\hat{g}(\hat{r}_{1/2}) = 0.5$ for $\hat{r}_{1/2}$. For the example with $w = 20$ m, $\hat{\rho} = 14.0$ m and $\hat{r}_{1/2} = 13.0$ m.

We know that the detection function was a half-normal for the example. Using that knowledge, closed form estimators are available and the analysis is simple to carry out by hand, provided the data are both ungrouped and untruncated. Using the results of Section 3.3.4,

$$\hat{\sigma}^2 = \sum_{i=1}^{n} \frac{r_i^2}{2n} = 94.81\,\text{m}^2$$

It follows that

$$\hat{h}(0) = \frac{2\pi}{\hat{\nu}} = \frac{1}{\hat{\sigma}^2} = 0.01055$$

and estimated density is

$$\hat{D} = \frac{144 \times \hat{h}(0)}{2\pi \times 30} = 0.00806 \text{ objects/m}^2, \text{ or } 80.6 \text{ objects/ha}$$

The effective radius of detection is estimated as $\hat{\rho} = \sqrt{2\hat{\sigma}^2} = 13.8\,\text{m}$, and the radius at which probability of detection is one-half is estimated by $\hat{r}_{1/2} = \sqrt{2\hat{\sigma}^2 \log_e 2} = 11.5\,\text{m}$. These estimates are in excellent agreement with the true values of $D = 79.6$ objects/ha, $\rho = 14.1\,\text{m}$ and $r_{1/2} = 11.8\,\text{m}$.

The results of Section 3.3 also yield variance estimates for this special case:

$$\widehat{var}[\hat{h}(0)] = \frac{4}{\sum_{i=1}^{n}(r_i^2 - 2\hat{\sigma}^2)^2} = 8.850 \times 10^{-7}$$

or

$$\widehat{se}[\hat{h}(0)] = 9.407 \times 10^{-4}$$

Thus

$$[cv\{\hat{h}(0)\}]^2 = \frac{\widehat{var}\{\hat{h}(0)\}}{\{\hat{h}(0)\}^2} = \frac{0.0009407^2}{0.01055^2} = 0.007951$$

Also, $[cv(n)]^2 = 0.010065$ from above, so that

$$\begin{aligned}\widehat{var}(\hat{D}) &= \hat{D}^2 \cdot \{[cv(n)]^2 + [cv\{\hat{h}(0)\}]^2\} \\ &= (80.6)^2[0.010065 + 0.007951] \\ &= 117.04\end{aligned}$$

and $\widehat{se}(\hat{D}) = 10.82$. The 95% log-based confidence interval is then [62.0, 104.7] objects/ha.

5.7.2 *Bootstrap variances and confidence intervals*

The bootstrap is a robust method, based on resampling, for quantifying precision of estimates. One circumstance in which the bootstrap is likely to be preferred is when the user wishes to incorporate in the standard error the component of variation arising from estimating the number of polynomial or Fourier series adjustments to be carried out. We recommend the following implementation.

Generate a bootstrap sample by selecting points with replacement from the k points recorded until the bootstrap sample also comprises k points. Repeat until B bootstrap samples have been selected. Typically, B will be around 1000. Density D is estimated from each bootstrap resample, and the estimates are ordered, to give $\hat{D}_{(i)}$, $i = 1, \ldots, B$. Then

$$\hat{D}_B = \frac{\sum_{i=1}^{B} \hat{D}_{(i)}}{B} \tag{5.10}$$

and

$$\widehat{var}_B(\hat{D}_B) = \frac{\sum_{i=1}^{B}(\hat{D}_{(i)} - \hat{D}_B)^2}{B-1} \tag{5.11}$$

while a $100(1-2\alpha)\%$ confidence interval for D is given by $[\hat{D}_{(j)}, \hat{D}_{(j')}]$, with $j = (B+1)\alpha$ and $j' = (B+1)(1-\alpha)$. It is convenient to select B so that j and j' are integer. Thus for $\alpha = 0.025$, one might select from the following values: 199, 239, 279, ..., 999. The estimate $\hat{D}$ calculated from the original data set is usually used in preference to the bootstrap estimate $\hat{D}_B$, with $se(\hat{D})$ estimated by $\sqrt{\widehat{var}_B(\hat{D}_B)}$. Applying this to the example with $B = 399$ (so that $(B+1)\alpha = 10$, an integer, for $\alpha = 0.025$), we take a sample of 30 points at random and with replacement from the 30 in the example data set. Suppose this yields the following points: 1, 1, 3, 5, 6, 6, 6, 8, 10, 10, 11, 12, 15, 15, 17, 17, 17, 18, 18, 20, 21, 22, 22, 25, 26, 26, 26, 28, 30, 30. The bootstrap sample therefore comprises each detection distance recorded at points 6, 17 and 26 three times, each distance recorded at points 1, 10, 15, 18, 22 and 30 twice, and each distance from points 3, 5, 8, 11, 12, 20, 21, 25 and 28 once. Those for remaining points are excluded. This bootstrap resample is analysed in exactly the same way as the actual sample, to yield an estimate $\hat{D}_1$. The exercise is repeated 399 times. (For real problems, we recommend larger numbers of resamples; $B = 999$ is adequate for most purposes.) The sample variance of these bootstrap estimates was 159.8, giving $\widehat{se}(\hat{D}) = \sqrt{159.8} = 12.6$ objects/ha. After ordering the bootstrap estimates, the 10th smallest value ($j = (B+1)\alpha = 10$) was found to be $\hat{D}_{(10)} = 53.6$ and the 10th largest value was $\hat{D}_{(390)} = 100.7$, giving an approximate 95% confidence interval for D of (53.6, 100.7) objects/ha. This compares with $\hat{D} \pm 1.96 \cdot \widehat{se}(\hat{D}) = (60.8, 100.4)$ objects/ha by the more traditional method. Assuming the distribution of $\hat{D}$ is log-normal and using eqn (3.72), we obtain the interval $(63.1, 102.9)$ objects/ha. Note that the lower limit is smaller for the bootstrap method. This is because cosine adjustments to the half-normal fit sometimes generated a fitted detection function with a flatter shoulder than that of the half-normal. If no adjustments to the half-normal fit are allowed, the bootstrap should duplicate the analytic method, except asymptotic normality is not assumed when setting confidence limits. Applying this with $B = 399$ gives $\widehat{se}(\hat{D}) = 10.8$ and an approximate 95% confidence interval for D of (62.5, 102.7), which is shifted slightly to the right of the symmetric analytic interval, reflecting the greater uncertainty in the upper limit, but agrees well with the interval calculated assuming the distribution of $\hat{D}$ is log-normal.

Variances of functions of the fitted density, such as $\hat{\rho}$ or $\hat{r}_{1/2}$, may be estimated using the methods of Section 3.3, or from the above bootstrap method, replacing the bootstrap estimate of density $\hat{D}_{(i)}$ by the appropriate estimate, such as $\hat{\rho}_{(i)}$ or $\hat{r}_{1/2_{(i)}}$. Adopting the analytic approach,

$$\frac{\partial \hat{\rho}}{\partial \hat{\sigma}^2} = \frac{1}{\sqrt{2\hat{\sigma}^2}}$$

so that

$$\widehat{se}(\hat{\rho}) = \sqrt{\frac{2\hat{\sigma}^6}{\sum_{i=1}^{n}(r_i^2 - 2\hat{\sigma}^2)^2}} = 0.61\,\text{m}$$

and

$$\widehat{se}(\hat{r}_{1/2}) = \widehat{se}(\hat{\rho}) \cdot \sqrt{\log_e 2} = 0.51\,\text{m}$$

By comparison, the bootstrap method yields $\widehat{se}(\hat{\rho}) = 1.12\,\text{m}$, with 95% confidence interval (12.62, 16.84) m, and $\widehat{se}(\hat{r}_{1/2}) = 0.93\,\text{m}$, with 95% confidence interval (10.51, 14.02) m. If no cosine adjustments are allowed, as above, we get $\widehat{se}(\hat{\rho}) = 0.66\,\text{m}$, with 95% confidence interval (12.55, 15.00) m, and $\widehat{se}(\hat{r}_{1/2}) = 0.55\,\text{m}$, with 95% confidence interval (10.45, 12.49) m. These results are in good agreement with the analytic results.

We noted earlier that the AIC value for the preferred analysis of the example data was almost the same as that using the uniform key with a single polynomial adjustment. However, the latter model gave an estimated density of 60.9 objects/ha, with 95% confidence interval [48.2, 77.0]. Thus the true parameter value, $D = 79.6$ objects/ha, is outside the confidence interval. The bootstrap option within Distance was implemented with $B = 199$ replicates, to obtain a variance for $\hat{h}(0)$ that allows for estimation of the number of polynomial terms required. It gave $\widehat{se}\{\hat{h}(0)\} = 0.000783$, compared with the analytic estimate of $\widehat{se}\{\hat{h}(0)\} = 0.000581$, which is conditional on a single term adjustment to the uniform key. Thus the variance is larger as expected, and the revised 95% confidence limit for D is [46.6, 79.3]. The true density is therefore still just outside the interval, probably because the uniform + polynomial model gives a negatively biased estimate of density for this data set. To attempt to improve the variance estimate corresponding to $\hat{D} = 60.8$, a component of variance corresponding to model selection uncertainty should be estimated. We do this by generating 199 bootstrap samples (preferably more for analyses of real data), and analysing each resample by the three models of Table 5.1 that gave competitive AIC values, namely uniform + cosine, uniform + polynomial and half-normal + Hermite polynomial. In each resample, the bootstrap estimate of density is taken to be the estimated density under the model with the smallest AIC. Under this rule, the uniform + cosine model was selected in 49 of the 199 replicates, the uniform + polynomial model in 92 and the half-normal + Hermite polynomial model in the remaining 58. The 95% percentile confidence interval was [48.0, 94.5] objects/ha, which is wider than the intervals obtained by assuming that the selected model is the correct model, and comfortably includes the true parameter value, $D = 79.6$.

In the above bootstrap implementations, the sampling unit was taken to be the individual point. This is valid if points are randomly distributed through the study area, and provides a good approximation if points are arranged as a regular grid. To reduce travel time between points, transect lines are sometimes defined, and counts are made at regular points along each line. If the spacing between lines is similar to the distance between neighbouring points on the same line, then the point may still be taken as the sampling unit. However, if separation between lines is large, then the line should be taken as the sampling unit. Thus lines are selected with replacement until the number of lines in the resample is equal to the number in the real sample, or, if the number of points per line is very variable, until the number of points in the resample is as close as possible to the number in the real sample. If a line is selected, the data from all points on that line are included in the resample.

5.8 Estimation when the objects are in clusters

If point transects are used for objects that are sometimes recorded in clusters during the survey period, the recording unit should be the cluster, not the individual object, and analyses should be based on clusters. In this section, various options for the analysis of clusters are considered. If it is assumed that (i) probability of detection is independent of cluster size and (ii) cluster sizes are accurately recorded, or alternatively that they are estimated without bias at all distances, then $E(s)$ may be estimated by the mean size of detected clusters, $\bar{s}$. Estimated cluster density is then

$$\hat{D}_s = \frac{n \cdot \hat{h}(0)}{2\pi k} \tag{5.12}$$

and estimated object density is

$$\hat{D} = \hat{D}_s \cdot \bar{s} = \frac{n \cdot \hat{h}(0) \cdot \bar{s}}{2\pi k} \tag{5.13}$$

Note that the formula for cluster density is identical to that for object density when the objects do not occur in clusters. The formula for the variance of $\hat{D}_s$ is also identical to that given for object density in Section 5.7.1. The variance of object density is now estimated by

$$\begin{aligned} \widehat{var}(\hat{D}) &= \hat{D}^2 \cdot \left\{ \frac{\widehat{var}(\hat{D}_s)}{\hat{D}_s^2} + \frac{\widehat{var}(\bar{s})}{\bar{s}^2} \right\} \\ &= \hat{D}^2 \cdot \left\{ \frac{\widehat{var}(n)}{n^2} + \frac{\widehat{var}[\hat{h}(0)]}{[\hat{h}(0)]^2} + \frac{\widehat{var}(\bar{s})}{\bar{s}^2} \right\} \end{aligned} \tag{5.14}$$

where

$$\widehat{var}(\bar{s}) = \frac{\sum_{i=1}^{n}(s_i - \bar{s})^2}{n(n-1)} \tag{5.15}$$

In practice, larger clusters often tend to be more detectable than small clusters, so that $E(s)$, and hence D, are overestimated. This is a form of size-biased sampling (Cox 1969; Patil and Ord 1976; Patil and Rao 1978; Rao and Portier 1985). Bias can be negative if the size of a detected cluster at a large distance from the observer tends to be underestimated. If either bias occurs, then the above method should be modified or replaced.

The simplest approach is based on the fact that size bias in detected clusters does not occur within a region around the point for which detection is certain. Hence, $E(s)$ may be estimated by the mean size of clusters detected within distance v of the point, where $g(v)$ is reasonably close to one, say 0.6 or 0.8. In the second method, a cluster of size s_i at distance r_i from the point is replaced by s_i objects, each at distance r_i. Thus, the sampling unit is assumed to be the object rather than the cluster. For the third method, data are stratified by cluster size (Quinn 1979, 1985). The selected model is then fitted independently to the data in each stratum. If size bias is large or cluster size very variable, smaller truncation distances are likely to be required for strata corresponding to small clusters. The fourth method estimates cluster density D_s conventionally, as does the first. Then, given the r_i, $E(s)$ is estimated by regression modelling of the relationship between s_i and r_i. All four approaches are illustrated in this section using program Distance. A fifth approach, in which cluster size is included as a covariate in the detection function, is addressed in Buckland *et al.* (in preparation).

The data used to illustrate the four methods were simulated from a half-normal detection function without truncation, in which the scale parameter σ was a function of cluster size:

$$\{\sigma(s)\}^2 = \sigma_0^2\left(1 + b \cdot \frac{s - E(s)}{E(s)}\right) \tag{5.16}$$

where $\sigma_0 = 30\,\text{m}$, $b = 0.75$ and $E(s) = 1.85$ for the population. (In Chapter 4, $\sigma(s)$ was assumed to be a linear function of s; for point transects, theoretical considerations suggest that it is more appropriate to assume $\{\sigma(s)\}^2$ is a linear function of s.) Cluster sizes s were generated by simulating values from the geometric distribution with rate $E(s) - 1$ and adding one, and a cluster of size s was detected with probability

$$g(r \mid s) = \exp\left[-\frac{r^2}{2\{\sigma(s)\}^2}\right]$$

The expected sample size was $E(n) = 96$, distributed between $k = 60$ points, with $var(n) = 2.65 \cdot E(n)$. True densities were $D_s = 283$ clusters/km^2 and $D = 1.85 \times 283 = 523$ objects/km^2. The bivariate detection function $g(r, s)$ is monotone non-increasing in r and monotone non-decreasing in s. The detected cluster sizes are not a random sample from the population of cluster sizes; the mean size of detected clusters $\bar{s}$ has expectation greater than $E(s)$.

A histogram of the untruncated distance data shows a rather long tail (Fig. 5.4). Truncation at 70 m deleted just under 10% of observations

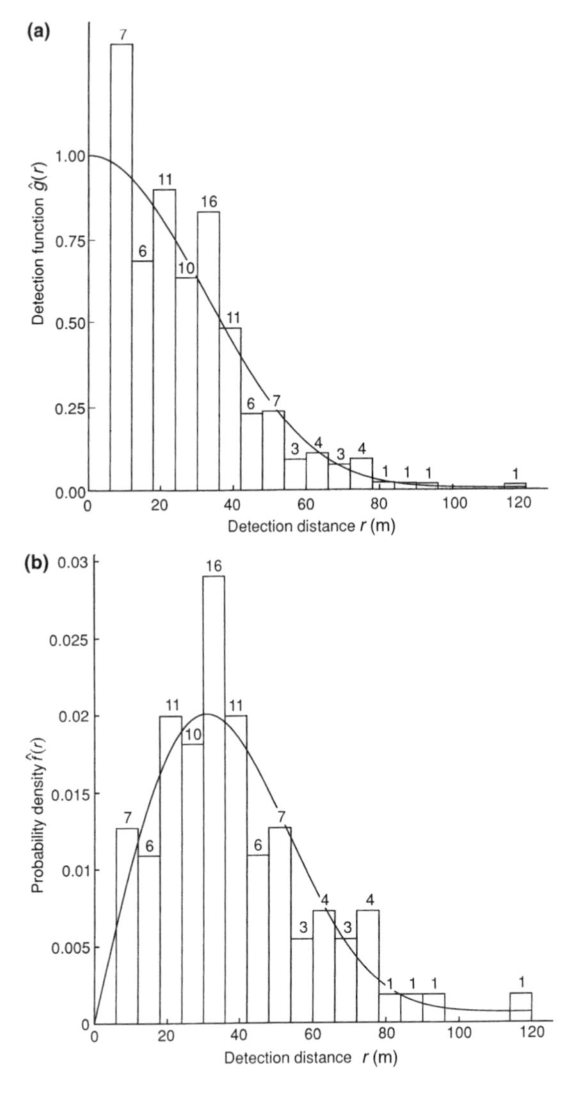

Fig. 5.4. Histograms of the example data using 20 distance categories for the case where cluster size and detection distance are dependent. The fit of a uniform + three-term cosine detection function to untruncated data is shown in (a), in which frequencies are divided by detection distance, and the corresponding density function is shown in (b).

(eight from 92), and allowed the data to be modelled more reliably. The same four models were applied as for Section 5.6: uniform + cosine, uniform + polynomial, half-normal + Hermite polynomial and hazard-rate + cosine. All four models fitted the truncated data well. AICs for the four models were 691.8, 693.7, 692.6 and 693.1, which marginally favours the uniform + cosine model. We therefore use it to illustrate methods of analysis of the example data.

The uniform + cosine model for the untruncated data required three cosine terms to adequately fit the right tail of the data (Fig. 5.4). By truncating the data, only a single cosine term is required (Fig. 5.5), and the size bias in the truncated sample of detected clusters is reduced. The fit of the model was good ($\chi_5^2 = 4.51$; $p = 0.48$). The estimated density of clusters was 258.1 clusters/km^2 ($\widehat{se} = 52.2$), compared with the true value of 283. The mean cluster size from the untruncated sample data was 2.293 ($\widehat{se} = 0.165$), which is biased high due to the size-biased sampling. The scatter plot of cluster size against detection distance (Fig. 5.6) shows wide scatter, but a significant correlation ($r = 0.272$). Truncation at $w =$ 70 m reduced this correlation to 0.180. Multiplying the density of clusters by the uncorrected estimate of mean cluster size from data truncated at 70 m ($\bar{s} = 2.202$; $\widehat{se}(\bar{s}) = 0.168$), the density of individuals is estimated as 574.6 objects/km^2 with $\widehat{se} = 115.1$ and 95% confidence interval [389.6, 847.5], which comfortably includes the true density of 523 objects/km^2.

5.8.1 *Standard method with additional truncation*

Observed mean cluster sizes and standard errors for a range of truncation distances are shown in Table 5.4. The detection function $g(r)$ was estimated using a truncation distance of w, while a truncation distance of v ($v \leq w$) was used to estimate mean cluster size. It seems that 70 m may be too large a truncation distance for unbiased estimation of mean cluster size, but an appropriate distance is difficult to determine, because mean cluster size does not stabilize as the truncation distance is reduced. Possible choices for truncation distance v range between 21.5 m, for which $\bar{s} = 1.650$, and 46.9 m, giving $\bar{s} = 2.030$. If strong size bias is suspected, a reasonable compromise might be $v = 27.0$ m, so that $\bar{s} = 1.806$ with $\widehat{se} = 0.199$. Replacing the estimates $\bar{s} = 2.202$ and $\widehat{se} = 0.168$ by these values, density is estimated as 471.2 individuals/km^2, with $\widehat{se} = 101.5$ and 95% confidence interval [310.4, 715.3]. In view of the difficulty in selecting v, and the sensitivity of the estimate to the choice, another approach seems preferable in this instance.

5.8.2 *Replacement of clusters by individuals*

One solution to size bias is to replace a cluster of size s_i by s_i objects at the same distance. There is then no requirement to estimate mean cluster size, and the problem is circumvented. However, another problem arises: the assumption that detections are independent is severely violated, invalidating analytic variance estimates and model selection procedures. Robust

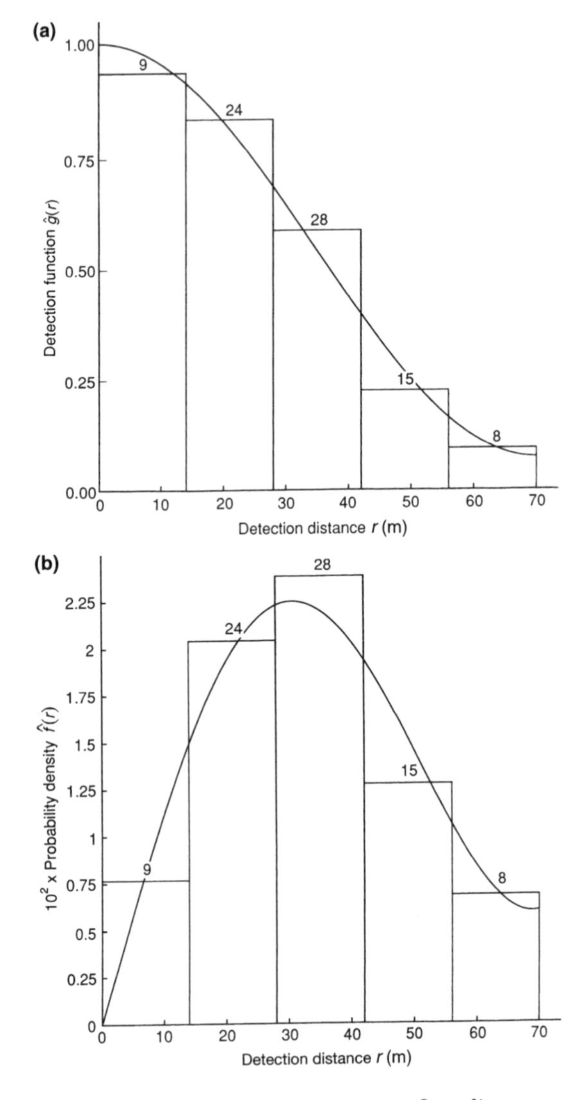

Fig. 5.5. Histograms of the example data using five distance categories and truncation at $w = 70$ m for the case where cluster size and detection distance are dependent. The fit of a uniform + one-term cosine detection function is shown in (a), in which frequencies are divided by detection distance, and the corresponding density function is shown in (b).

methods for variance estimation avoid the first difficulty, but model selection is more problematic. One solution is to select a model taking clusters as the sampling unit, then refit the model (with the same series terms, if any) to the data with object as the sampling unit. Adopting this strategy,

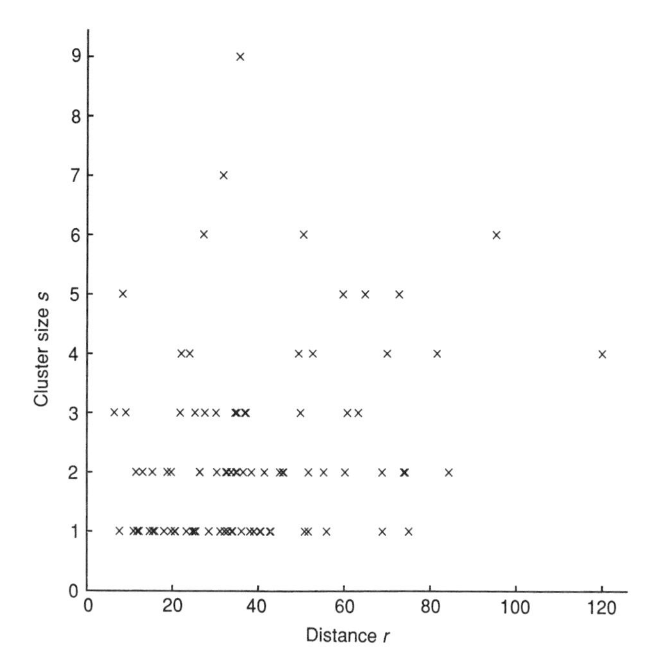

Fig. 5.6. Scatterplot of the relationship between cluster size and detection distance. The correlation coefficient is 0.272 ($w = \infty$).

Table 5.4. Observed mean cluster sizes and standard errors for various truncation distances v. Probability of detection at the truncation distance for cluster size estimation, $\hat{g}(v)$, was estimated from a uniform + one-term cosine model with $w = 70$ m (Fig. 5.5) for $v \leq 70$ m, and from a uniform + three-term cosine model with $w = 120$ m (Fig. 5.4) for $v = 120$ m

Truncation distance v (m)	n	$\bar{s}$	$\widehat{se}(\bar{s})$	$\hat{g}(v)$
120.0	92	2.293	0.165	0.005
70.0	84	2.202	0.168	0.07
46.9	67	2.030	0.183	0.30
36.8	52	2.135	0.228	0.50
31.9	38	2.079	0.243	0.60
27.0	31	1.806	0.199	0.70
21.5	20	1.650	0.232	0.80
14.9	10	2.000	0.422	0.90

Table 5.5. Summary of results for different stratification options. The model was uniform with cosine adjustments; distance data were truncated at $w = 70$ m. True $D = 523$ objects/km^2

Cluster sizes	Sample size after truncation	Effective search radius (m)	AICc	$\hat{D}$	$\widehat{se}(\hat{D})$	95% confidence interval for D
All	84	41.3	691.8	574.6	115.1	(387.8, 851.4)
1	35	39.2	283.3	120.8	24.7	
2–9	49	43.9	407.7	413.5	100.5	
All			691.0	534.3	103.5	(365.6, 781.1)
1	35	39.2	283.3	120.8	24.7	
2	24	41.6	197.6	147.1	42.2	
3–9	25	46.0	211.6	256.1	80.9	
All			692.5	524.0	94.5	(367.9, 746.2)

a uniform + one-term cosine model was fitted to the distance data truncated at 70 m, and the following estimates obtained. Number of objects detected, $n = 185$. Estimated density, $\hat{D} = 526.2$ objects/km^2, with analytic $\widehat{se} = 104.6$ and 95% confidence interval [355.3, 779.1]. These estimates are lower than those obtained assuming cluster size is independent of distance, and the point estimate is appreciably closer to the true density of 523 objects/km^2. Average cluster size can be estimated by the ratio of estimated object density (526.2) to estimated cluster density (258.1), giving 2.039.

5.8.3 *Stratification*

Stratification by cluster size can be an effective way of handling size bias. For the example data, if two strata are defined, one corresponding to individual objects and the other to clusters ($\geq$ two objects), sample sizes before truncation are 36 and 56, respectively. If the second stratum is split into clusters of size two and clusters of more than two individuals, the respective sample sizes in the three strata before truncation are 36, 27 and 29. The data were analysed for both choices of stratification.

Results are summarized in Table 5.5. As for the line transect example in the previous chapter, no precision is lost by stratification, despite the small samples from which $f(0)$ was estimated. The estimated densities are lower than that obtained by assuming cluster size is independent of detection distance, as would be expected if size bias is present. Both stratifications yield similar estimated densities, and they bracket the estimate obtained by the previous method. The true density is 523 objects/km^2, very close to both estimates. For the case of two strata, the AICc values summed across strata (691.0) is smaller than the AICc value for the case without stratification (691.8). This suggests that stratification by cluster size is preferable to no stratification. If the second stratum is further subdivided, the sum of the AICc values increases to 692.5, suggesting that there are insufficient data to support more than two strata.

In an analysis stratified by cluster size, mean cluster size may be estimated by a weighted average of the mean size per stratum, with weights equal to the estimated density of clusters by stratum. Alternatively, $E(s)$ may be estimated as overall $\hat{D}$ from the stratified analysis divided by $\hat{D}_s$ from the unstratified analysis. For two strata, this yields $\hat{E}(s) = 534.3/258.1 = 2.070$, and for three strata, $\hat{E}(s) = 524.0/258.1 = 2.030$. Both estimates are rather higher than the true mean cluster size of 1.85.

5.8.4 *Regression estimator*

Average cluster size can be estimated from a regression of cluster size on estimated detection probability. This procedure estimates the average cluster size for clusters close to the centreline, where detection is assumed to be certain, so that size bias is not present. The loss in precision in correcting for size bias using regression is generally small. The method of regressing $z_i = \log_e(s_i)$ on $\hat{g}(x_i)$ (Section 3.5.4), applied to the example data, yields $\hat{E}(s) = 1.749$ and $\widehat{se}\{\hat{E}(s)\} = \sqrt{\widehat{var}\{\hat{E}(s)\}} = 0.124$. The corresponding density estimate is 456.3 individuals/km^2, with $\widehat{se} = 90.4$, $cv = 19.8\%$ and 95% confidence interval [310.6, 670.4]. The estimate $\hat{E}(s)$ is close to the true parameter value of 1.85. The resulting density estimate (456.3) is low relative to the true density (523), although the confidence interval comfortably includes the true value.

5.9 Assumptions

The assumptions of point transect sampling are discussed in Section 2.1. There has been considerable confusion on whether objects must be assumed to be randomly distributed, both in the literature and among biologists. If objects are distributed stochastically independently from each other, but with variable rate depending on location, then the assumption that points rather than objects are randomly located suffices unless the rate shows extreme variation over short distances (of the order of a typical detection distance). If the rate can change appreciably in a short distance or if the presence of one object greatly increases the likelihood that another object is nearby (thus violating the assumption that detections are independent events), then given random placement of points, reliable estimation may still be possible provided robust variance estimation methods are used and provided that the results of goodness of fit and likelihood ratio tests (which will tend to give spurious significances) are viewed with suspicion. The more serious the departure from random, independent detections, the larger the sample size required to yield reliable analyses. Robust empirical or resampling methods should always be used for estimating the variance of sample size, as described in Section 5.7, to guard against the effects of clustered detections. The most extreme departures from a random distribution of objects are when the objects occur in well-defined clusters. In such cases, the above problems are avoided by taking the cluster rather than the object to be the sampling unit (although the clusters themselves may have a clumped spatial distribution). Strict random placement of points can be modified. For example, stratification of the study area allows sampling intensity to vary between strata, or a regular grid of points may be randomly superimposed on the area. Use of a regular grid allows the biologist to control the distance between points.

Surveys should be designed to minimize departures from the assumption that probability of detection at the point is unity ($g(0) = 1$). For example, the assumption is likely to be more reasonable for songbirds if the recording time at each point is long (giving each bird time to be detected) or if surveys are carried out in early morning, when detectability may be an order of magnitude higher (Robbins 1981; Skirvin 1981). We do not concur with the argument that early morning should be avoided when carrying out point transects. The reasoning behind it is that bird detectability varies rapidly during the first hour or two of daylight. Although detectability may vary less later in the day, it will also be lower, and densities of some species may be appreciably underestimated. Whenever possible, survey work should be carried out when detectability is greatest, and survey design should be given careful consideration to allow for variation in detectability. Models that are robust to variable detectability (pooling robust) should be used to analyse the data.

Time of season also determines whether it is reasonable to assume that probability of detection at the point is unity. For multiple species studies, it may be necessary to carry out surveys more than once, say early and late in the season. For any given species, the data collected closest to the time that it is most detectable can then be used. For many songbirds, it may be practical to survey only territorial males.

For point transect sampling, we consider that it is necessary to assume that the detection function has a shoulder because we believe that reliable estimation is not possible if it fails, although small departures from the strict mathematical requirement that $g'(0) = 0$ need not be serious. Unlike line transect data, only a very small proportion of point transect distances is close to zero, because the area covered close to the point is small. Thus, there is a case for designing surveys to ensure that $g(r) = 1$ out to some predetermined distance. If there is an area about the point for which detection is perfect, then different point transect models will tend to give more consistent estimation. When $g'(0) = 0$ but $g''(0) < 0$, the stronger criterion of an area of perfect detection fails. Methods based on squaring detection distances (Burnham *et al.* 1980: 195) and the method due to Ramsey and Scott (1979, 1981b) may then perform poorly. Even when the criterion is satisfied, but the distance up to which detection is certain is close to zero, such methods can be poor.

The mathematical theory assumes that random movement of objects does not occur. In line transect sampling, random movement prior to detection can be tolerated provided average speed of objects is appreciably less than (i.e. up to about one-third of) the speed of the observer (Hiby 1986). The problem is more serious for point transects, for which the observer is stationary. Bias occurs because probability of detection is a non-increasing function of distance from the point, so that objects moving at random are more likely to be detected when closer to the point, leading to overestimation of object density. As noted above, the assumption that $g(0) = 1$ is more plausible if recording time at each point is large, but bias arising from random object movement increases with time at the point; thus recording time at each point is a compromise, and is typically 5–10 min for songbird surveys. Even so, substantial upward bias in point transect estimates of density is common, and we advocate a 'snapshot' approach. In this, the observer strives to locate objects close to the point at a single snapshot moment. Typically, this moment might be a few minutes after arrival at the point, to give the objects (usually birds) time to settle and the observer time to ascertain locations and identities of objects. The observer remains at the point for a few minutes after the snapshot moment, to relocate objects detected before the snapshot moment, to allow estimation of their positions at the snapshot moment. The extra time is also used to estimate distances (preferably with the aid of a laser range finder), and to identify any unidentified object. If the location of an object at the snapshot moment is not known, or cannot be estimated with reasonable certainty, its detection distance is not recorded.

Response to the observer may take the form of movement towards or away from the observer, or of a change in the probability of detection of the object. Movement towards the observer has a similar effect on the data as random movement, and leads to overestimation of density (Fig. 5.7). Movement away from the observer tends to give rise to underestimation (Fig. 5.7), as does a decrease in detectability close to the point, if this is sufficient to violate the assumption that $g(0) = 1$. An increase in detectability, as when birds 'scold' the intruder, is generally helpful. However, if birds also move in response to the observer, or if females are seldom detected except very close to the point, the detection function might be difficult to model satisfactorily. The effects of response to the observer have been considered by Wildman and Ramsey (1985), Bibby and Buckland (1987) and Roeder *et al.* (1987).

Bibby and Buckland considered two 'fleeing' models. In the first, each object was assumed to maintain a minimum distance (its 'disturbance radius' r_{d}) between itself and the observer. The radius was allowed to vary from object to object, and was assumed to follow a negative exponential distribution. The detection function was assumed to be half-normal. If the data were to be analysed using a binomial half-normal model (Section 6.6.1) with the division between near and far sightings set at $c_1 = 30\,\text{m}$, and if 50% of detections would fall within c_1 in the absence of evasive behaviour, Bibby and Buckland calculated that the bias in $\hat{D}$ (evaluated by numeric integration) would be -9% when the mean disturbance radius was 10 m, -20% for 15 m, -30% for 20 m and -55% for 40 m. In this case, bias might be deemed 'acceptable' ($<10\%$ in magnitude) if the mean disturbance radius was of the order of one-third the median detection distance or less. In their second fleeing model, many objects close to a sample point become undetectable, because they either leave at the approach of the observer, moving beyond the range of detection, or take to cover, remaining silent until the observer has departed. The probability that an object at distance r is undetectable was modelled as half-normal. In otherwise identical circumstances to the first model, bias in $\hat{D}$ was found to be -24% when the point at which 50% of objects become undetectable was 10 m, -44% for 15 m, -61% for 20 m and -88% for 40 m. They concluded that species for which the second model applied were unsuitable for surveying by the point transect method, but considered that the first model, for which bias was less severe, would apply to most species of woodland songbird that show evasive behaviour.

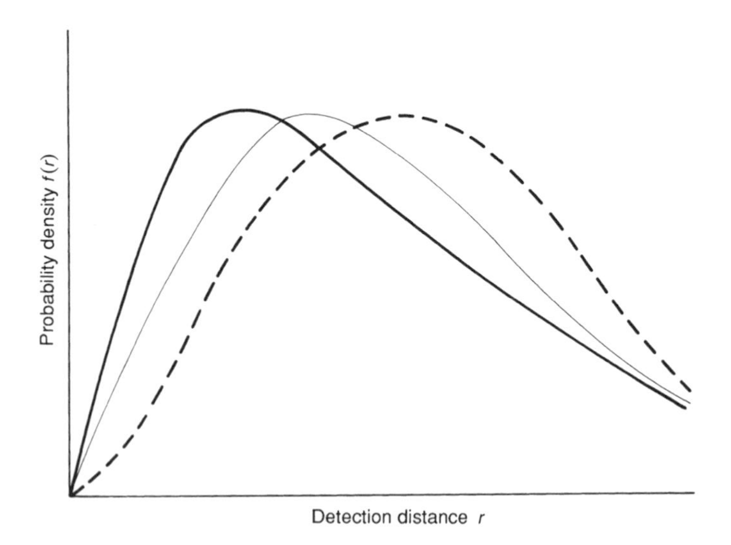

Fig. 5.7. Plots of the real probability density function (——), the apparent function when there is movement away from the observer (— — —), and the apparent function when there is movement towards the observer (——). Estimated density of birds is proportional to the slope of the curve at zero distance from the point, so that density is overestimated when there is movement towards the observer and underestimated when movement is away from the observer.

Roeder *et al.* (1987) also considered two models for disturbance, in which the probability of disturbance was exponentially distributed, being one at distance zero. They then simulated data in which a 'disturbed' object either moved exactly 10 m away from the observer (Model 1) or hid (Model 2). Their conclusions, based on analyses using the method due to Ramsey and Scott (1979, 1981b), the Fourier series method on squared distances and an order statistic method, were consistent with those reported above. In cases where there is an area of perfect detectability well beyond any effects arising from evasive behaviour, Wildman and Ramsey (1985) showed that their method is still valid under a model in which objects move away from the observer, and can be modified if objects close to the observer are known to hide.

The paucity of observations close to the point has been termed a 'doughnut' or 'donut' effect. This effect can result from birds moving away from the point as the observer approaches. Wildman and Ramsey (1985) used data on the omao or Hawaiian thrush (*Phaeornis obscurus*) as a good example of this. In some instances, a poor choice of model can lead to erroneous identification of a doughnut; the empirical distribution function of detection distances is useful for assessing whether a doughnut really exists.

Occasionally, ornithologists have misinterpreted the histogram of the frequency of detections plotted as a function of linear distance (e.g. Fig. 5.1b) as being evidence of a paucity of detections near the point (Rosenstock 1998). Based on this misinterpretation, they have concluded that analysis methods for distance sampling could not perform well because the assumption involving movement was seriously violated. However, such histograms have few observations near the point merely because the sampled area near the point is very small relative to that sampled at larger distances.

Distances are assumed to be measured without error (or to be assigned to defined distance intervals without error), but the assumption is less problematic than for line transects in two respects. First, only the observer-to-object distance is required for modelling. This is often easier to measure or estimate than the perpendicular distance of the object from a transect line, especially if a detected object is not visible or audible, or has moved, by the time the observer reaches the closest point on the transect to it, or if densities are high, so that the observer may need to keep track of several detections simultaneously. Second, to reduce the problems inherent in estimating perpendicular distances for line transect sampling, sighting distances and sighting angles are often recorded. Effort tends to be concentrated ahead of the observer, so that measurement errors in the angles often give rise to recorded angles, and hence calculated perpendicular distances, of zero. Such data are notoriously difficult to model. Point transect data do not exhibit this problem; small observer-to-object distances are seldom recorded as zero, and few small distances occur, as the area surveyed close to a point is small.

In songbird point transect surveys on Arapaho National Wildlife Refuge, locations of detected birds were marked, and were later measured to the nearest decimetre (Knopf *et al.* 1988). Such accuracy is not usually possible; for example, up to 90% of detections are purely aural in woodland habitats (Reynolds *et al.* 1980; Scott *et al.* 1981b; Bibby *et al.* 1985), so that the location of the bird must be estimated. Bias in distance estimation should be avoided. If distances are consistently overestimated by 10%, densities are underestimated by $100(1 - 1/1.1^2) = 17\%$; if they are underestimated by 10%, densities are overestimated by $100(1/0.9^2 - 1) = 23\%$. Bias in line transect density estimates would be smaller (9% and 11%, respectively). Provided distance estimation is unbiased on average, small measurement errors are not problematic. Permanent markers at known distances are a valuable aid to obtaining unbiased estimates, and laser range finders are effective over typical songbird detection distances, at least when the habitat is sufficiently open to use them. Scott *et al.* (1981b) suggest that optical range finders are accurate to ±1% within 30 m, and to ±5% between 100 and 300 m, whereas trained observers are accurate to ±10–15% for distances to birds that can be seen. Laser range finders now far surpass optical range finders for accuracy, offering estimation to the nearest metre from around 20 m up to several hundred metres.

If most objects are located aurally, then the assumption that an object is not counted more than once from the same point may be problematic. If, for example, a bird calls or sings at one location, then moves unseen by the observer to another location, and again vocalizes, it is likely to be recorded twice. Training of observers, with warnings about more problematic species, can reduce such double counting. In some surveys, points are sufficiently close that a single bird may be recorded from two points. Although this violates the independence assumption, it is of little practical consequence.

5.10 Summary

Relative to line transects, relatively few distances are recorded close to zero distance in point transect surveys. Thus estimation of the central parameter ($h(0)$ for point transects and $f(0)$ for line transects) is more difficult, and model selection more critical. This was seen for the first example, where estimation was satisfactory if the correct model was selected, but if the uniform + polynomial model was selected, underestimation occurred, even though the model selection criteria indicated that the model was good. One of the contributory factors to this result was that the detection function used to generate the data was the half-normal, which does not have an area of perfect detection around the point, even though it has a shoulder. Expressing this mathematically, $g''(0) \neq 0$, even though $g'(0) = 0$. If field methods are adopted that ensure an area of perfect detectability, estimation is more reliable, and different models will tend to give similar estimates of density. The hazard-rate key is best able to fit data that show a large area over which detection is perfect, because the hazard-rate detection function can fit a wide, flat shoulder. It performed relatively badly on the example data sets largely because it tended to fit a flat shoulder to the simulated data, which were generated from the half-normal, which possesses a rounded shoulder.

To estimate densities reliably from point transect sampling, design and field methods should be carefully determined, following the guidelines of Chapter 7, and the data should be checked for recording and transcription errors. Histograms of the distance data are a useful aid for gaining an understanding of any features or anomalies, and give an indication of how much truncation is likely to be required. Several potential models should be considered, and model selection criteria applied to choose between them. Special software is essential if efficient, reliable analysis is to be carried out. Variance estimation methods should be chosen for their robust properties; the model that gives the most precise estimate is not the best model if either the estimate is seriously biased or the variance estimate ignores significant components of the true variance. A strategy for data analysis is outlined in Section 2.5.

Systematic error in estimated distances must be avoided. Observer training is essential if data quality is not to be compromised (Chapter 7). If more than one observer collects the data, analyses should be attempted that stratify by observer, to detect observer differences. The importance of this is illustrated in Section 8.6. It may prove beneficial to stratify analysis by other factors, such as species, location, habitat, month, year, or any factor that has a substantial impact on detection probabilities. AIC may be used to determine which factors affect detection, thus reducing the amount of stratification and increasing parsimony. If the factor is ordinal or a continuous variable, it might enter the analysis as a covariate, so that its effect on detectability is modelled.

If objects occur in clusters, the location of the centre of each cluster and the number of objects in the cluster should be recorded. If clusters occur but are not well-defined, the observer should record each individual object, and its location, and use robust variance estimation methods. It is also useful to indicate which detected objects were considered to belong to the same cluster, so that a comparative analysis can be carried out by cluster. The location of the cluster can then be determined by the analyst, by calculating the geometric centre of the recorded locations.

The checklist of stages in line transect analyses given at the end of Chapter 4 may also be used for point transect analyses. In that checklist, replace 'line' by 'point', 'perpendicular distance' by 'detection distance' and 'Section 4.*' by 'Section 5.*'. Also, the rule of thumb for selecting a truncation point w for detection distances is that $\hat{g}(w) \doteq 0.10$, or less satisfactorily, that roughly 10% of observations are truncated (Section 5.3).

5.11 Exercises

1. (a) In a point transect survey, detection distances are estimated, with some degree of rounding to the nearest 10 m. Distances are truncated at $w = 110$ m; probability of detection at this distance is roughly 0.10. Apart from the rounding errors in recorded distances, all main assumptions are met. One analyst chooses to analyse the distances ungrouped, whereas a second chooses to group the data into eight intervals with cutpoints at 0, 15, 25, 35, 45, 55, 65, 85 and 110 m. They argue over the relative merits of the analyses. Do you think this decision is likely to have much effect on estimated density $\hat{D}$ or its variance?

(b) If detection distances were grouped in the field into these eight intervals, does this necessarily mean that the assumption that distances are correctly recorded is violated? Explain.

(c) Which of the following schemes for grouping the data into eight intervals is best? Which is worst? Explain your answer.

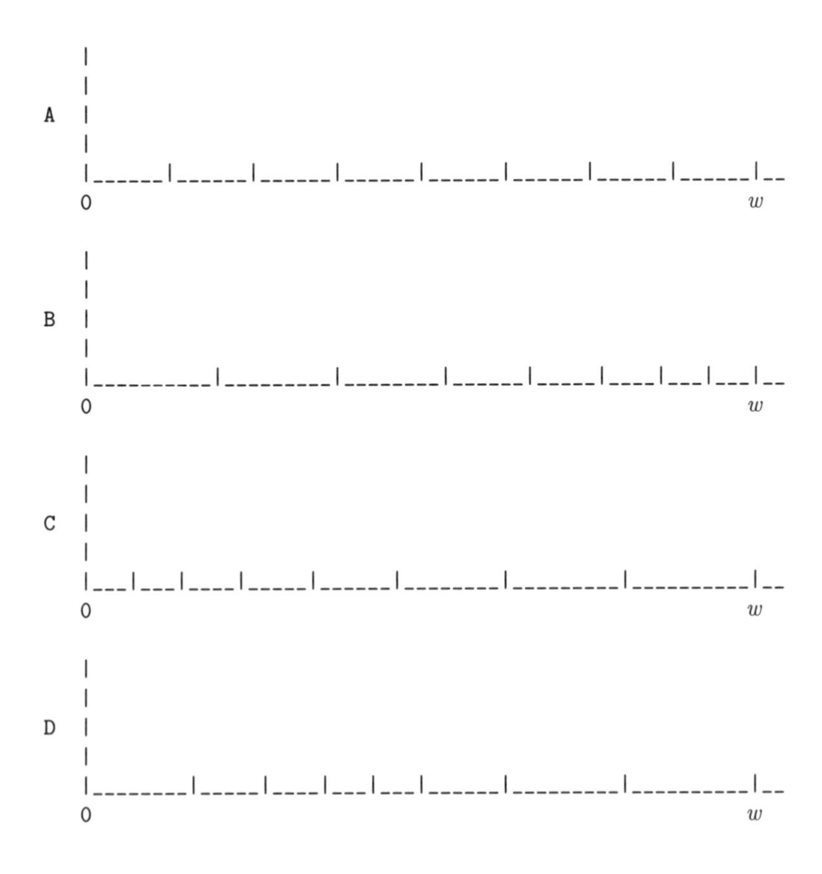

2. If a point transect survey is to be conducted on each of the following species, with the mode indicated, which assumption (if any) is most likely to be violated?

(a) Red foxes ('snapshot' method).

(b) A small songbird that is typically active and mobile within its territory (five-minute count).

(c) A scarce tree species within rainforest (observer remains at the point until satisfied that all visible trees of that species have been detected and their distances from the point measured by laser range finder).

(d) A large gamebird in open habitat ('snapshot' method).

3. In point transect sampling, bird density D is estimated as

$$\hat{D} = \frac{n\,\hat{h}(0)}{2\pi k}$$

where n is the number of birds detected, k is the number of points sampled, and

$$h(0) = \lim_{r \to 0} \frac{f(r)}{r} = f'(0)$$

r is distance from the point, and

$$f(r) = \frac{r\,g(r)}{\int_0^\infty s\,g(s)\,ds}$$

is the density function of detection distances.

(a) If the detection function is given by $g(r) = \exp(-a\,r^2),\ 0 \le r < \infty$, show that the maximum likelihood estimator of $h(0)$ is

$$\hat{h}(0) = \frac{2n}{\sum_{i=1}^{n} r_i^2}$$

where r_i is the ith recorded distance.

(b) Hence estimate density as birds/ha for the following data. (One hectare= $10000\,\text{m}^2$.)

```
k=10 points

Detection distances (m) at each point:

Point 1: 12, 35;  point 2: 10, 25, 30, 45;  point 3: 8, 15, 50;
point 4: 20, 20, 30, 35, 45, 60;  point 5: no birds recorded;
```

`point 6: 20, 25, 30, 30, 40; point 7: 30, 40, 45, 45;`
`point 8: 25; point 9: 15, 25, 30, 40, 40, 50; point 10: 35.`

(c) Estimate the variance of n for these data, and test whether the birds can be assumed to be distributed randomly through the surveyed habitat.

4. For the following three point transect data sets, the number of detection distances is tabulated by distance interval. For each set, plot a histogram of the data, and comment on whether there is evidence of failure of the assumptions of point transect sampling. Where there is such evidence, what kind of failure might have generated such a data set?

Data set A

Interval (m)	0-15	15-25	25-35	35-45	45-65	65-90
Frequency	48	31	34	29	31	24

Data set B

Interval (m)	0-15	15-25	25-35	35-45	45-65	65-90
Frequency	29	52	59	46	39	23

Data set C

Interval (m)	0-15	15-25	25-35	35-45	45-65	65-90
Frequency	9	35	67	55	88	33

5. The histograms below show the fit of the hazard-rate model to ungrouped point transect data on house wrens. Using the same group intervals as in the histograms, the observed frequencies of perpendicular distances are

Interval endpoint (m)	7.5	12.5	17.5	22.5	27.5	32.5	42.5	62.5	92.5
Frequency	76	139	114	117	109	97	104	72	11

(a) Does the model appear to provide an adequate fit to these data?

(b) Explain why two plots are used to show the fit of the model to point transect data, whereas only one is used for line transect data.

(c) Discuss the strengths and limitations of the following methods for selecting between models: goodness-of-fit tests; likelihood ratio tests; AIC.

(d) Summarize the options within Distance that can be used to improve the fit of the hazard-rate model to the wren data. Assess which options are most likely to prove useful on these data.

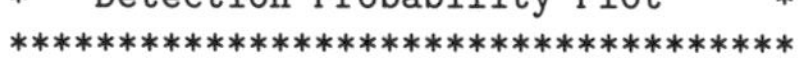
****
* Probability Function Estimation *
* Detection Probability Plot *
****

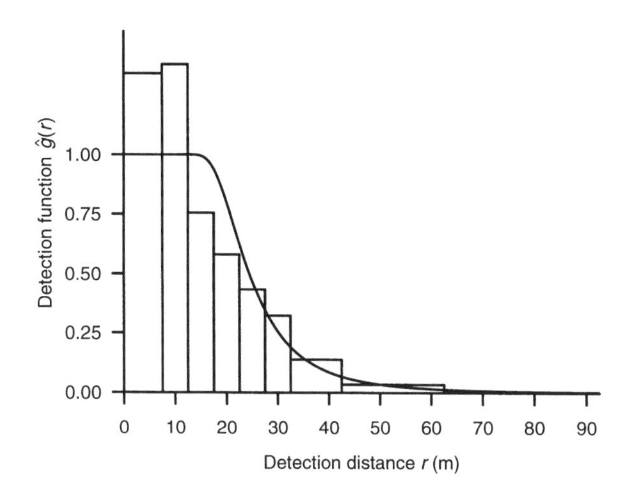
Detection function $\hat{g}(r)$
1.00
0.75
0.50
0.25
0.00
0
10
20
30
40
50
60
70
80
90
Detection distance r (m)

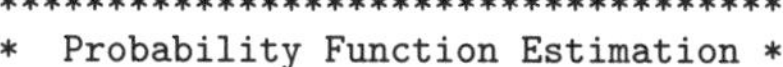
****
* Probability Function Estimation *
* Probability Density Plot *
****

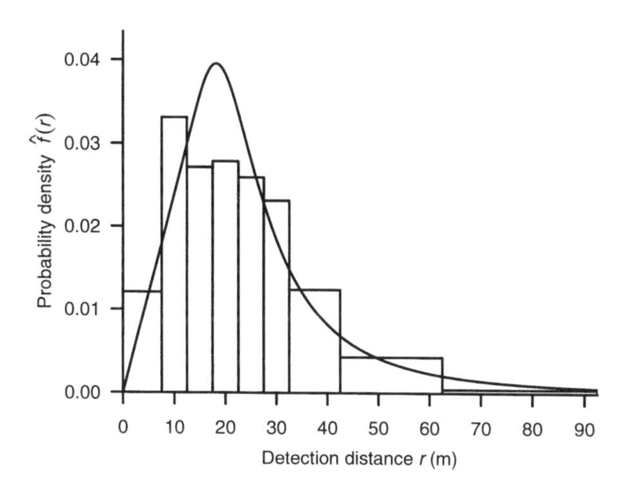
Probability density $\hat{f}(r)$
0.04
0.03
0.02
0.01
0.00
0
10
20
30
40
50
60
70
80
90
Detection distance r (m)

6

Related methods

6.1 Introduction

In this chapter, we describe distance sampling methods that are closely related to the standard line and point transect sampling of Chapters 4 and 5. We also consider models that do not fit into the key + adjustment formulation of earlier chapters. The material on these other models is not exhaustive, but is biased towards recent work, and models that may see future use and further methodological development. Most of the older models not described here are discussed in Burnham *et al.* (1980).

A theme that arises in several sections of this chapter is the use of multipliers (Section 3.1.4) to widen the applicability of distance sampling methods. These methods fall naturally under the heading of 'indirect methods' because they estimate animal density indirectly, instead of the direct approach of standard distance sampling. They include dung and nest surveys (Section 6.2), surveys of objects that are not continuously available for detection (Section 6.3), cue count surveys (Section 6.4), and a method for estimating abundance of fast-moving objects (Section 6.5). Although the multipliers for each method represent completely different quantities, they share the characteristic that they all convert an estimate of the abundance of something produced by the animals ('cues') into an estimate of abundance of the animals themselves. In the case of dung and nest surveys, the 'cue' is the dung or nest. For objects that move fast much of the time, the 'cue' is the object when it is stationary. For objects that are not continuously available for detection, and for cue count surveys, the cue might be the animal itself when it is available for detection, or evidence or presence of the animal, such as a blow in the case of whales.

Indirect methods are useful when cues produced by the animals are more suited to distance sampling methods than the animals themselves. We estimate the cue density and then use knowledge of the expected number of cues per animal to convert this into an estimate of animal density. Let $\hat{R}$ be the estimated cue density (number of cues per unit area) obtained by standard distance sampling methods. Given an estimate (c, say) of the mean number of cues per animal that are available to the observers during the time they conducted the survey, animal density is estimated by $\hat{R}/c$.

To obtain an estimate of the mean number of cues per animal that are available to the observers during the time they conducted the survey, we must estimate mean cue production rate (the number of cues produced per animal per time period). If the cues persist for any length of time, we must also estimate mean cue lifetime or decay rate.

If the survey involves observing the production of instantaneous cues, only an estimate, $\hat{\eta}$, of the mean number of cues produced per animal per unit time is required to estimate animal density from cue density. If observers searched for time T, an estimate of the number of cues per animal available to the observers is $c = \hat{\eta}T$.

If the survey is effectively an instantaneous survey of persistent cues (such as dung), then either an estimate of the mean lifetime of cues present at the time of the survey is also required, or c must be estimated directly. Given an estimate, $\hat{\xi}$, of the mean cue lifetime, an estimate of the number of cues per animal that are available to the observers at a single point in time is $c = \hat{\eta}\hat{\xi}$.

6.2 Dung and nest surveys

6.2.1 *Background*

In some circumstances, when a direct distance sampling survey of a population is unlikely to be effective, it may be possible to estimate the density of some object produced by the animals of interest. If both the production rate and the disappearance rate of these objects can be estimated, then it is possible to estimate the number of animals that produced the objects. One example is deer in forest habitat. In a direct survey, deer may move away from the observer before they can be detected. It is then more reliable to carry out line transect surveys to estimate the density of dung pellet groups, which can be converted to estimated animal density if estimates of defecation rate and decay rate can be obtained (Marques *et al.* in press). Other animals for which dung surveys can be effective include elephants (Fay 1991; Fay and Agnagna 1991; Barnes *et al.* 1995; Plumptre and Harris 1995; Walsh and White 1999), cape buffalo (Plumptre and Harris 1995), foxes, large cats and antelope. Similarly, surveys of apes seldom yield sufficient sample size for reliable line transect analysis, and so the density of nests is estimated. Given estimates of the rate at which new nests are made, and of the mean rate of decay, again animal density can be estimated (Plumptre and Reynolds 1996, 1997; Plumptre 2000). Note one key difference between these surveys and direct surveys: in direct surveys, an estimate of abundance at the time of the survey is obtained, whereas for dung and nest surveys, the final abundance estimate is an average over a time period corresponding roughly to the mean time to decay for the dung or nests.

6.2.2 *Field methods*

In line transect surveys of dung, detection distances are typically very short, perhaps mostly within a metre or two of the line if there is significant ground cover. Because the line is not always well-defined, observers therefore have a tendency to record a high proportion (sometimes 30–50%) of detections as on the line, as they have no meaningful point from which to measure. Such data cannot be analysed reliably. Given a team of two people, one solution to this problem is for one person to pull or lay a cord or cable along the ground, while the other detects the dung and measures its distance from the cord or cable. This and other solutions are discussed in Section 7.8.1.

Elephant dung or dung that comprises pellet groups may cover a significant area of ground. In these cases, it is essential that distances are measured from the line to the centre of the dung; if distances are measured to the nearest edge of the dung, many distances will be recorded as zero, and again, reliable estimation is not possible.

In areas of high dung density, it may prove more cost-effective to carry out strip transect surveys. These entail a simple count of the number of dung piles, or pellet groups, within a strip centred on the line and of predefined width. The observer should search the full width of the strip thoroughly, to ensure that detectability does not fall off close to the strip's edges, and there should be a rigorous criterion for determining whether detected dung is within the strip; otherwise, observers have a tendency to record marginal cases as in the strip, leading to bias. Field trials may be needed to determine a suitable width for the strip, or to compare the efficiency of strip transects with that of line transects.

For nest surveys, distance estimation is seldom problematic, except that nests may occur in loose clusters. In this circumstance, it is generally better to record the distance to each nest individually, and analyse the data as described in Sections 3.5.6 and 4.8.2.2. This is because clusters of nests are often not well-defined, so that measuring to the centre of the cluster can be problematic. Sometimes in these circumstances, the distance to the nearest object(s) or to the object first detected is used as the location of the nest cluster. This biases distances downwards, and hence generates upward bias in the density estimate. A more minor issue for distance estimation in nest surveys when those nests may be high above the ground is that the required distance is the horizontal displacement of the nest from the line, which may be difficult to estimate.

6.2.3 *Analysis*

For ease of presentation, we assume that the surveyed objects are dung piles.

6.2.3.1 *Estimating equations*

If there is no stratification, analysis is straightforward. We first estimate dung density R using standard methods. Thus from eqn (3.1),

$$\hat{R} = \frac{n}{a \cdot \hat{P}_a} \tag{6.1}$$

For line transects, this yields (see eqn (4.7))

$$\hat{R} = \frac{n \cdot \hat{f}(0)}{2L} \tag{6.2}$$

and for point transects (from eqn (5.6))

$$\hat{R} = \frac{n \cdot \hat{h}(0)}{2\pi k} \tag{6.3}$$

If we now divide estimated dung density by $\hat{\xi}$, the estimated mean time to decay (in days say) of dung (where time to decay is the reciprocal of the decay rate), then we obtain $\hat{G}$, an estimate of the dung production per day per unit area:

$$\hat{G} = \frac{\hat{R}}{\hat{\xi}} \tag{6.4}$$

Finally, we need to divide $\hat{G}$ by $\hat{\eta}$, the estimated daily production of dung by one animal (number of dung piles per day) to obtain $\hat{D}$, our estimate of animal density:

$$\hat{D} = \frac{\hat{G}}{\hat{\eta}} = \frac{\hat{R}}{\hat{\xi}\hat{\eta}} \tag{6.5}$$

This is simply line transect sampling with multipliers (Section 3.1.4), and the above eqn is essentially eqn (3.16) with $\hat{\gamma} = 1/\hat{\xi}$ and with the term in $\hat{E}(s)$ omitted. The variance of animal density is

$$\widehat{var}(\hat{D}) = \hat{D}^2 \cdot \left\{ \frac{\widehat{var}(\hat{R})}{\hat{R}^2} + \frac{\widehat{var}(\hat{\xi})}{\hat{\xi}^2} + \frac{\widehat{var}(\hat{\eta})}{\hat{\eta}^2} \right\} \tag{6.6}$$

where $var(\hat{R})$ is obtained from the line transect analysis of the dung distance data.

Estimation becomes more complex when there are strata, corresponding for example to geographic blocks or habitats. If strata correspond to habitats, decay rate and $\hat{f}(0)$ might both be expected to vary by stratum, but defecation rate might be assumed constant across strata. Examples of complex stratified analyses are given by Marques *et al.* (in press).

6.2.3.2 *Estimating decay rates*

Decay rates of dung typically vary spatially and by season. It is important therefore to design an experiment or observational study for estimating decay rate. For valid analysis, the mean time to decay of the dung piles that are on the ground at the time of the line transect survey is required. If that survey is of short duration, the most cost-effective way to estimate decay rate is the following.

First establish criteria for deciding whether a given dung pile is considered to have decayed. These criteria should be exactly the same in the decay rate experiment as in the line transect survey. Next determine a time period over which the majority (say 90% or more) of dung piles can be expected to decay. Commence regular searches for fresh dung this amount of time ahead of the line transect survey. If typical mean times to decay are say 10–20 days, the search should be restricted to dung no more than 24-h old. This time period can be extended for longer mean times to decay, but experiments may be required to assess reliability in determining whether dung is fresh. Mark each fresh dung pile detected, so that it can later be relocated. There should be at least five or six visits to the study site, approximately evenly spaced in time, between the first visit and the eventual line transect survey. The number of fresh dung piles located at each visit need not be large, as the major source of variability in the final abundance estimate is likely to be encounter rate, and the contribution to overall variance of the estimated mean time to decay is typically small (Marques *et al.* in press). However, adequate sample size is needed to allow reliable modelling of the decay rate; a minimum of 50 dung piles over the duration of the experiment might be sufficient. Searches for fresh dung piles should be carried out in a range of environments representative of the study area. Ideally, this will entail visits to several locations selected at random within the area.

Having established marked dung piles, only one subsequent visit is required to each dung pile. This should be conducted at the time of the line transect survey. The recorded data are binary: 1 if the dung pile does not meet the decay criteria at this visit, and 0 if it does, or can no longer be located. A logistic regression can then be carried out on these data, with time lapsed between marking and revisiting the dung pile as the covariate. If the logistic curve is at, or very close to, unity for zero elapsed time, the cumulative distribution function of time to decay is given by one minus the fitted logistic curve. Mean time to decay is then estimated by differentiating the distribution function to give the probability density function (pdf) $f(t)$ say of time to decay, from which mean time to decay is estimated by numerical integration: $\int_0^T t\, f(t)\, dt$ where T is some arbitrarily large elapsed time (corresponding to the maximum plausible time to decay). If the logistic curve is not close to unity at time zero, some transformation of the elapsed times may be required prior to carrying out the logistic regression.

In practice, additional covariates, such as habitat variables at the location of the marked dung pile or rainfall, might be included in the logistic regression. This both improves and complicates estimation of mean time to decay; the methods of Buckland *et al.* (1999), although developed for a different purpose, may be used. Transformations of the covariate of interest, elapsed time in our case, are also addressed in that paper.

Typically, decay rates are estimated by making repeat visits to a marked dung pile until it is judged to have decayed (Plumptre and Harris 1995). Its 'lifetime' is then recorded (at least approximately), and survival models used to estimate mean time to decay. However, this estimates mean time to decay of fresh dung present at the start of the experiment. What we require is the mean time to decay of any dung present at the end of the experiment, when the line transect survey is carried out. If decay rates are independent of date, these two quantities are equivalent. However, if seasonal variation occurs, or is suspected, we recommend against this strategy. If line transect surveys are carried out over an extended time period, then it may be necessary to make repeat visits to marked dung piles, so that the mean time to decay can be estimated for different times of the year.

Similar considerations apply for estimating mean time to disappearance of nests, for example for monitoring ape numbers.

6.2.3.3 *Estimating defecation or nest production rate*

Defecation rates may be estimated through direct observation of individuals in the population of interest. This is often impractical, in which case captive animals, if available, might be studied. In this case, conditions should be as close as possible to those for the population to be surveyed. Another option with captive animals is to place them in a large, natural enclosure that has no dung at the outset of the study. Dung piles are then counted after a given period of time, which is sufficiently short that none have decayed, from which the defecation rate is calculated as

$$\frac{\text{number of dung piles}}{\text{number of animals} \times \text{number of days in enclosure}}$$

Again, similar strategies may be adopted if the surveyed objects are nests.

6.2.4 *Assumptions*

We again assume that dung piles are being surveyed. Similar considerations apply for nests. For estimating dung pile density, the standard assumptions of line transect sampling apply (Section 2.1). To convert that density to animal density, further assumptions are required.

First it is important to note that conceptually, for direct distance sampling methods, animal density at the instant of the survey is being estimated. Thus no assumption of closure of the population is required. Of course, in practice, the survey cannot be conducted in an instant of time, and closure is effectively assumed for the duration of the survey. By contrast, in indirect surveys, the estimate of density corresponds to the time period over which dung piles, present at the time of the line transect survey, were deposited. This adds no new problems if the population is closed throughout that period. If it is not, then the method estimates an average of the density through that time period, weighted by the dung decay curve, with greatest weight given to the period immediately preceding the line transect survey, for which there has been insufficient time for any dung piles to have decayed.

In deer surveys for example, it may be easy to use direct methods to estimate numbers in open habitat. It is tempting therefore to use indirect methods, which require additional assumptions and resources, only in closed habitat, where direct surveys might not be possible. However, movement between open and closed habitat can then be problematic. The direct surveys estimate the number of animals in open habitat at the time of the survey. This number may vary by season, or between day and night. The indirect surveys estimate average numbers of animals in closed habitat over a period of some weeks or months preceding the survey of dung. We cannot therefore assume in such a study that total abundance is reliably estimated by the sum of estimated abundances in the two habitats.

Second, we must assume that the dung piles monitored are a representative sample of the dung piles deposited over the period of the decay rate experiment. If decay rates are constant across animals, habitats, weather conditions and season, this assumption is not required. In practice, heterogeneity can be substantial, so that survey design is important. In the presence of heterogeneity, we must estimate the mean time to decay of all dung piles deposited prior to the line transect survey of dung piles, as a function of time between deposition and the survey. If we can identify a random sample of dung piles at each of several time points in the lead up to the line transect survey, with the earliest time point such that at least 90% of dung piles will have decayed by the time of the survey, then we simply need to record which of these dung piles are still present at the time of the survey. The methods of Section 6.2.3.2 then allow reliable estimation of the required mean decay rate. The design of the decay rate experiment should therefore seek to ensure that the fresh dung piles identified at each time point are representative, for example by searching for dung piles using a standard search pattern at random locations throughout the survey region.

Third, we must assume that we have an unbiased estimate of the deposition rate of dung piles, together with a reliable estimate of its precision. This requires that an independent random sample of animals of reasonable size (ideally around 20) from the study population be monitored. This is seldom feasible, so that measures to ensure that the estimate is as representative as possible should be taken. Precision of the estimate is not critical, provided variation between individual animals is not large. However, bias in the estimate, caused for example by using captive animals on a different diet, and with different behaviour patterns, from the population of interest, can be problematic.

6.3 Line transect surveys for objects that are not continuously available for detection

6.3.1 *Periods of detectability interspersed with periods of unavailability*

Consider line transect surveys of objects that are only visible at well-spaced periods in time, so that $g(0) < 1$. For example, some species of whale dive for prolonged periods, followed by periods at the surface. Suppose detected whales are only recorded if they are at the surface at the time they come abeam of the observation platform, and their perpendicular distance is estimated at that time. Then a conventional line transect analysis yields an estimate of the product of the density of whales with the proportion of whales at the surface at any given time. If that proportion can be estimated, then so can population abundance. This strategy is of little use on slow-moving platforms such as ships, since most detected whales will have dived, or moved in response to the vessel, by the time the vessel comes abeam. However, it can be very successful for aerial surveys. Its disadvantage is that further survey work must be carried out to estimate the proportion of whales at the surface at a given time. This is done by monitoring individual whales over prolonged periods. Possible problems are that it may not be possible to monitor sufficient whales for sufficiently long periods; monitored whales may be affected by the presence of the observer, and may spend an atypical amount of time at the surface; if whales go through periods of short dives followed by longer dives, most of the monitored sequences may be short-dive sequences, since whales are more likely to be lost if they dive for a longer period; whales that habitually spend more time at the surface are more likely to be detected and monitored; it can be difficult to define exactly what is meant by 'at the surface', especially if monitoring of individual whales is done from a surface vessel, and the line transect surveys from the air. Hain *et al.* (1999) used airships (blimps) to estimate surface and dive times of right whales in coastal waters of the southeastern United States, thus ensuring better comparability with aircraft survey data.

If a whale at the surface is considered to be a cue, then the conventional survey gives an estimate of cue density at a single point in time ($\hat{R}$). An estimate of the proportion of time whales are on the surface is an estimate of the expected number of cues per animal at the time the observer passes the animal (c). Hence the animal density estimate $\hat{D} = \hat{R}/c$.

Another example is desert tortoise, which spend much of their time underground. A conventional line transect survey may be conducted to estimate the number of animals above ground at the time of the survey, and simultaneously, a survey can be conducted, for example using radio-tagged animals, to estimate the proportion that are above ground. This allows estimation of total abundance, taking the multiplier c of Section 3.1.4 to be the estimated proportion above ground. If data are collected for several sites and years, together with relevant covariates such as date, time of day, temperature and precipitation, then c may be modelled using logistic regression. We are then able to predict c from the model, so that annual telemetry studies at every site are not needed.

6.3.2 *Objects that give discrete cues*

When objects give only discrete cues to indicate their presence, such as whale blows or bird song, there is the danger that detection on the trackline will be uncertain ($g(0) < 1$). This danger can be reduced by reducing the speed of the observer (line transect sampling), or increasing time at the point (point transect sampling), so that objects close to the line or point are certain to give a cue while in detection range. However, this strategy increases the upward bias in density estimates if the objects are mobile.

Cue counting (Section 6.4) is a simple and natural approach for such populations. If this approach is not feasible, then methods for surveys in which detection on the trackline is uncertain may be used. Most methods rely on having two observation platforms that operate independently, or at least with one-way independence, with the primary platform ignorant of detections made by the second platform. Methods for estimating object density generally rely on determining or estimating which objects or cues are detected from both platforms – so-called duplicate detections. For methods based on identification of duplicate objects, it is necessary to separate the area of search of the two platforms if detection cues are discrete. This is because an animal that gives a particularly visible cue within a common search area will tend to be detected by both platforms, whereas one that does not give a cue within the search area will be missed by both platforms, thus violating the requirement of independent detections from the two platforms. By separating the areas of search, as in Buckland and Turnock (1992) and Borchers *et al.* (1998a), the assumption of independence is more plausible. Another solution is to use models for duplicate cues, for which the assumption of independent detections is more plausible, especially if the effects of covariates on cue detection are modelled (Schweder 1974, 1977, 1999; Schweder and Høst 1992; Schweder *et al.* 1999). It is also easier to identify whether a cue is seen by both platforms than whether an animal is, as the different platforms may see different cues from the same animal. The disadvantage of cue-based methods is that cue rate must be estimated. The problems in estimating cue rate for whales are largely as noted in Section 6.3.1 for estimating the proportion of time spent at the surface.

For double-platform surveys of objects that give discrete cues, a few general recommendations on field procedures can be made. First, all detected cues from a given object should be recorded, even if object-based methods are used, as this eases the task of assessing which detections are duplicates. To facilitate this goal further, the exact time of each cue should be noted, preferably using a computerized recording system. Methods are generally sensitive to the judgement of which detections are duplicates, so every attempt should be made to minimize uncertainty, and the uncertainty should be reflected in estimated variances (Schweder *et al.* 1991). Ancillary data such as animal behaviour, cluster size and weather should be recorded for each detection, to allow the analyst to use stratification or covariate modelling to reduce the impact of heterogeneity on estimation. The pooling robustness property, so useful when detection on the line is certain, is lost when such detection is uncertain, hence the need to model heterogeneity (Borchers *et al.* 1998a,b; Schweder 1999).

Double-platforms methods are described in detail by Buckland *et al.* (in preperation). They are useful not only for objects that provide only discrete detection cues but also for surveys in which it is not possible to provide enough search effort to ensure that all objects on the line are detected, even if they offer a continuous detection cue.

Models for discrete cues might also be useful for line or point transect surveys of songbirds in which detection relies primarily on hearing the song. By modelling the rate at which birds of a given species sing, as a function of variables such as time of day, season and habitat, bird densities might be reliably estimated from distance sampling surveys of their cues. This is a possible area of research, which might extend the applicability of distance sampling to more species. For example some species are very mobile between songbursts, and their abundance may be overestimated by as much as an order of magnitude by conventional point transect sampling, in which the observer counts for several minutes at each point. Other species sing rather infrequently, at least at certain times of the day or year, and their abundance may tend to be underestimated by conventional methods.

6.4 Cue counting

6.4.1 *Introduction*

Cue counting is suitable when multiple discrete (instantaneous) cues are observed during a survey. Because cues are instantaneous, no estimate of cue lifetime is required; only estimates of cue production rate per animal are needed to convert estimates of cue density to estimates of animal density.

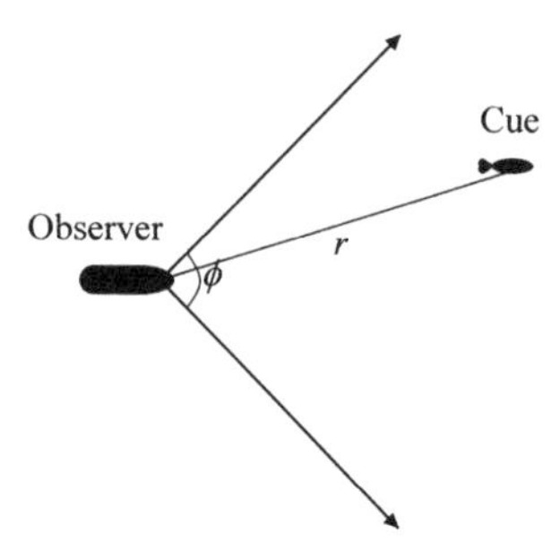

Fig. 6.1. In cue counting, the observer scans a sector of angle ϕ ahead of the vessel. Distances are estimated to detected cues within that sector. Cues outside the sector are not recorded.

Cue counting was developed for estimating whale numbers (Hiby, 1982, 1985; Hiby and Hammond 1989). It has very similar design considerations as line transect sampling – and in fact is sometimes carried out simultaneously with line transect sampling – yet theoretically it is much more closely related to point transects. It is a simpler alternative to the use of double-platform methods for estimating the abundance of objects that give discrete detection cues (Section 6.3.2). An observer scans a sector ahead of the viewing platform – usually an airplane or the bow of a ship – and records the distance to each detected cue (Fig. 6.1). The cue is usually defined to be a whale blow. Cues are recorded irrespective of whether the whale was previously detected, and it is not necessary to estimate pod (cluster) size. The method yields estimates of cue density, which can only be converted into whale density by estimating the cue or blow rate, η, from separate surveys. Cue density is estimated much as object density is estimated from point transect data. The observer records only radial distances. Perpendicular distances are not needed, and angles only determine whether a cue is within or outside the observation sector. To estimate cue rate η, individual whales are followed, and the observed rate is used as an estimate of the cue rate for the whole population. This is the main weakness of the approach, as relatively few whales can be monitored for sufficiently long periods to obtain reasonable cue rate estimates. Further, the sample of whales used to estimate η may not exhibit typical cue rates; for example whales with high cue rates are more likely to be detected and less likely to be 'lost' before an estimate can be obtained than whales with a low cue rate (an example of size-biased sampling), and whales monitored over a long time period may change their cue rate in response to the vessel.

6.4.2 *Density estimation*

Suppose cues are recorded out to a distance w, within a sector of angle ϕ (Fig. 6.1). For cue counting, as for point transect sampling, the area of a ring of incremental width δr at distance r from the observer is proportional to r. It follows that $f(r) = 2\pi r g(r)/\nu$, where $\nu = 2\pi \int_0^w r g(r)\,dr$. Given that a cue occurs in the sector of area $c \cdot a$, where $c = \phi/2\pi$ is the proportion of the circle of area $a = \pi w^2$ that is counted, let the probability that it is seen be P_a. Then this probability is $\nu/(\pi w^2)$. Thus $a \cdot P_a = \nu$, which holds as $w \to \infty$. We assume that all cues very close to the observer are seen ($g(0) = 1$). Eqn (3.15), with $\hat{E}(s) = 1$, yields the following estimate of cue 'density' per unit time (i.e. the number of cues per unit area per unit time):

$$\hat{D}_c = \frac{2\pi n}{\phi \hat{\nu} T} \tag{6.7}$$

where n is the number of cues recorded in time T. The constant T is the total time that the observer is searching (i.e. 'on effort'), and corresponds to the line transect parameter, L. If the cue rate is estimated as $\hat{\eta}$ cues per unit time per animal, then estimated whale density is

$$\hat{D} = \frac{2\pi n}{\phi \hat{\nu} T \hat{\eta}} \tag{6.8}$$

Note that this can be written as

$$\hat{D} = \frac{\hat{R}}{c} \tag{6.9}$$

where

$$\hat{R} = \frac{2\pi n}{\phi \hat{\nu}} \tag{6.10}$$

is the estimated density of cues that occurred during the survey period (of length T), and

$$c = \hat{\eta} T \tag{6.11}$$

is the estimated number of cues generated per animal during the time period T. As for point transects,

$$\hat{\nu} = \frac{2\pi}{\hat{h}(0)} \tag{6.12}$$

where

$$h(0) = \lim_{r \to 0} \frac{f(r)}{r} \tag{6.13}$$

so that

$$\hat{D} = \frac{n \cdot \hat{h}(0)}{\phi T \hat{\eta}} \tag{6.14}$$

The value of $\hat{h}(0)$ may be obtained by modelling the recorded distances to cues, as if they were distances from a point in a point transect survey. Distance has a cue count option to carry out this analysis (below).

6.4.3 *Assumptions*

As noted previously, a valid estimate of the cue rate (and of the variance of this estimate) is assumed to be available. Thus the study to estimate cue rate is assumed to monitor a random sample of whales from the population. There should be no bias towards whales with high cue rates, and the period that a given whale is monitored should not be terminated when its cue rate changes, for example when a sperm whale dives for a prolonged period, having been monitored at the surface. The survey vessel should not affect the cue rate of monitored animals.

The definition of what constitutes a 'cue' should be consistent between the cue rate experiment and the cue count survey. For whales that have a very visible blow, this assumption is easily met. Greater care is required for other species.

Because successive cues from the same whale, or cues from more than one whale in a pod, may be counted, the distances are not independent observations. This does not invalidate the method, but analytic variances should not be used. The bootstrap, applied by taking say cruise legs as the sampling unit, provides valid variance estimation.

Line transect sampling of whale populations is beset with problems of how to estimate $g(0)$, especially for aerial surveys, where a whale may be below the surface while it is in range of the observer, and for species such as sperm whales, which typically dive for around 40 min at a time. Cue counting does not require that all whales on the centreline are detected. Instead, it assumes that all cues occurring immediately ahead of the observer are seen. Thus, of those on the centreline, only whales that are at the surface when the vessel passes are assumed to be detected with certainty. In practice, whales may show vessel avoidance, so that the recorded number of cues very close to the vessel is depressed. Because the area surveyed close to the vessel is small, the effect of vessel avoidance might be expected to be small, unless avoidance occurs at relatively large distances. If avoidance is suspected, the distance data may be left-truncated. This solution should prove satisfactory provided the effects of vessel avoidance only occur well within the maximum distance for which the probability of detecting a cue is close to unity.

If cues immediately ahead of the vessel might be missed, double-platform methods similar to those described for line transect surveys in Buckland *et al.* (in preparation) may be used. Cue counting has the advantage over methods that require determination of whether the same animal (or animal cluster) is seen from both platforms in that it is easier to identify whether a single cue is seen from both platforms, for example by recording exact times of cues. The two platforms may see different cues from the same animal, so that 'duplicate detections' are less reliably determined for animal-based methods than for cue-based methods.

Even if estimation of $g(0)$ is not a major concern, consideration should be given to using double-platform methods, to allow estimation of the size of errors in distance estimation. Random error in distance estimation can result in substantially positively biased estimates of density from point transect methods, and cue count methods in particular. With line transect surveys, the expected number of animals in the population at all perpendicular distance intervals is the same; provided errors are typically smaller than the shoulder width of the detection function (Chen 1998; Section 7.4.3), bias arising from underestimating some distances from the line tends to cancel with bias arising from overestimating other distances. With point transect and cue count methods, by contrast, the expected number of animals in any distance interval increases as distance increases. Random error in distance estimation therefore tends to result in a net transfer of animals towards distance zero. If the proportion of animals erroneously allocated to a nearer distance interval is the same as that erroneously allocated to a further distance interval, more animals will move in from the further interval than will move out from the nearer interval – because there are more animals in the further interval. This has the effect of increasing the slope at the origin of the density of observed distances ($h(0)$). Because $h(0)$ is a multiplier in the density estimation equation, the extent to which it is increased is the extent to which the density estimate is positively biased. This can be very substantial if distances are estimated imprecisely.

Figure 6.2 shows an example of the effect on the distribution of estimated radial distances, and on the resulting estimate of $h(0)$, of multiplicative normal random distance estimation error with a coefficient of

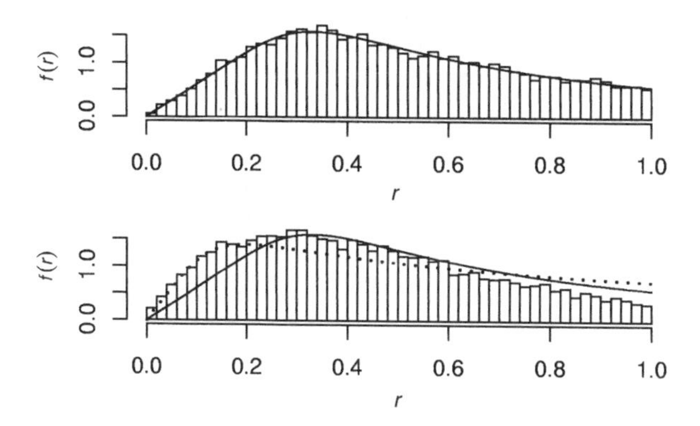

Fig. 6.2. The top plot shows the distribution of true radial distances of observed animals and the true underlying $f(r)$. The bottom plot shows the distribution of estimated radial distances of observed animals and the estimated $f(r)$ (dashed line), together with the true $f(r)$. Note that the slope at the origin of the former is substantially larger than the slope at the origin of the latter.

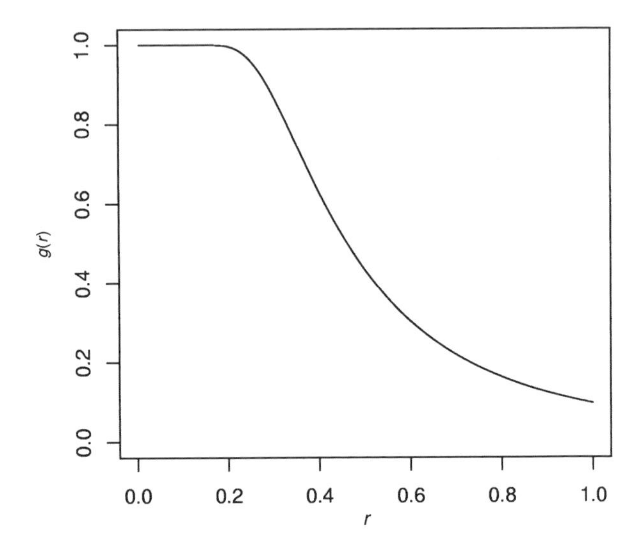

Fig. 6.3. The true detection function from which observations in Fig. 6.2 were generated.

variation of 35%, when the true detection function is the hazard-rate function shown in Fig. 6.3 and the estimated detection function is also a hazard-rate function. The detection function was estimated treating the distances as exact; bias may be less if data are binned into distance intervals. The hazard-rate fit to the recorded distances in the bottom plot is good out to a distance of about 0.1 but not beyond. In practice one might include adjustment terms to improve the fit beyond 0.1, but this will not alter $h(0)$ substantially. This simulation with 100 000 animals results in a positive bias in density of approximately 85%. Bias becomes more severe as the coefficient of variation of distance errors increases, and as the true detection function becomes more 'spiked'.

Hiby *et al.* (1989) describe a method of estimating and incorporating random errors in distance estimation to avoid bias of this sort.

6.4.4 *Example*

Cue counting has been used in aerial surveys to estimate fin whale densities near Iceland (Hiby *et al.* 1984) and in shipboard surveys to estimate whale densities in the North Atlantic (Hiby *et al.* 1989) and minke whale densities in the Antarctic (Hiby and Ward 1986a,b; Ward and Hiby 1987). We use here data from Hiby and Ward (1986a) to illustrate the method. Annual surveys of Southern Hemisphere minke whales have been carried out since the 1978–79 season. The first attempt to use cue counting during shipboard surveys occurred on the 1984–85 cruise. Hiby and Ward considered that cues close to the vessel were under-represented, possibly because whales showed vessel avoidance behaviour or because blows close to the vessel were under-recorded by observers. We therefore analysed the data both with no left-truncation and with left-truncation at 0.4 n.m. (nautical mile). The data were right-truncated at 3 n.m. Under the hazard-rate model, frequencies at distances less than 0.4 n.m. are not significantly below expected frequencies, and truncation makes little difference; the only anomaly is the relatively high frequency at 0.8–1.0 n.m. (Fig. 6.4), which may be chance fluctuation, or, more likely, preferential rounding to that distance interval. Hiby and Ward (1986a) appear to have interpreted these data too pessimistically, suggesting that detections close to the vessel are too few because (1) blows are less visible at short distances, (2) whales show vessel avoidance behaviour or (3) observers did not appreciate the need to record all cues at short distances. Because successive cues are not independent, goodness of fit tests are likely to give spurious significant results. If they are carried out regardless for the hazard-rate model, they are not significant at the 5% level, so Hiby and Ward's conclusion that the data cannot be analysed seems unduly pessimistic. Data sets collected more recently suggest that the method performs adequately.

The fits of the hazard-rate model to the data both with and without left-truncation are shown in Fig. 6.4. In these trials, both blows and sightings of the body of the whale were counted as cues. Hiby and Ward (1986a) estimated the cue rate at 34.98 cues per whale per hour ($\widehat{se} = 4.74$). Supplying this estimate to Distance, together with an estimate of time on effort of 35.8 h (430 n.m. divided by an average speed of around 12 knots), yields an estimated density of 0.24 whales/n.m.^{2} from untruncated data and 0.26 whales/n.m.^{2} from the truncated data. The goodness of fit statistics are $\chi^2_6 = 11.7$ and $\chi^2_3 = 7.3$, respectively. The p-values for the goodness of fit tests are invalidated by the lack of independence between successive cues from the same animal or animal cluster. Similarly, the analytic estimates of variance are invalid. Without the raw data, it is not possible to apply either the bootstrap or the empirical method to obtain valid variance estimates, because cue counts are not given by cruise leg in Hiby and Ward. In Fig. 6.5, the fits of the Fourier series model to these data, with and without left-truncation, are shown. It yields an estimated density of 0.24 whales/n.m.^{2} without truncation and 0.31 whales/n.m.^{2} with truncation, with respective goodness of fit statistics of $\chi^2_7 = 18.9$ and $\chi^2_3 = 9.1$, indicating a worse fit than the hazard-rate model. Again the p-values corresponding to these statistics are invalid, and we do not present them. The flatter shoulder of the hazard-rate model enables it to fit the counts at short distances more closely. The estimate of density from a line transect survey carried out at the same time as the cue rate trial was 0.37 whales/n.m.^{2}.

6.5 Distance sampling surveys for fast-moving objects

6.5.1 *Line transect surveys*

We consider here the problem of estimating abundance of seabirds from shipboard ('at sea') surveys. The solutions suggested here are potentially relevant to other line transect surveys of fast-moving objects.

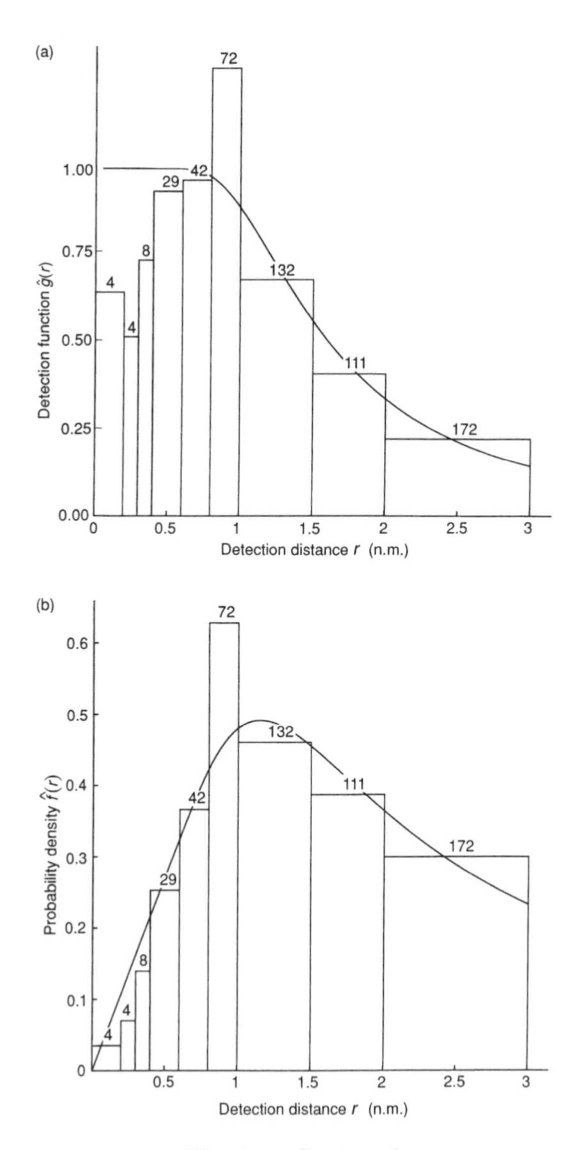

Fig. 6.4. *Continued*

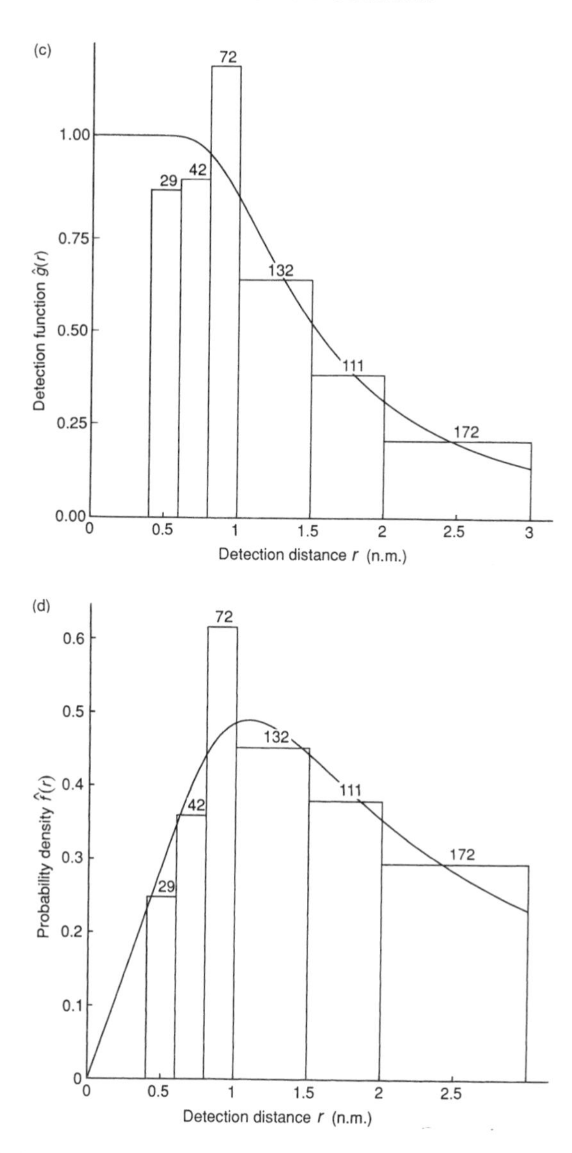

Fig. 6.4. (a)–(d) Histograms of the cue count data. Also shown are the fits of the hazard-rate model to the data without left-truncation (a and b) and with left-truncation (c and d). The fitted detection functions are shown in (a) and (c) and the corresponding density functions in (b) and (d).

Tasker *et al.* (1984) reviewed methods for estimating abundance of seabirds from ship-based transects. They noted that most methods, including line transect sampling, were positively biased, due to the presence of birds in flight. As stated in Section 4.9.3, any movement of objects should be slow relative to the speed of the observer (Hiby 1986). This is clearly not the case for shipboard surveys of seabirds in flight.

Tasker *et al.* (1984) proposed three methods of counting seabirds. Only the first is designed to yield estimates of bird density: strip counts are carried out of birds on the sea, whilst instantaneous counts are made

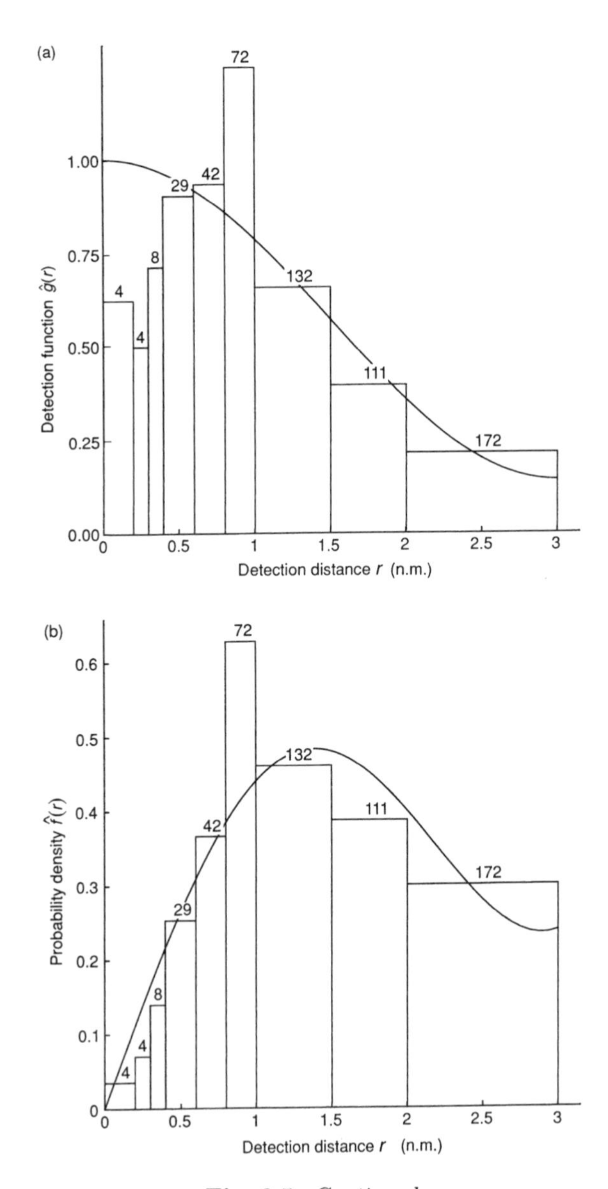

Fig. 6.5. *Continued*

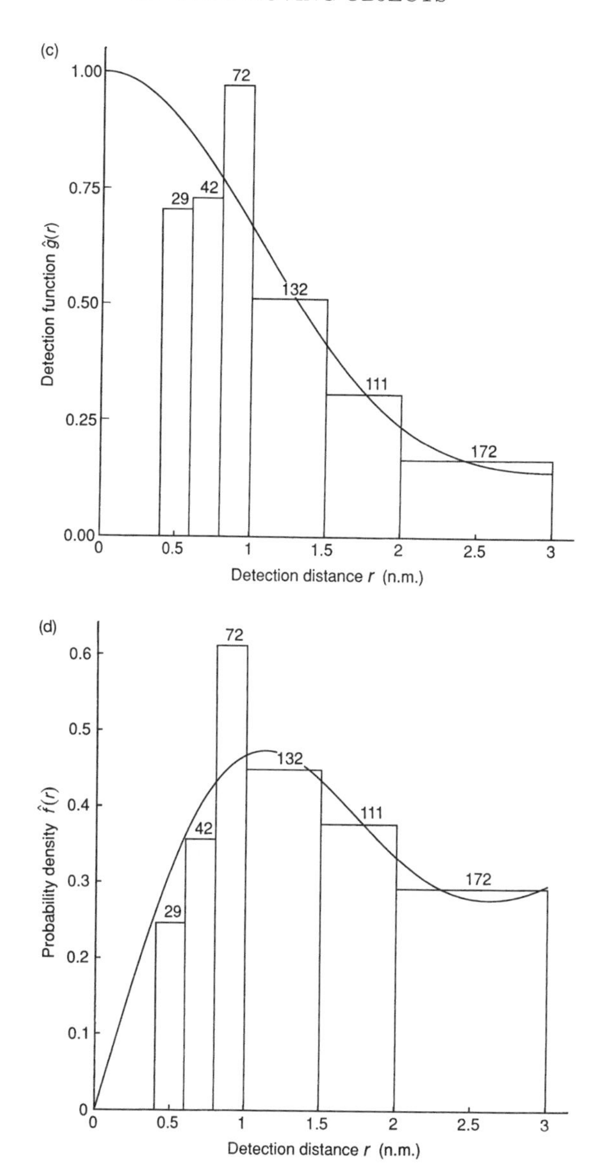

Fig. 6.5. (a)–(d) Histograms of the cue count data. Also shown is the one-term Fourier series fit to the data with no left-truncation (a and b) and the two-term fit to the left-truncated data (c and d). The fitted detection functions are shown in (a) and (c) and the corresponding density functions in (b) and (d).

periodically of birds in flight within a strip of defined width and length. Both counts are then treated as if they are quadrat or strip counts. Further procedures were proposed for birds moving across the bows of the ship and for birds associated with the ship.

In practice, it is difficult to carry out instantaneous counts within a predefined area. For smaller species, it may not be possible to detect all birds present. For all species, it is difficult to estimate whether a detected bird is within the defined area at the instant of the count, and there may be a tendency to include birds where there is doubt, which would generate positive bias in the density estimate. The method is also inefficient for scarce species, because birds seen, but not within the defined area at the instant of the count, are ignored.

Spear *et al.* (1992) also used strip counts, but included birds in flight in their standard counts. They then adjusted their counts, to take account of both speed and direction of birds in flight.

Another solution is to use conventional strip or line transect sampling (depending on whether densities are high or low) to estimate abundance of birds on the sea. Separate surveys can then be conducted, for example using radio tags, to estimate the proportion of time spent on the sea. The abundance estimates are divided by this estimated proportion. The closer the proportion is to one, the greater the precision of this method. The same approach is sometimes used in aerial surveys of whales, in which only those at the surface when abeam are recorded (Section 6.3.1).

If an animal is considered to be giving a detection 'cue' while it is on the sea surface, then the conventional survey gives an estimate of cue density ($\hat{R}$). An estimate of the proportion of time birds or whales are on the surface is an estimate of the expected number of cues per animal at the time the observer passes the animal (c). Hence the animal density estimate is $\hat{D} = \hat{R}/c$.

For scarce species that spend much of their time in flight, a line transect method of analysing detections of birds in flight would be advantageous. This is possible if (a) birds do not respond to the observation platform and (b) birds pass within detection range of the observer at most once during coverage of a given line. Whenever a flying bird (or flock) is detected, wait until it comes abeam of the observation platform, and only then record its position. For line transects, its perpendicular distance is estimated at this point; for strip transects, the bird is recorded only if it is within the strip when it comes abeam. The bird is not recorded if it is lost from view before it comes abeam. If it alights on the water, it is included as a bird on the sea, and analysed separately from those birds recorded as in flight. Having obtained separate density or abundance estimates for resting/feeding and for flying birds, sum the two estimates.

If birds are known to respond to the observation platform, but only when quite close to it, the above procedure may be modified. Determine the smallest distance d beyond which response of flying birds to the platform is likely to be minimal. Instead of waiting for the bird to come abeam, its position is now recorded when its path intersects with a line perpendicular to the transect a distance d ahead of the platform. For this procedure to work, probability of detection at distance d, $g(d)$, should equal, or be close to, one. In this circumstance, flying birds that are first detected after they have crossed the line will have mostly intersected it at relatively large perpendicular distances, and can be ignored.

Hashmi (in preparation) developed a similar method for estimating abundance of migrating species of seabird, for when birds travel perpendicular to the transect line, as occurs when birds pass through a strait, while the observer crosses the strait. In this circumstance, it is more effective to estimate the distance of the bird from the observer when the bird crosses the transect line, rather than the abeam line, and Hashmi's method is based on these distances.

6.5.2 *Point transect surveys*

The simplest solution to the problem of fast-moving objects in point transect sampling is to define a 'snapshot' moment. The observer will typically spend a few minutes identifying what is present before the snapshot moment, and a few more after the snapshot moment verifying locations and distances (Section 5.9). In the case of bird surveys, sometimes birds in flight will be detectable at the snapshot moment. Their position at that moment should be estimated as accurately as possible; setting a watch to beep, so that the moment is well-defined, may help. Flying birds not detectable at that moment are excluded.

Point counts to a fixed radius can be biased by records of birds in flight, even if the snapshot method is adopted, because of the tendency to record flying birds as inside that radius when in doubt. By contrast in point transect sampling, the observer does not have to decide whether to count the bird or not; rather, the distance is estimated if the bird is detectable at the snapshot moment.

Marsden (1999) used point transect sampling to estimate densities of parrots and hornbills. He analysed only data from stationary birds in these surveys. He then used the presence of vantage points to watch birds, and record the proportion of time spent in flight. This allows densities to be corrected for birds in flight, in much the same way as noted above for line transect surveys of seabirds. Typically, such corrections are much smaller than might be anticipated from the proportion of birds recorded in flight during point transect surveys. For example, Marsden (1999) noted that percentages of birds in flight recorded during ten-minute counts at points ranged from 29% to 77%, whereas subsequent monitoring of birds gave upward corrections to the stationary bird density estimates of between 2% and 19%. This reflects the substantial over-representation of flying birds in point transect data if the snapshot method is not adopted.

6.6 Other models

6.6.1 *Binomial models*

Binomial models are a special case of multinomial models, the theory for which is given in Section 3.3.2. We examine them briefly, since closed-form estimators are available for some underlying models for the detection function; these are sometimes used as indices of abundance, to assess change in abundance with habitat (Section 8.8) or over time.

Line and point transect methods sometimes provide a quick and inexpensive alternative to census methods for generating population abundance indices of songbirds. In areas of thick cover, the observer may rely heavily on aural detection, with perhaps fewer than 10% of detected birds visible. Difficulty in both locating the bird and moving through vegetation make measurement of each detection distance impractical, and the disturbance would also cause many birds to move or change their behaviour. Bibby *et al.* (1985) stated:

'Recording the distance at which each bird was detected would have been desirable but was not practicable when so many were heard and not seen. Overcoming this difficulty might have risked swamping the observer's acuity for other birds when an average of about nine birds was recorded at each five-minute session. A single decision as to whether or not each bird was within 30 m when first detected was easier to achieve in the field and sufficient to permit estimates of density.'

Sometimes, therefore, birds are simply recorded according to whether they are within or beyond a specified distance c_1. To help classify those birds close to the dividing distance, permanent markers may be positioned on trees or bushes at distance c_1. Only single-parameter models may be fitted to such data, and it is not possible to test the goodness of fit of any proposed model. The data may be analysed using the multinomial method for grouped data. Because there are only two groups (with the second cutpoint $c_2 = \infty$), the sampling distribution is binomial. As for the models of Chapter 3, numerical methods will be required in general, but below we consider the half-normal binomial model for point transects (Buckland 1987a), for which analytic estimates are available.

Define

$$g(r) = \exp\{-(r/\sigma)^2\}, \quad 0 \leq r < \infty \tag{6.15}$$

Then

$$\nu = 2\pi \int_0^\infty r g(r)\, dr = \pi\sigma^2 \tag{6.16}$$

and

$$f(r) = \frac{2r}{\sigma^2} \cdot \exp\Big\{ -\Big(\frac{r}{\sigma}\Big)^2 \Big\} \tag{6.17}$$

Given the binomial likelihood, simple algebra yields the maximum likelihood estimate

$$\hat{h}(0) = \frac{2\pi}{\hat{\nu}} = \frac{2}{c_1^2} \cdot \log_e\Big(\frac{n}{n_2}\Big) \tag{6.18}$$

where n is the number of birds detected and n_2 is the number beyond distance c_1.

Thus eqn (5.6) yields

$$\hat{D} = \frac{n \cdot \hat{h}(0)}{2\pi k} = \frac{n \cdot \log_e(n/n_2)}{c_1^2 \pi k} \tag{6.19}$$

where k is the number of plots sampled. Suppose for the example of Fig. 5.1 that distances had been recorded simply as to whether they were within or beyond 15 m. Then $c_1 = 15$ m, $k = 12$, $n = 55$ and $n_2 = 13$, yielding $\hat{D} = 0.00935$ birds/m^2, or 93.5 birds/ha.

The asymptotic variance of $\hat{h}(0)$ is

$$\widehat{var}\{\hat{h}(0)\} = \frac{4(1/n_2 - 1/n)}{c_1^4} \tag{6.20}$$

so that the variance of $\hat{D}$ may be estimated using the methods of Chapter 5. For the example, we obtain $\widehat{se}(\hat{D}) = 22.0$ birds/ha, compared with 17.6 birds/ha when exact distances are analysed.

Two measures of detectability are $r_{1/2}$, the point at which the probability of detection is one half, and ρ, the effective radius of detection. Further algebra yields

$$\hat{r}_{1/2} = \sqrt{\frac{2 \cdot \log_e 2}{\hat{h}(0)}}, \tag{6.21}$$

with

$$\widehat{var}(\hat{r}_{1/2}) = \frac{2 \cdot \log_e 2 \cdot (1/n_2 + 1/n)}{c_1^4 \cdot \{\hat{h}(0)\}^3} \tag{6.22}$$

while

$$\hat{\rho} = \sqrt{\frac{2}{\hat{h}(0)}}, \tag{6.23}$$

with

$$\widehat{var}(\hat{\rho}) = \frac{2 \cdot (1/n_2 + 1/n)}{c_1^4 \cdot \{\hat{h}(0)\}^3} \tag{6.24}$$

Thus for the example we have $\hat{r}_{1/2} = 10.4\,\text{m}$, with $\widehat{se}(\hat{r}_{1/2}) = 1.1\,\text{m}$, and $\hat{\rho} = 12.5\,\text{m}$, with $\widehat{se}(\hat{\rho}) = 1.3\,\text{m}$.

The efficiency of this binomial point transect model for estimating density relative to the half-normal model applied to ungrouped detection distances (Ramsey and Scott 1979; Buckland 1987a) is typically around 65–80% (Buckland 1987a). This loss is relatively small; more serious is that robust models with more than a single parameter cannot be used on binomial data, and there is no information from which to test whether the form of the half-normal model is reasonable.

Buckland (1987a) also derived analytic results for a linear binomial model for point transects, and found that density estimates for a variety of species are similar under the two models (Section 8.8). In practice, a detection function is unlikely to be approximately linear, so we give just the half-normal model here. The linear model had been considered earlier by Järvinen (1978), but only partially developed.

The choice of c_1 requires some comment. For the half-normal model, the value that minimizes the variance of $\hat{h}(0)$, of $\hat{r}_{1/2}$ and of $\hat{\rho}$ is $c_1 = 1.78/\sqrt{h(0)} = 1.26\sigma$, which implies that roughly 80% of detections should lie within c_1. Buckland (1987a) finds that estimation is more robust when around 50% lie within c_1. Further, for simultaneously monitoring several species, an average value of 80% across all species may mean that few or no birds of quiet or unobtrusive species are detected beyond c_1. A practical advantage in selecting a smaller value for c_1 than the optimum is that the observer will be more easily able to determine whether a bird is within or beyond c_1.

As a safeguard, two cutpoints, c_1 and c_2, with $0 < c_1 < c_2 < \infty$, might be used, so that the sampling distribution is trinomial. The data could be analysed using the results for the general multinomial distribution in Section 3.3.2, but detection functions with at most two parameters could be used. Another option would be to use the above binomial model, first using cutpoint c_1, then using cutpoint c_2. If the two density estimates differed appreciably, this might be an indicator that the model is not robust. Otherwise the two estimates might be averaged. For surveys of several species, the first cutpoint might be used for quieter, more unobtrusive species, and the second for louder, more obvious species.

Järvinen and Väisänen (1975) developed three binomial models for line transects, in which the detection function was assumed to be linear, negative exponential or half-normal. The last of these is the most plausible, but a closed-form estimator is not available for it. Program Distance allows the user to implement this model using numeric methods. Otherwise, its limitations and advantages are very similar to those of the binomial half-normal model for point transects, described above.

Although the goodness of fit of a binomial model cannot be tested, the homogeneity of the binomial data can. Suppose each line or point transect is assigned to one of R groups, which might, for example, be R geographic regions or woods. Then an $R \times 2$ contingency table analysis may be carried out, where the frequencies are n_{ij}, $i = 1, \ldots, R$, $j = 1, 2$. Then n_{i1} is the total count within distance c_1 for group i, and n_{i2} is the total count beyond c_1. If a significant test statistic is obtained, then there is evidence that the detection function varies between groups. This method extends in the obvious way to multinomial models.

A variety of now outdated line transect methods is given in Burnham *et al.* (1980). In particular, a nonparametric binomial method once thought to have promise is the Cox method, derived by Eberhardt (1978a). We no longer recommend this method. However, as a matter of intellectual curiosity, we derived the analogue of the Cox method for point transect sampling. A linear detection function is assumed, so the model is very similar in concept to the linear binomial point transect model of Buckland (1987a). The difference is that the Cox method assumes that data are truncated at a distance for which the linear detection function is non-zero, whereas the method of Buckland, in common with the linear line transect model of Järvinen and Väisänen (1975), assumes data are untruncated in the field; an estimate of the point at which probability of detection becomes zero is provided by the model.

Let the distance data be grouped with the first two cutpoints being c_1 and c_2. Let the corresponding counts in these two intervals be n_1 and n_2 with total $n = n_1 + n_2$. Let k be the number of points sampled. The Cox estimator is derived by assuming that $g(r) = 1 + b \cdot r$ is an adequate model over the range $0 \leq r \leq c_2$. (It would be better to assume a quadratic form, $g(r) = 1 + b \cdot r^2$, but we use Eberhardt's formulation.) Of course the parameter b is negative. Based on just the counts in these first two intervals, we can get an estimate of b and hence an estimator of density, D:

$$\hat{D} = \frac{n}{k\pi c_2^2}\left[\frac{(c_2/c_1)^3 \cdot (n_1/n) - 1}{(c_2/c_1) - 1}\right] \tag{6.25}$$

In the simple case of the two intervals having equal widths, Δ (i.e. $c_1 = \Delta$ and $c_2 = 2\Delta$), the result reduces to

$$\hat{D} = \frac{7n_1 - n_2}{4k\pi\Delta^2} \tag{6.26}$$

In Burnham *et al.* (1980: 169), the general Cox case was given for line transects. The $\hat{f}(0)$ in their publication is not in the same form as used here for $\hat{h}(0)$. For comparison we provide the results below.

Cox estimator for point transects

$$\hat{h}(0) = \frac{2}{c_2^2}\left[\frac{(c_2/c_1)^3 \cdot (n_1/n) - 1}{(c_2/c_1) - 1}\right] \tag{6.27}$$

Cox estimator for line transects

$$\hat{f}(0) = \frac{1}{c_2}\Big[\frac{(c_2/c_1)^2 \cdot (n_1/n) - 1}{(c_2/c_1) - 1}\Big] \tag{6.28}$$

For this same context of two cutpoints and truncation at c_2, if we assume the detection function has the generalized form $g(y) = 1 + b \cdot y^p$ for $0 \leq y \leq c_2$ and where p is a known integer ≥ 1, then relevant results for point and line transects are

$$\hat{h}(0) = \frac{2}{c_2^2}\Big[\frac{(c_2/c_1)^{2+p} \cdot (n_1/n) - 1}{(c_2/c_1)^p - 1}\Big] \tag{6.29}$$

and

$$\hat{f}(0) = \frac{1}{c_2}\Big[\frac{(c_2/c_1)^{1+p} \cdot (n_1/n) - 1}{(c_2/c_1)^p - 1}\Big] \tag{6.30}$$

Corresponding theoretical sampling variances are

$$var\{\hat{h}(0)\} = \frac{4}{c_2^4}\Big[\frac{(c_2/c_1)^{2+p}}{(c_2/c_1)^p - 1}\Big]^2 \cdot \frac{p_1(1-p_1)}{n} \tag{6.31}$$

and

$$var\{\hat{f}(0)\} = \frac{1}{c_2^2}\Big[\frac{(c_2/c_1)^{1+p}}{(c_2/c_1)^p - 1}\Big]^2 \cdot \frac{p_1(1-p_1)}{n} \tag{6.32}$$

where $p_1 = E(n_1)/n$, and is estimated by $\hat{p}_1 = n_1/n$.

6.6.2 *Estimators based on the empirical cdf*

Emlen (1971, 1977) developed a non-mathematical approach for line transect analysis of songbird data. He assumed that a characteristic proportion of birds of any species will be detected in the surveyed area $2wL$. He called this proportion the coefficient of detectability. This corresponds to the parameter P_a defined in Section 3.1. The method typically uses data from only two to four distance categories, and an estimator of the product $E(n) \cdot f(0)$ is determined from a smoothed frequency histogram of perpendicular distances. The density estimate from this model is probably best considered as a rough index of relative abundance, and is not recommended (Burnham *et al.* 1980: 164). However, the concept has been extended to develop the following estimators, based on the empirical cumulative distribution function of distances.

Ramsey and Scott's (1979, 1981b) point transect model was similar to Emlen's line transect model, but with a more formal statistical basis. Suppose cutpoints are defined at distances $c_0 = 0, c, 2c, \ldots, kc$, so that the truncation point $w = kc$. Let $A(0, i)$ be the area of the circle of radius ic, and let $A(i, j)$ be the area of the annulus with inner radius ic and outer radius jc. Thus, $A(i,j) = \pi c^2(j^2 - i^2)$. Let $n(i,j)$ be the number of birds counted within the annulus. Then the corresponding density $D(i,j)$ may be estimated by

$$\hat{D}(i,j) = \frac{n(i,j)}{A(i,j)} \tag{6.33}$$

The value i is chosen to be the smallest value such that the likelihood of differing densities within the areas $A(0,i)$ and $A(i,j)$ is at least four times the likelihood of equal densities for all $j > i$. That is, i is the smallest value that satisfies

$$[\hat{D}(0,i)]^{n(0,i)} \cdot [\hat{D}(i,j)]^{n(i,j)} \geq 4 \cdot [\hat{D}(0,j)]^{n(0,j)} \quad \text{for all } j > i \tag{6.34}$$

Having calculated i,

$$\hat{D} = \frac{n(0,i)}{A(0,i)} \tag{6.35}$$

with

$$\widehat{var}(\hat{D}) = \frac{\hat{D}}{A(0,i)} \tag{6.36}$$

This variance estimate is only valid if birds are randomly distributed throughout the study area, so is likely to be poor, but an empirical estimate may be obtained, for example using the bootstrap. The density estimate is only valid if there is an area of perfect detectability, assumed to extend out to distance ic.

Wildman and Ramsey (1985) developed the 'CumD' estimator, which is similar to the above, but estimates the distance out to which detection is certain without the need to group the data. The estimator is defined for both line and point transects, and observations are transformed to detection 'areas', defined as $a = 2Lx$ for line transects and $a = \pi r^2$ for point transects. These areas are ordered from smallest to largest, giving $a_1 \leq a_2 \leq \cdots \leq a_n$, with empirical distribution function $F_n(a)$. Let $j(0) = a_{j(0)} = 0$, and let $j(1)$ be the largest integer such that

$$d_1 = \frac{j(1)}{a_{j(1)}} = \max\left\{\frac{j}{a_j} \mid j \geq \sqrt{n}\right\} \tag{6.37}$$

Then for $m = 2, 3, \ldots$ let $j(m)$ be the largest integer such that

$$d_m = \frac{j(m) - j(m-1)}{a_{j(m)} - a_{j(m-1)}} = \max\left\{\frac{j - j(m-1)}{a_j - a_{j(m-1)}} \mid j > j(m-1)\right\} \tag{6.38}$$

Straight lines linking the points $[a_{j(m)}, j(m)/n]$, $m = 0, 1, 2, \ldots$, form a convex envelope over $F_n(a)$, which is the isotonic regression estimate of

$F(a)$ (Barlow *et al.* 1972) and yields an estimated detection function that is a non-increasing function of distance from the line or point. The slopes d_m are average estimates of density of detections within strips or annuli of increasing distance from the line or point. These equate to estimates of object density if all objects within a given strip or annulus are detected. Likelihood ratio tests of the equality of density between the first region and the next $m-1$, $m = 2, 3, \ldots$, are used to provide a stopping rule. The smallest value of m, m^* say, is chosen such that the null hypothesis is rejected, and all objects within the corresponding distance a_{j^*}, where $j^* = j(m^*)$, are assumed to have been detected, yielding an estimate of density.

This innovative and intuitively appealing approach is computationally inexpensive and easily programmed. However, probability of detection should be at or close to unity for some distance from the line or point for the method to yield good estimates; because of the paucity of sightings close to a point, this distance must be appreciably greater for point transects than for line transects for comparable performance. Also, estimation of a distance up to which all objects are detected through hypothesis testing causes the method to underestimate density when sample size is small, as there are too few data for the tests to have much power. Bias in the method is therefore a strong function of sample size, at least for small samples. Investigation of how large sample size should be for the method to perform well would be useful. Wildman and Ramsey (1985) show that it must be very large if the true detection function, expressed as a function of area, is negative exponential. For point transects, this corresponds to the half-normal detection function, when expressed as a function of distance, and bias is still of the order of 23% for a sample size of 10 000.

6.6.3 *Estimators based on shape restrictions*

Johnson and Routledge (1985) developed a nonparametric line transect estimator based on shape restrictions for which they found 'a general improvement in efficiency over existing estimators.' The method has not seen wide use, although software TRANSAN is available (Routledge and Fyfe 1992).

The density function $f(x)$ is constrained to be non-negative and to integrate to unity. In addition, Johnson and Routledge added the monotonicity constraint that $f(x)$ must be monotonic non-increasing and the shape constraint that $f(x)$ must have a concave shoulder, followed by a convex tail, separated by a single point of inflection. The range of concavity must be determined, or guessed, by the user, and it is suggested that the percentage of detections that fall within the point of inflection might be as high as 90% or below 50%.

The parameter $f(0)$ is estimated by grouping the distance data, and using the frequencies in a histogram estimator of $f(x)$. Let h_i represent the height of the ith histogram bar, $i = 1, \ldots, u$. Find the adjusted heights, $\tilde{h}_i$, by minimizing

$$\sum_{i=1}^{u} (h_i - \tilde{h}_i)^2 \tag{6.39}$$

subject to the imposed constraints. Then $f(0)$ is estimated by $\hat{f}(0) = \tilde{h}_1$.

Johnson and Routledge used a bootstrap approach to quantify precision, but one that is more sophisticated than the general purpose bootstrap described in Section 3.6.4. For the single-parameter case, a guess can be made of say the lower $100(1 - 2\alpha)\%$ confidence limit for $f(0)$, and bootstrap resamples generated. The parameter $f(0)$ is estimated from each resample, and if the proportion exceeding the estimate from the true data is greater than α, then the current guess of the limit is estimated to be too large, and is reduced. The process is repeated until the true limit is located with adequate precision. Johnson and Routledge suggested that a more efficient search procedure might be developed. A general algorithm for evaluating confidence limits in this way for single-parameter problems, which updates the estimated limit after each resample, was given by Buckland and Garthwaite (1990). Its optimal properties were noted by Garthwaite and Buckland (1992).

Johnson and Routledge based their conclusions on estimator performance on a simulation study, in which the Fourier series and half-normal models were compared with the shape restriction estimator. If the procedures recommended in earlier chapters were implemented, more severe data truncation would have been carried out for some of their simulations, and the Fourier series and half-normal models would have been rejected as inappropriate in some. However, the shape restriction method proved to be robust to choice of truncation point and to the true underlying detection function, and therefore merits further investigation and development.

6.6.4 *Kernel estimators*

6.6.4.1 *Line transect sampling*

There are several methods for fitting probability densities using kernels (Silverman 1986). They are based on the concept of replacing a data point (a distance from the line here) by a probability distribution, centred on that point. This is done for all observations, and the distributions are combined, to provide the estimated density function:

$$\hat{f}(x) = \frac{\sum_{i=1}^{n} \phi(x - x_i; h)}{n} \tag{6.40}$$

where $x_i, i = 1, \ldots, n$ are the recorded distances, and the kernel function ϕ is itself a pdf, whose variance is controlled by the smoothing parameter h.

Typically, we might choose $\phi(z;h)$ to be a normal distribution with respect to z with mean zero and standard deviation h.

Seber (1986) suggested the use of kernel estimators of the pdf of distances from the line in line transect sampling. Buckland (1992a) compared the kernel estimator of Silverman (1982) with the Hermite polynomial model for fitting the deer data from survey 11 of Robinette *et al.* (1974). To force the algorithm to fit a symmetric density, differentiable at zero (and hence possessing a shoulder), each distance x was replaced by the two distances, x and $-x$. The optimum choice of smoothing parameter for a normal distribution with standard deviation σ, i.e. $h = 1.06\sigma n^{-0.2}$, yielded a comparable but less smooth fit to the data than the Hermite polynomial model. The kernel method is less computer-intensive than the methods recommended here, but the kernel estimate of $f(0)$ is sensitive to the choice of window width, especially if there is rounding in the data (Buckland 1992a). Barabesi and Fattorini (1994) showed that automated bandwidth selection leads to large bias in estimates of $f(0)$, whereas estimates corresponding to interior points ($f(x)$ where $0 \ll x \ll w$) have low bias. They advised use of orthogonal series estimators of $f(0)$ in preference to kernel estimators. A further disadvantage of the kernel method is that it is problematic to include more than one or two covariates, in contrast with the methods of Section 3.7.2. One advantage of the kernel method is that any given observation has only a local effect on the fitted density. For parametric or semiparametric methods, if the model fails to fit the tail of the distribution well, its fit at zero distance may be adversely affected.

Chen (1996a) evaluated a standard kernel estimator of $f(x)$, based on a Gaussian kernel, and developed a reliable analytic method for estimating confidence intervals. His approach is therefore computationally competitive, and he concluded that, for the range of cases in his simulation study, coverage was better than for the Fourier series model.

Chen (1996b) also developed a kernel method for bivariate line transect data, comprising distances x and cluster sizes s. He defined $\beta(0) = \int_0^\infty s f(0,s)\, ds$, where $f(0,s)$ is the bivariate density $f(x,s)$ evaluated at $x = 0$, then used bivariate kernel methods to estimate $f(x,s)$ and hence $\beta(0)$. The variance of $\hat{\beta}(0)$ was estimated using a bootstrap method. Mack and Quang (1998) adopted a similar approach, but rather than model the bivariate density $f(x,s)$, they used univariate kernels to estimate the quantities $\int_0^M s^j f(0,s)\, ds$ and $\int_0^M s^j f'(0,s)\, ds$, where $j = 0,1,2$ and M is the largest possible cluster size. They derived estimators for cluster density, individual density and mean cluster size, and the corresponding variances, as functions of these quantities. The approach treats the discrete variable cluster size as if it were continuous; a discrete version might be useful for populations in which cluster size is typically small.

Chen (1999) further developed kernel methods for double-platform surveys, in which detection on the trackline is not certain. The methods allow detection probabilities to be modelled as a function of cluster size, and potentially of additional covariates.

Mack *et al.* (1999) developed kernel methods for the case that the detection function does not satisfy the shape criterion (Section 3.4.1.3). However, this is likely to be of limited practical use, as hazard-rate modelling of the detection process (Section 6.6.5) suggests that, when detection on the line is certain, models should satisfy the shape criterion even if the hazard function is sharply spiked. Further, even if it is plausible that the true detection function is spiked, estimation of such a curve is prone to bias when the true model is not known.

Brunk (1978) developed a kernel method based on orthogonal series, an approach which is a close parallel to the key + adjustment formulation, especially if the adjustment terms are orthogonal to the key, as for the Hermite polynomial model. Buckland (1992a) found that Brunk's method gave unstable estimation relative to the Hermite polynomial model when the data were simulated from a markedly non-normal distribution.

Gerard and Schucany (1999) showed that kernel methods for modelling the detection function are more reliable if local bandwidth selection procedures are used. Barabesi (2000) used a semiparametric approach, based on a half-normal key function whose parameters are chosen on the basis of a local kernel-smoothed likelihood function. A simulation study suggested that this approach is competitive relative to the Hermite polynomial model and to the kernel methods of Barabesi and Fattorini (1994) and Gerard and Schucany (1999).

6.6.4.2 *Point transect sampling*

Quang (1993) proposed a method based on kernel techniques. As in his line transect developments, he assumed that perfect detectability occurs somewhere, but not necessarily at zero distance; that is, that $g(r) = 1$ for some value of $r \geq 0$. Given $g(0) = 1$, it was noted in Chapter 3 that density estimation could be reduced to estimation of $h(0) = \lim_{r \to 0} f(r)/r$. Under Quang's formulation, this generalizes, so that the maximum value of the ratio $h(r) = f(r)/r$ over all r must be estimated. He used the kernel method with a normal kernel (Silverman 1986) to estimate $f(r)$ and hence $h(r)$.

The concept of maximizing $h(r)$ is also applicable to series-type models. Suppose a model is selected whose plotted detection function increases with r over a part of its range, thus indicating that objects close to the point are evading detection, either by fleeing or by remaining silent. Then $\hat{h}(0)$, which is the estimated slope of the density at zero, $\hat{f}'(0)$, may be replaced by the maximum value of $\hat{h}(r) = \hat{f}'(r)$ in the point transect equation for estimated density (Section 3.6.1).

The kernel method of Mack and Quang (1998), referenced in the previous section, is also applicable to point transect sampling. On the basis of simulations and of analyses of real data sets, the authors conclude that their approach is a viable alternative to other existing estimators.

6.6.5 *Hazard-rate models*

The hazard-rate model implemented in Distance is one of a family of derived models. Instead of specifying a form for the detection function $g(y)$, a functional form for the 'hazard' of detection is specified, from which the detection function is derived. The 'hazard' measures the instantaneous risk of detection of an object as the observer approaches along the line (line transects) or as time passes (point transects). For objects that give discrete detection cues, such as songbirds (cue is a burst of song or a call) or whales (cue is a blow or surfacing), the risk of detecting an individual cue can be modelled, giving rise to discrete hazard-rate models. Continuous and discrete hazard-rate models are described in Buckland *et al.* (in preparation).

6.7 Distance sampling surveys when the observed area is incompletely covered

A conceptual distinction can be made between a survey in which every object in the observed (or covered) area is potentially detectable and one in which a proportion of objects is undetectable due to their location with respect to the line or point, even though they are within the truncation distance w. For most purposes, we need draw no distinction between these two cases; in the second case, the fitted detection function reflects both the fall-off in detectability with distance and the decreasing proportion of objects that are potentially detectable. We need merely to assume that all objects close to the line or point are detectable, so that $g(0) = 1$ (unless we adopt methods that do not require this assumption; see Buckland *et al.* in preparation). If we assume $g(y)$ is a non-increasing function of y, we also need to assume that the proportion of objects potentially detectable is a non-increasing function of y.

In some circumstances, the above effect may be sufficiently extreme to render modelling of a 'composite' $g(y)$ unreliable. This may occur for example in line transect surveys when the habitat is predominantly open but with patches of dense vegetation, such as forest plantations. The methods described here are applicable in those circumstances, but the density estimate obtained applies only to the open habitat. Essentially, the areas of dense cover must be assumed to be outside the study area. (Those areas might be surveyed by other means, for example using indirect dung surveys, Section 6.2. However, dung surveys measure average usage over the typical time period for dung to decay, whereas direct surveys estimate

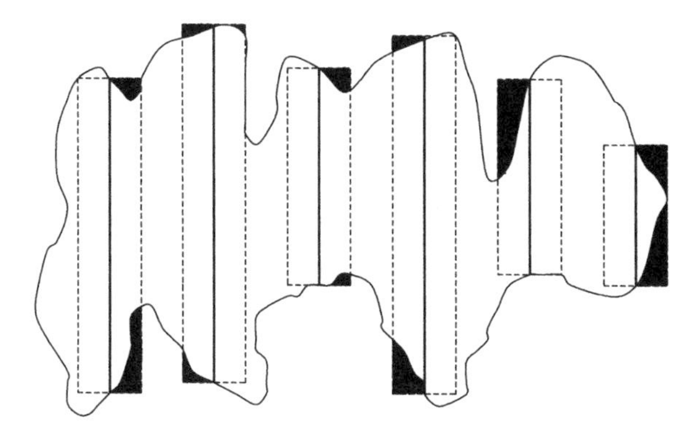

Fig. 6.6. A line transect survey in which a substantial proportion of the surveyed strips lies outside of the study area (black shading). Conventional analysis will yield a valid estimate of density, but the estimated detection function will drop more quickly than it should, because the area surveyed within the strips that is also within the study area decreases with distance from the line.

abundance at the time of the survey, so care must be taken.) A better example of when the approach described below is useful is point transect sampling in which detection relies on the object being visible, such as night surveys of red foxes or rabbits with spotlights. In that circumstance, the terrain or an obstacle such as a building or a hedge may lead to parts of the surveyed area being hidden from the observer. The blind areas around each point can be mapped, allowing application of the methods described here. Object density is assumed to be independent of whether or not the area is blind.

Another circumstance in which these methods are potentially useful is when the study area is small, so that a significant proportion of the surveyed strips or circles lies outside the boundaries of the study area (Fig. 6.6). The lines or points fall within the boundaries by design, and the proportion of surveyed area falling within the boundaries can be determined as a function of distance from the lines or points, for example using a Geographic Information System.

We define $e(y)$ to be the proportion of area that is available to be surveyed, expressed as a function of distance from the line or point. By mapping the areas that cannot be surveyed, $e(y)$ can be measured. It is evaluated for each line or point, and averaged across all lines or points for which a common detection function is assumed. We assume that $e(0) = 1$. For conventional distance sampling, $e(y) = 1$ for $0 \leq y \leq w$.

Previously, for line transect sampling we had the relationship (eqn (3.6))

$$f(x) = \frac{g(x)}{\int_0^w g(x)\, dx} \tag{6.41}$$

This is now replaced by

$$f(x) = \frac{g(x)e(x)}{\int_0^w g(x)e(x)\,dx} \tag{6.42}$$

For point transect sampling, we had previously (eqn (3.8))

$$f(r) = \frac{rg(r)}{\int_0^w rg(r)\,dr} \tag{6.43}$$

We now have

$$f(r) = \frac{rg(r)e(r)}{\int_0^w rg(r)e(r)\,dr} \tag{6.44}$$

Because $e(y)$ is a known, measured quantity, no new estimation problems arise, and software implementing the methods of Chapter 3 may be readily modified.

We stress that circumstances in which the modifications of this section are likely to prove worthwhile are not common. On occasion, there may be interest in modelling $g(y)$, having adjusted for the effect of blind areas. If this is not of interest, conventional analysis will yield a fitted detection function that estimates the product $g(y)e(y)$ in the above notation. If $e(y)$ is smooth and well-behaved, modelling of the product using standard methods should give reliable estimation of object density and abundance.

6.8 Trapping webs

The estimation of population size (N) from capture data is usually formulated as a capture–recapture problem (e.g. White *et al.* 1982). There, traps are positioned, often at intersections of a rectangular grid, and animals are captured, marked, and released for possible recapture on a subsequent trapping occasion. If the trapping grid is enclosed or the trapped area samples the entire area of interest, then density = number/area can be estimated. However, the usual case is that an area surrounding the trapping grid contains animals that are subject to being captured and thus the effective area being sampled is larger than the area of the grid. One might naïvely estimate density as $\hat{N}/A_g$, where A_g = the area covered by the trapping grid. However, density is then overestimated because grid area is smaller than the area actually sampled by the traps. This problem has been well known for over half a century (Dice 1938). The use of a trapping web (Anderson *et al.* 1983) is an attempt to reformulate the density estimation problem into a distance sampling framework, where density is estimated directly, rather than separately estimating population size and effective area (but see Wilson and Anderson 1985a for an alternative).

6.8.1 *Survey design and field methods*

Trapping webs are a special case of point transect sampling useful in estimating density of animal populations where 'detection' is accomplished by trapping. Animals are trapped in live, snap-trap or pitfall traps. Mist nets or other devices can be employed. Such trap devices are placed in a 'web' such that the density of traps is highest near the centre (Fig. 6.7), thus attempting to ensure that $g(0) = 1$. Within the study area, k webs are laid out according to some randomized design. The number of webs should usually be at least 15 or 20.

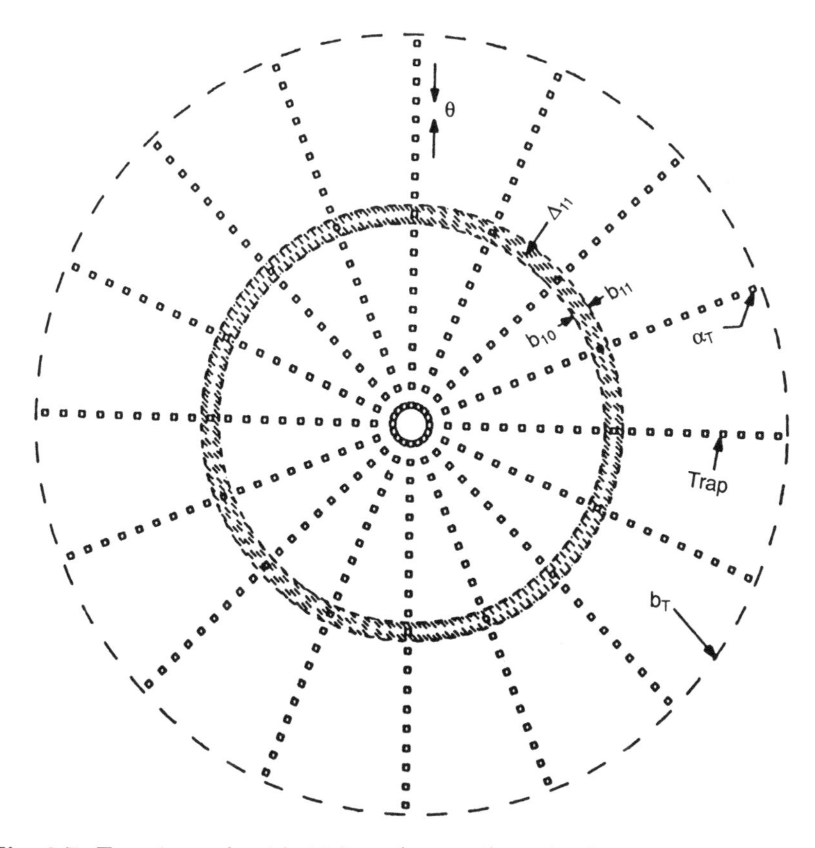

Fig. 6.7. Trapping web with 16 lines ($m = 16$), each of equal length α_T and 20 traps per line ($T = 20$), giving in total 320 traps. The traps are at distances $\alpha_1, \ldots, \alpha_T$ from the centre of the web. The points along each line, halfway between traps, are denoted by b_i, $i = 0, 1, \ldots, T$, where $b_0 = 0$ is the centre and b_T is the boundary of the web, just beyond the last trap. Captures in the 11th ring of traps are assigned to the shaded ring Δ_{11}, which has area $c_{11} - c_{10}$, where $c_i = \pi \cdot b_i^2$.

The web design consists of m lines of equal length, α_T, and each of T traps, radiating from k randomly chosen points. A useful rule of thumb is to ensure that $m \times T \geq 100$ for each web. The traps are located along each of the m lines, usually (but not necessarily) at some fixed distance interval θ, starting at distance $\alpha_1 = \theta/2$. Points b_i are defined along each line, halfway between traps, for $i = 1, 2, \ldots, T$, with $b_0 = 0$ representing the web centre, and b_T the boundary of the web beyond the last trap. The traps are then at distances $\alpha_i = i\theta - 0.5$ for $i = 1, 2, \ldots, T$, and the b_i are at distances $i\theta$ for $i = 0, 1, 2, \ldots, T$.

Thus, traps are placed in rings of increasing radius from the web centre at equal distances along the m lines (Fig. 6.7). All captures in the ith ring of traps are considered to be detections of objects at distance α_i from the centre of the web. The distance data are analysed as grouped data. That is, the total number of captures arising from the ring of traps at distance α_i are treated as grouped data over the interval from distance b_{i-1} to b_i. The total area of the web out to interval i is $c_i = \pi b_i^2$ and the area trapped by the ith ring of traps is then $\Delta_i = c_i - c_{i-1}$. Generally, only first captures (removal data) are recorded and used in the estimation of density. This procedure reduces the impact on estimation of heterogeneity in trap response due to trap-happy or trap-shy animals.

As with point transects, movement of objects through time will result in objects being overrepresented by the traps near the centre of the web, thus leading to overestimation of population density. Placing additional traps near the centre of the web may exacerbate this overestimation, and is not now recommended. Use of at least $m = 8$ lines per web is suggested, and 10, 12 or even 16 might be considered. Guidelines for the number of traps are less well defined, although a practical objective is to obtain a total sample (pooled over all webs) of trapped animals of at least 60–80, and preferably around $n = 100$. A pilot study using 100–150 traps may often lead to insight on the number required to achieve an adequate sample size. Animals do not have to be marked, unless they are to be returned to the population and thereafter ignored in future samples ('removal by marking'). Simple marking with a felt-tip pen will often suffice.

Trap spacing remains to be studied and we offer only the guideline that traps be spaced along the lines at a distance roughly equal to half the home range diameter of the species being studied. Wilson and Anderson (1985b) suggested 4.5–8-m spacing for mice, voles or kangaroo rats. The field trapping can be done over sequential nights (or days) until it seems clear that no new animals are being caught near the centre. Alternatively, if at the centre of the web most animals that have been marked and released have subsequently been recaptured, then one might conclude that sufficient trapping occasions have been carried out.

Trapping webs can be established using a stake at the web centre and a long rope with knots to denote the trap spacing. Then, the investigator can travel in a circle laying out traps in roughly straight lines radiating from the centre. However, it is not important that the traps are on perfectly straight lines. Disturbance of the site should be minimized while traps are being placed in the sampled area. In multiyear surveys, the location of each trap is often marked by a numbered metal stake. We recommend that recaptures of released animals are recorded, and that each trap has a unique number, allowing captures to be assigned to traps. These data allow assumptions to be assessed, and additional analytic methods, such as bootstrap sampling within a web, to be implemented.

6.8.2 *Assumptions*

Analytic theory for the trapping web is an application of point transect sampling theory and the general assumptions apply. The three major assumptions of distance sampling are slightly restated here for the trapping web:

1. All animals at the centre of the web are captured at least once during the t occasions ($g(0) = 1$). That is, trapping continues until evidence exists that no new animals are being caught near the centre of the web.
2. During the trapping period, animals move over distances that are small relative to the size of each web. Thus, movement through the web is not allowed. Trap spacing is an important consideration and is species-dependent, taking account of the size of home ranges or 'territories'.
3. Each trap is placed correctly in the proper distance category. This assumption is trivial even if the trap spacing has been somewhat imprecise.

Assumption 1 is critical but can be monitored by examining the number of new individuals trapped near the web centre over trapping occasions. Animals near the centre, or whose area of activity includes the web centre, should be captured with probability one. However, if substantial movement occurs over the t occasions (assumption 2), animals that are initially away from the centre of a web may move, eventually to be caught where the trap density is highest. This situation is analogous to point transect sampling where the observation period is long and birds move around the study area. Such movement causes detections near the point to increase and leads to positive bias in the estimator. Bias may be worse if animals are attracted to the point or web centre; this is expected if baited traps are used.

6.8.3 *Estimation of density*

The basic data are the number of first captures in traps in ring j of web i on trapping occasion l, n_{ijl}, where $i = 1, 2, \ldots, k$, $j = 1, 2, \ldots, T$ and $l = 1, 2, \ldots, t$. Pooling the data over t occasions, the data can be summarized as n_{ij}, where

$$n_{ij} = n_{ij1} + n_{ij2} + \cdots + n_{ijt} \tag{6.45}$$

Hence n_{ij} is the number of animals trapped in the jth ring of the ith web. Let the total sample size be $n = \sum_i \sum_j n_{ij}$. Then density can be estimated by

$$\hat{D} = \frac{n \cdot \hat{h}(0)}{2\pi k} \tag{6.46}$$

where the estimate $\hat{h}(0)$ is obtained through standard point transect methods (Chapter 5). The estimator of the sampling variance is

$$\widehat{var}(\hat{D}) = \hat{D}^2 \cdot [\{cv(n)\}^2 + \{cv[\hat{h}(0)]\}^2] \tag{6.47}$$

If the population is distributed randomly (i.e. Poisson), then Wilson and Anderson (1985a) recommend $[cv(n)]^2 = 1/n$. Generally, some degree of spatial aggregation can be expected, and $[cv(n)]^2 = 2/n$ or $4/n$ might then be more appropriate. If the number of replicate webs, k, is sufficient, it is preferable to estimate the sampling variance of n empirically (Section 3.6.2).

Data analysis is similar to the general theory for point transects, including model selection and inference issues. The challenge with the trapping web is to collect trapping data that mimic the assumptions of point transect sampling and analysis theory. The use of only initial captures is advantageous as behavioural response to capture is eliminated. However, theory is being developed to allow recaptures to be used in the estimation of density, to improve precision. The usual problems concerning capture heterogeneity are absent in the trapping web approach, if the three assumptions are met.

Excessive animal movement near the web centre is problematic. The density of traps near the web centre is high relative to that near the edge of the web. Thus, even if animal movement is random, there is a tendency to trap animals near the web centre, regardless of their original location. If the trap spacing is too small, the problem is made worse and overestimation will likely result. If the animals tend to move in home ranges that are small relative to the size of the web and the trap spacing, then the trapping web may perform well. Alternatively, if animals move somewhat randomly over wide areas in relation to the size of the web and the trap interval chosen, then overestimation may be substantial (see the darkling beetle example, below).

6.8.4 *Monte Carlo simulations*

Wilson and Anderson (1985b) performed a Monte Carlo study to investigate the robustness of density estimation from trapping web data. Their simulations mimicked small mammal populations whose members were allowed to move in defined home ranges. Home range was simulated from bivariate normal, bivariate uniform and bivariate U-shaped distributions, and from a 'random excursion' model. More details are given by Wilson and Anderson (1985b). A 4-ha area was simulated, 320 traps were positioned in a two-dimensional plane, and animal density was set at two levels, 100/ha and 25/ha. Home range centres were allowed to be spatially random (Poisson), or clumped at three levels of aggregation. Three average probabilities of first capture were simulated at 0.09, 0.16 and 0.24, and these probabilities were allowed to vary by time (trapping occasion), behaviour (trap-shy or trap-happy) and heterogeneity (individual variability); this is model M_{tbh} in Otis *et al.* (1978: 43). Trapping was simulated for six, eight and ten occasions.

The Monte Carlo results indicated that the combination of a trapping web design and a point transect estimator of density was quite robust. The procedure had typically low bias under a wide variety of realistic situations. Confidence interval coverage was lower than the nominal level, due in part to the use of the Fourier series estimator (Buckland 1987a). The method was recommended in cases where the capture probability was >0.16 and the number of trapping occasions was at least six. In some extreme situations (e.g. the random excursion model, with low capture probabilities and a clumped spatial distribution), the bias was in the 20–30% range, which might still be less than traditional capture–recapture estimators.

The trapping web is easy to implement in the field. No unique marks or tags are required, and several different types of trap can be used. The results of Wilson and Anderson (1985b) and Parmenter *et al.* (in preparation) indicated that the trapping web was very promising as an alternative to standard capture–recapture methods. The work of Parmenter *et al.* (1989), summarized in Section 6.8.6, was carried out as a field test of the method where the true density was known.

6.8.5 *A simple example*

Anderson *et al.* (1983) presented an example of trapping web data from a 4.8-km area south of Los Alamos, New Mexico, where *Peromyscus* spp. were trapped for $t = 4$ nights on a web very similar to that of Fig. 6.7, with trap spacing of $\theta = 3\,\text{m}$. The mice were captured in baited live traps and marked using a monel metal tag placed in one ear. Only initial captures were used in the analysis; animals were thus 'removed by marking'. No unmarked mice were caught in the inner area (out to ring 7) on the fourth night and only two new captures were made in this area on the third night. This was taken as evidence that the probability of capture near the web centre was one. A plot of the histogram indicated that mice from beyond the web were being attracted to the baited traps in the web, as the number of captures in rings 19 and 20 (i.e. n_{19} and n_{20}) was somewhat higher than expected. Thus, the distance data were truncated to exclude the two outer rings. This left 76 'detections' in 18 distance groups for analysis; frequencies were 1, 1, 0, 6, 2, 2, 3, 2, 4, 7, 4, 5, 8, 6, 7, 6, 7 and 5, respectively (Anderson *et al.* 1983). Note the lower frequencies in the inner rings, where the area sampled is small relative to that in the outer rings.

Two models were fitted to these data: half-normal and hazard-rate, each with cosine adjustment parameters. No adjustment parameters were required and the AIC values were similar for the two models (424.18 and 425.76, respectively; $\Delta = 1.58$). Both models fit the data well as judged by the χ^2 goodness of fit tests ($\chi^2 = 13.2$ with 16 df and 13.6 with 15 df, respectively). Density is estimated at 97.8 mice/ha ($\widehat{se} = 21.3$) under the half-normal model and 86.1 mice/ha ($\widehat{se} = 12.7$) under the hazard-rate model. These estimates compare with 76 animals trapped on the web of area 0.97 ha, suggesting that most animals were caught. Only one web was sampled, hence no inference to a larger area is justified in this simple example. In fact, the above standard errors obtained from Distance assume that animals are distributed over a larger area according to a homogeneous Poisson process; valid inference applied to the web alone could be drawn by assuming that sample size $n \sim \text{binomial}(N, P_a)$, where $N = D \cdot a$ is the total number of animals in the area $a = 0.97$ ha sampled by the web. Then $var(n) = NP_a(1 - P_a)$, estimated by $\widehat{var}(n) = n(1 - \hat{P}_a)$. Then

$$\hat{D} = \frac{n}{a \cdot \hat{P}_a} \tag{6.48}$$

and

$$\widehat{var}(\hat{D}) = \hat{D}^2 \left[\frac{\widehat{var}(n)}{n^2} + \frac{\widehat{var}(\hat{P}_a)}{\hat{P}_a^2} \right] \tag{6.49}$$

For the half-normal model, this yields a standard error for $\hat{D}$ of 18.8 mice, and for the hazard-rate model, 8.5 mice.

6.8.6 *Darkling beetle surveys*

Populations of two species of ground-dwelling darkling beetles were studied in a shrub-steppe ecosystem in southwestern Wyoming to field test the validity of the trapping web on a series of known populations (Parmenter *et al.* 1989). These beetles (10–30-mm body length) attain natural densities so great (>2000 beetles/ha) that relatively small plots could be surveyed and still have test populations of reasonable size. These beetles are wingless and could be contained by low metal fences. They are easily marked on their elytra with coloured enamel paint, and are relatively long-lived, allowing longer periods of trapping and increased capture success. Pitfall traps were made from small metal cans (80–110 mm), and the web was surrounded by a metal enclosure wall. Traps were placed along 12 lines, each 11-m long, with 1-m trap spacing along the lines. Nine additional traps were placed at the centre of the web, giving 141 in total.

Beetles were captured, marked with enamel and released. These marked beetles constituted the population of known size that was subsequently sampled using the trapping web design. Surveys were done in two different years and different colours were used to denote the year. Several subpopulations, each of known size, were released, allowing analyses to be carried out both separately and in combination, to test the method on a wide range of densities. Additional details were given in Parmenter *et al.* (1989).

Overall the method performed quite well, yielding a correlation coefficient between $\hat{D}$ and D of at least 0.97 for each of four models for $g(r)$. The negative exponential model performed better than the Fourier series, exponential power series and half-normal models. The data exhibited a spike near the web centre, almost certainly caused by considerable movement of beetles. All models were fitted after transforming the distance data to areas, a procedure that is no longer recommended. In summary, the results of these field tests were certainly encouraging.

Reanalysis of the darkling beetle data using current theory and program Distance provided a less optimistic impression in that density was substantially overestimated, leading to important insights. The first is that traps were too closely spaced along the lines; trap spacing should have been greater to compensate for the wide area over which beetles of this species move. Second, the beetles had no 'home range' and thus tended to wander widely in relation to the size of the web, which is a function of trap spacing. The trapping web design was envisioned for use with animals that have some form of home range or 'territory'. Third, it is clear that some random movement may result in too many animals being trapped near the web centre. The problem can also arise in bird surveys where some random movement results in the detection of too many birds near the point. This condition leads to a spiked distribution and overestimation of density. Clearly, the additional nine traps placed near the web centre aggravated this problem.

If the data are spiked, one might analyse the distance data using some left-truncation to eliminate the high numbers trapped near the web centre. Alternatively, one could constrain $\hat{g}(r)$ to be a low order, slowly decreasing function that does not track the spiked nature of the data, but this solution is rather arbitrary and may be ineffective. More experience is needed with sampling from populations of known size to understand better trap spacing and appropriate modelling. Still, this application of point transect sampling has advantages over capture–recapture.

6.9 Point-to-object and nearest neighbour methods

The term 'distance sampling' has been used by botanists in particular to describe methods in which a random point or object is selected, and distances from it to the nearest object(s) are measured. A discussion of these methods was given by Diggle (1983: 42–4). In the simplest case, the distance y to the nearest object is measured; y is a random variable with pdf $f(y)$. However, there is no detection function; the nearest object will be detected with probability one. This is very different from the distance sampling from which this book takes its title, for which there is also a sample of distances y with pdf $f(y)$. The two pdf's can be very similar mathematically, but they are conceptually very different. It is the concept of a detection function that distinguishes the distance sampling of this book. Hence we do not describe point-to-object and nearest neighbour methods in detail here.

For point-to-object and nearest neighbour methods, if the distribution of objects is random, then object density is estimated by

$$\hat{D} = \frac{kn}{\pi \sum_{j=1}^{k} \sum_{i=1}^{n} r_{ij}^2} \tag{6.50}$$

where k is the number of random points or objects, n is the number of point/object-to-object distances measured at each point or object, and r_{ij} is the distance of the ith nearest object to the jth random point or object, $i = 1, \ldots, n; j = 1, \ldots, k$.

When the distribution of objects is overdispersed (i.e. aggregated), density is underestimated if distances are measured from a random point, and overestimated if distances are measured from a random object. An average of the two therefore tends to have lower bias than either on its own. Diggle (1983) listed three *ad hoc* estimators of this type.

Some authors have used point-to-object distances only, together with a correction factor for non-Poisson distribution (Batcheler 1975; Cox 1976; Warren and Batcheler 1979), although Byth (1982) showed by simulation that the approach can perform poorly.

Nearest neighbour and point-to-object methods have been used primarily to measure spatial aggregation of objects, and to test the assumption that the spatial distribution is Poisson. Their sensitivity to departures from the Poisson distribution is useful in this context, but renders the methods bias-prone when estimating object density. Except in special cases, such as estimating the density of forest stands (Cox 1976), we do not recommend these methods for density estimation. Their disadvantages are

- All objects out to the nth nearest to the selected point or object must be detected. In areas of low density, this may require considerable search effort.
- It can be time-consuming to identify which are the n nearest objects, and at lower densities, it may prove impractical or impossible to determine them.

- The effective area surveyed cannot be easily predicted in advance, and is highly correlated with object density; a greater area is covered in regions of low object density. By contrast, good design practice in line and point transect surveys ensures that area covered is independent of object density within strata, leading to more robust estimation of average object density.

Point transects and point-to-object methods may both be considered as generalizations of quadrat counts. In both cases, the quadrat may be viewed as circular. For point transects, the area searched, $a = k\pi w^2$, is fixed (and possibly infinite), but the observer is not required to detect all objects in that area. For point-to-object methods, the number of objects n to be detected from each point is fixed, but the radius about each point is variable; all objects within that radius must be detected.

6.10 Exercises

1. The British Trust for Ornithology's Breeding Bird Survey is conducted using line transect methods. Lines are walked in each of a large number of 2-km squares, selected according to a stratified random scheme. Observers record data on every species encountered. To keep the methodology simple for the large number of volunteers that participate, detected birds are assigned to one of three distance categories. For small birds in a woodland habitat, only two categories are used, so that binomial models are required to estimate abundance.

   Suppose for a given species, n birds are detected from lines of total length L, of which n_0 are within distance x_0 of the line. Assume that probability of detecting a bird at distance x is given by $g(x) = \exp(-\lambda x)$, $0 \leq x < \infty$.

   (a) Given that a bird is detected, what is the probability that it is within x_0 of the line? (Express your answer as a function of λ.)

   (b) Hence write down the probability function $p(n_0)$ of n_0.

   (c) Show that the maximum likelihood estimator of λ is given by $\hat{\lambda} = -(1/x_0)\log_e(1 - n_0/n)$.

   (d) Hence obtain an estimator of bird density, D.

   (e) Describe how you would estimate variance of your density estimate (i) analytically and (ii) by a computer-intensive method.

   (f) Is the above model a reasonable choice for $g(x)$? Explain your answer, and suggest an alternative model that might be better.

   (g) Discuss the advantages and disadvantages of collecting data in just two distance intervals. You should consider both practical field issues and any problems that such a survey creates for the data analyst.

(h) In the pilot survey for the above scheme, point transect surveys, with the same distance categories, were also tested. Review the pros and cons of the above scheme relative to one based on point transects.

2. The figure shows the locations and sizes of dung piles detected on four transects (the lines down the centre of the plots) in the same survey region. The circles represent the size of the detected dung.

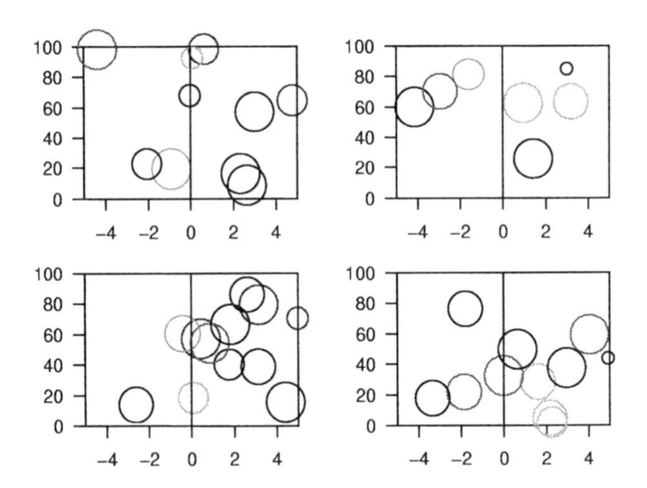

(a) Estimate the perpendicular distances from the transect line to the centre of each dung pile by eye.

(b) Use program Distance to estimate $f(0)$.

(c) If the estimated dung deposition rate is 20 per animal per day with coefficient of variation 8%, and the mean number of days for dung to disappear is 130 with coefficient of variation 10%, estimate animal density.

(d) Repeat parts (a) and (b), but instead of estimating distances to the centre of each dung pile, estimate the distance from the line of the closest edge of each dung pile. (Record dung piles that intersect the line as zero.) Hence estimate the bias in the estimate of animal density that would result if the field worker erroneously recorded these distances.

3. A point transect analysis of cues detected on an aerial cue count survey of minke whales gives an estimate of $h(0)$ of 60 with a coefficient of variation of 27%. Observers searched the 180° ahead of the aircraft on the survey. Estimates of survey time and numbers of cues detected by stratum are shown in the table below. The stratum labelled 'All' shows data pooled over all strata. The animals occur singly.

Table 1. Survey time (in hours), number of cues and stratum areas

Stratum	T	Cue count	Surface area (nm^2)
All	36	126	43 500
1	8	71	3 500
2	4	10	4 000
3	14	25	12 500
4	3	5	7 000
5	7	15	16 500

(a) Because whales give multiple cues in the time they are visible, cue detections are not independent. Suggest how you might obtain an estimate of the coefficient of variation (cv) of $\hat{h}(0)$ that does not rely on the assumption of independence. (The cv for $\hat{h}(0)$ given above was obtained using such a method.)

(b) Estimate cue abundance and its cv, both within each stratum and over all strata.

(c) Minke cue rate has been estimated to be 50 cues per whale per hour ($cv = 15\%$). Use this to estimate the number of cues per animal available to the observers in each stratum. Hence obtain a point estimate of minke whale abundance and its cv in each stratum and in the whole survey region.

7
Study design and field methods

7.1 Introduction

The analysis methods presented in Chapters 3–5 depend on proper field methods, a valid survey design and adequate sample size. This chapter presents broad guidelines for the design of a distance sample survey and outlines appropriate field methods. In general, a statistician or quantitative person experienced in distance methods should be consulted during the initial planning and design of the study. Just as important is the need for a pilot study, which can provide rough estimates of the encounter rate n/L (line transect sampling) or n/k (point transect sampling) and associated variances, from which refined estimates are obtained of n and of L or k for the main study. A pilot study also tests operational procedures and provides training for participants. It can also provide insights into how best to meet the key assumptions.

Careful consideration should be given to the equipment required to allow collection of reliable data. This may include optical or laser range finders, binoculars with reticles, angle boards or rings, a camera, a compass and various options for an observation platform, which might vary from none (other than one pair of feet) to a sophisticated aircraft or ship, or even a submersible (Fig. 7.1).

The primary purpose of material presented in this chapter is to ensure that the critical assumptions are met. Considerable potential exists for poor field procedures to ruin an otherwise good survey. Survey design should focus on ways to ensure that three key assumptions are true: $g(0) = 1$, no movement in response to the observer prior to detection and accurate measurements (or accurate allocations to specified distance categories). If the population is clustered, it is important that cluster size be determined accurately. In addition, a minimum sample size (n) in the 60–80 range and $g(y)$ with a broad shoulder are certainly important considerations. Sloppiness in detecting objects near, and measuring their distance from, the line or point has been all too common, as can be seen in Section 8.4. In many line and point transect studies, the proper design and field protocol have not received the attention required.

Fig. 7.1. Line transect sampling can be carried out from several different types of observation platform. Here, a two-person submersible is being used to survey rockfish off the coast of Alaska. Distances are measured using a small, hand-held sonar gun deployed from inside the submersible.

Traditional strip transects and circular plots should be considered in early design deliberations. However, if there is any doubt that all objects within the strip or circle are detected, then distances should be taken and analysed (Burnham and Anderson 1984). The trade-offs of bias and efficiency between strip transects and line transects have been addressed (Burnham *et al.* 1985). Other sampling approaches should also be considered by reviewing the compendium of alternatives described by Seber (1982). New methods, such as adaptive sampling (Thompson 1990; Pollard and Buckland 1997), are occasionally developed making the updated reviews of Seber (1986, 1992) and Schwarz and Seber (1999) valuable references. A common alternative for animals is capture–recapture sampling, but Shupe *et al.* (1987) found that costs for mark–recapture exceeded those for walking line transects by a factor of three in rangeland studies in Texas. (Mark–recapture studies can of course provide additional information, such as estimates of survival probabilities.) Guthery (1988) presented information on time and cost requirements for line transects of bobwhite quail.

If all other things were equal, one would prefer line transect sampling to point transect sampling. More time is spent sampling in line transect surveys, whereas more time is often spent travelling between and locating sampling points in point transect sampling (Bollinger *et al.* 1988). In addition, it is common to wait several minutes prior to taking data, to allow the animals (usually birds) time to readjust to the disturbance caused by the observer approaching the sample point. For large study areas, point transect sampling becomes more advantageous if the travel between points can be done by motorized vehicle, or if the points are established along transect lines, with fairly close spacing (a form of cluster sampling) rather than a random or systematic sample of points throughout the area. Line transect sampling efficiency is further improved relative to point transect sampling because it is objects on or near the line or point that are most important in distance sampling. The objects seen at considerable distances (i.e. distances y such that $g(y)$ is small, say less than 0.1) from the line or point contain relatively little information about density. In point transect surveys, the count of objects beyond $g(r) = 0.1$ may be relatively large because the area sampled at those distances is so large. By contrast, relatively few objects are recorded close to the point.

Point transect sampling is advantageous when terrain or other variables make it nearly impossible to traverse a straight line safely while also expending effort to detect and record animals. Multi-species songbird surveys in forest habitats are usually best done using point transect sampling. Point transects may often be more useful in patchy environments, where it may be desirable to estimate density within each habitat type; it is often difficult to allocate line transects to allow efficient and unbiased density estimation by habitat. One could record the length of lines running through each habitat type and obtain estimates of density for each habitat type (Gilbert *et al.* 1996). However, efficiency may be poor if density is highly variable by habitat type, but length of transect is proportional to the size of habitat area. Additionally, habitat often varies continuously, so that it is more precisely described at a single point than for a line segment. Detection may be enhanced by spending several minutes at each point in a point transect, and this may aid in ensuring that $g(0) = 1$. Remaining at each point for a sufficient length of time is particularly important when cues occur only at discrete times (e.g. bird calls). However, some individuals may move into the sample area if the observer remains at the point too long, generating upward bias in estimates of density. In terrestrial line transect surveys, the observer may need to stop periodically to search for objects.

7.2 Survey design

Survey design encompasses the selection of a target sample size to achieve a desired level of precision (Section 7.2.2) and placement (allocation) of lines or points (Section 7.2.1). However, before a survey can be designed adequately, the study population and objectives must be clearly defined. Unfortunately, this is often overlooked and the objectives are not met because they were not considered in the formulation of the design. What biological population is being sampled? What is the study area and time frame? Is the population geographically and demographically closed over the study area and time frame? What is the study objective – estimation of abundance, comparison of densities across habitats, trend analysis? These seem like fairly simple questions and often they are, but they should not be overlooked and they should be answered definitively before proceeding with a survey design.

The population of interest must be clearly defined in time and space. What will be sampled? Where will it be sampled? When will it be sampled? Is the population all birds in a particular marsh or county? Or is it only breeding territorial males? Is the area occupied by the population of interest likely to change over the duration of the study? Answers to these types of questions can affect how the survey is designed and what data are collected. Before selecting a method of transect placement, the study area must be clearly delimited. A good map or aerial photo of the study area is nearly essential in planning a survey. In defining the study area, it is important to consider whether the population is closed during sampling and between sampling occasions. If animals leave and enter the study area during sampling, the time frame and the temporal allocation of transects should be considered.

The study objectives must also be clearly defined to decide how much sampling is needed and how sampling is allocated temporally and spatially. The smallest temporal and spatial scales of interest should be specified and the level of precision needed at each scale should be defined to allow selection of an adequate sample size. Whilst we recommend choosing a sufficient line length or number of points to detect a minimum of 60–80 observations, we also stress that the sample size should be determined based on the level of precision that is needed to achieve the research or management objectives. For monitoring programmes with annual sampling, provided protocol is held constant and the same observers are used, it may prove feasible to assume that the detection function remains constant over time, in which case the objectives might be attainable with appreciably less than 60–80 observations per year.

In choosing an adequate sample size, the researcher should decide at what scale it is reasonable to assume that detection probability is nearly constant. Detection probability often varies with topography, habitat type, season, observer, and various other factors. Pooling and model robustness will accommodate these differences if detection probability and density are estimated at the same scale. However, if the sample size is inadequate to estimate detection probability at the scale of interest and detection probability is estimated at a larger scale than density, the density estimates at the smaller scale may be unreliable if detection probability varies at the smaller scale. This is particularly important for comparative studies (e.g. differences in density related to habitat). If the sample size for any comparative unit (e.g. habitat type) is insufficient to estimate detection probability, it will not be possible to adjust for those differences, and true density differences will be confounded with differences in detection probability. For example, in a comparative study of bird density in woodland and grassland habitats, it would be essential to have a sufficient sample size in each habitat to estimate the very different detection probabilities that would likely occur in those habitats. The development of methods to incorporate covariates into the estimation of detection probability (Buckland *et al.* in preparation) should reduce the sample size needed to estimate habitat-specific detection functions.

If the study objective is to make seasonal or annual comparisons of abundance or density, it is important to consider whether the area inhabited by the population changes over time. For example, marine coastal seabirds may shift offshore in response to changes in upwelling or coastal freshwater runoff. By using a larger survey area with stratification, with only a marginal increase in sampling, it may be possible to guard against misinterpretation of distributional shifts as temporal changes.

Although it is no guarantee for a successful survey, clear definition of the population and objectives is an important first step that will improve the chance for success, and help clarify and refine the choices in the survey design process. In the following sections, we give some broad guidelines for transect layout schemes and estimation of sample size requirements. We cannot address every possible design issue, so we strongly recommend that a statistician or quantitative person experienced in distance sampling methods should be consulted during the initial planning and design of the study.

7.2.1 *Transect layout*

In distance sampling, as in all forms of sampling, it is important to use an objective method for sample selection. We discuss five concepts that should be addressed in selecting a line or point transect layout: (1) replication, (2) randomization, (3) sampling coverage, (4) stratification and (5) sampling geometry.

Replication (i.e. multiple lines or points) should be used in any survey design. At a minimum, 10–20 replicate lines or points should be surveyed to provide a basis for an adequate variance of the encounter rate and a reasonable number of degrees of freedom for constructing confidence intervals (Section 3.6). If replicate lines or points are resampled for bootstrapping, even more replicates are desirable.

Some form of random probability sampling should be used in the line and point transect selection. Subjective judgement should not play a role here. Transects should not be deliberately placed along roads or trails, as these are likely to be unrepresentative. Transects following or parallel to ridges, hedgerows, powerlines or stream bottoms are also likely to be unrepresentative of the entire area. We strongly recommend against biasing samples towards such unrepresentative areas except within a stratification framework. Transects placed subjectively (e.g. 'to avoid dense cover' or 'to be sure the ridge is sampled') are poor practice, and should always be avoided.

In a strict sense, our estimates of the sampling variance of encounter rate are only valid if the lines or points are randomly and independently located (at least within strata). However, the precision of a systematic sample is often superior to random sampling (although our estimate of this variance assumes that the sample is random, so fails to exploit this feature of a systematic sample). There is no compelling reason to use completely random placement of lines or points. A systematic grid of lines or points with a random starting position is quite sufficient as long as the systematic spacing does not coincide with a regular spatial feature. If there are smooth spatial trends in the large-scale density over an area, then systematically placed lines or points are better than random placement, because they give better spatial coverage. Ideally, the analysis would fit these trends by some means and derive the variance from the model residuals (Burdick 1979). This topic is addressed in Buckland *et al.* (in preparation).

Standard distance sampling analysis methods assume that all portions of the study area (ignoring stratification for the moment) have an equal probability of being included in the sample (i.e. equal coverage probability, equal to $2\mu L/A$ for line transects and $k\nu/A$ for point transects). Designs with non-uniform coverage probability are valid as long as all portions of the study area have a non-zero coverage probability (Hiby and Hammond 1989). Valid analysis then requires computation of coverage probabilities for the observations, which can be rather tedious if they depend on the geometry of the survey area and transect selection. Non-uniform coverage probability designs are an area of current research (Buckland *et al.* in preparation) and the potential advantages and drawbacks are uncertain. They have been useful in situations in which it was not possible to achieve uniform coverage probability because the survey area as defined by an ice edge was not known *a priori* (Hiby and Hammond 1989).

Spatial stratification of the study area should be considered to improve precision. Another benefit of stratification is that the study area is divided into smaller areas, for which managers may want separate abundance estimates, or which may provide survey blocks of a more manageable size. Stratification is an easy way to develop a non-uniform coverage probability design that can be analysed by standard distance sampling methods.

Within each stratum, coverage is uniform, but coverage probability usually varies across strata. Optimal allocation of sample size and sampling effort across strata is described in Section 7.2.2.3. Exercise caution in developing strata and avoid over-stratifying the study area. Stratification will only yield modest gains in precision unless the density differences are large, and there is always the potential to achieve less precise estimates if the density during the survey does not match the estimates used for the sampling allocation. This is a particular concern for surveys in which animals are not tied to particular regions and may shift their distribution (e.g. marine surveys). Another danger of stratification is that too little effort is allocated to, or too few detections are obtained from, a low density stratum, so that reliable design-based estimation of abundance in the stratum cannot be achieved.

It may be necessary to stratify the study area if abundance estimates are required by sub-area, corresponding for example to habitat types or management blocks. The formulae of Section 7.2.2.1 should be used to ensure an adequate sample size for each stratum. If little is known about the density in each stratum of interest, the sampling effort should be allocated in proportion to the size of the stratum. Sampling can also be partitioned into temporal strata (e.g. monthly or seasonal).

Post-stratification can be used in some cases. For example, the individual lines or points can be repartitioned by habitat type, based on a large-scale aerial photograph on which line or point locations are drawn accurately, or on habitat data gathered at the time of the survey. Thus, estimates of density by habitat type can be made. For example, Gilbert *et al.* (1996) used a Geographic Information System (GIS) together with aerial photographs for the long-term nesting studies of waterfowl at the Monte Vista National Wildlife Refuge.

Sampling geometry is a more significant issue for line transect sampling than point transect sampling, and relates to the configuration and orientation of the set of lines. The best choice of sampling geometry will depend on the survey region, logistics and efficiency, knowledge of density gradients or patterns and their interplay with the other sampling concepts. In most cases, there is no benefit in resampling the same region within one survey, so designs that permit overlapping transects should usually be avoided; this requirement limits the number of possible layouts. Also, designs that require extensive and time-consuming travel between transect lines or points are inefficient. The size and number of replicate lines will depend on the target L and the desired number of replicates.

Several options exist for the layout of individual lines in a line transect survey. A favoured and practical layout for many types of survey is a systematic design using parallel transects, with a random first start (e.g. Figs 1.3 and 1.5). Transects extend from boundary to boundary across the study region and are usually of unequal length unless the region is rectangular. Transects are normally spaced at a sufficient distance to avoid detecting an object from two neighbouring transects, although this is not usually critical unless sampling a line changes the animal distribution at neighbouring, as yet unsampled, lines. If a systematic sample is considered undesirable, but the possibility of having two lines very close together is to be avoided, an option is to divide the study area into a series of contiguous strips of width $2w$. Select k of these at random (where k is chosen to give approximately the predetermined total line length, L), and establish a line transect in the centre of each selected strip. Thus, transects would be parallel, but the spacing between transects would be unequal.

A parallel line design will provide uniform coverage probability unless the connecting lines between the parallel lines are also surveyed. We recommend against that practice. If it is adopted, it should incorporate a stratification scheme to account for the unequal coverage probability at the edges of the study region. In that situation, the lines would have to terminate at a random distance within each edge stratum to achieve equal coverage probability.

A system of parallel lines can be inefficient or costly if the travel time between lines is large, or the observation platform expensive. An alternative is a series of connected line segments that form a continuous zig-zag or saw-tooth design (Fig. 7.2), for which no time is lost traversing between lines. This design is particularly useful for aerial and ship surveys where travel time is costly, although in aerial surveys, the time spent travelling between lines can be beneficial for observer rest and rotation. Except in rectangular survey regions, saw-tooth designs should not use a fixed angle for determining the position of successive segments. A design based on waypoints along the boundaries of the region, equally spaced with respect to distance along the major axis of the design (Fig. 7.2), gives more nearly

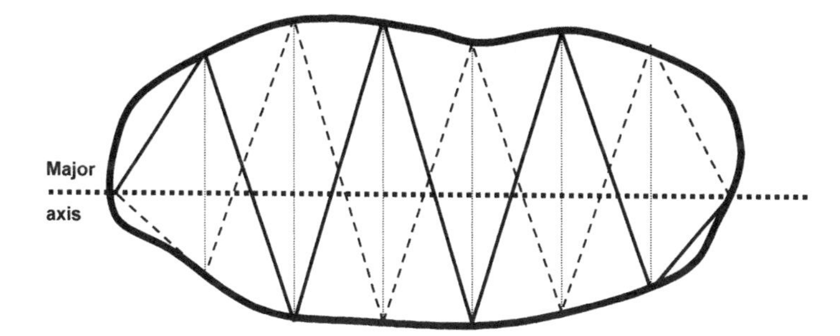

Fig. 7.2. A saw-tooth design based on waypoints at the study area boundaries, equally spaced with respect to distance along the major axis. The dashed saw-tooth shows a return route, should sufficient effort be available, designed to maximize spatial coverage.

even coverage probability. The method can be modified, using curved transects in non-rectangular regions, to achieve uniform coverage, provided the survey region is convex. If it is not, a possibility is to define a convex hull around the survey area, and generate the design within it. (Conceptually, the smallest convex hull enclosing the survey area may be found by cutting out a map of the area and stretching an elastic band around the resulting shape.) Segments of effort that fall outside the survey area are then discarded. If this leads to too much inefficiency, due to gaps in the survey effort, another option is to stratify the area into convex (or nearly convex) sub-areas, and allocate effort to sub-area in proportion to its size. These issues are discussed further in Buckland *et al.* (in preparation).

For many land-based studies, it is not feasible to extend transect lines across the entire study area. In this case, a similar strategy can be adopted as for point transect sampling: a regular grid of points can be superimposed on the area, and a short transect positioned passing through each point. This approach can also be used for a point transect design in which clusters of points are sampled at each location; the points can be located at equally spaced locations along the transect. In such a design, there is often a practical advantage to having a closed circuit such as a rectangle for each short transect (Fig. 7.3). The observer can then start from the most easily accessed point on any given circuit, and because the start point is also the endpoint, less time is required to return, after completion of the transect. This design may be advantageous where a system of roads exists on the study area, giving ready access to any circuit that intersects a road. Having selected random points, or a grid of random points, for the design, the circuit should be located according to a predetermined rule. For example the selected point might give the southwest corner of a rectangle that is orientated north-south. Many parts of central and western North America have roads on a one-mile grid, 'section lines', making this design easy to implement in the field.

When designing a survey based on circuits, the issue of edge effects must be considered. Two options are illustrated in Fig. 7.3. In Fig. 7.3a, a buffer zone has been defined around the study area so that points outside the area can be sampled if the circuit based on that point extends into the area. For circuits that intersect the boundary, only the segments within the study area are surveyed. For large study areas, this will entail on occasion a long journey just to cover a negligible segment of transect. In this case, a simple rule is to discard circuits that are less than 50% inside the study area. This alone would lead to undersampling of the edge of the study area, where densities may differ. This bias can be removed by having a rule that increases retained incomplete circuits (which are more than 50% inside the area) up to the same amount of effort as complete circuits. A simple way to do this is to shift the circuit until it just falls inside the study area, but this too generates some bias away from sampling the edge. Another solution,

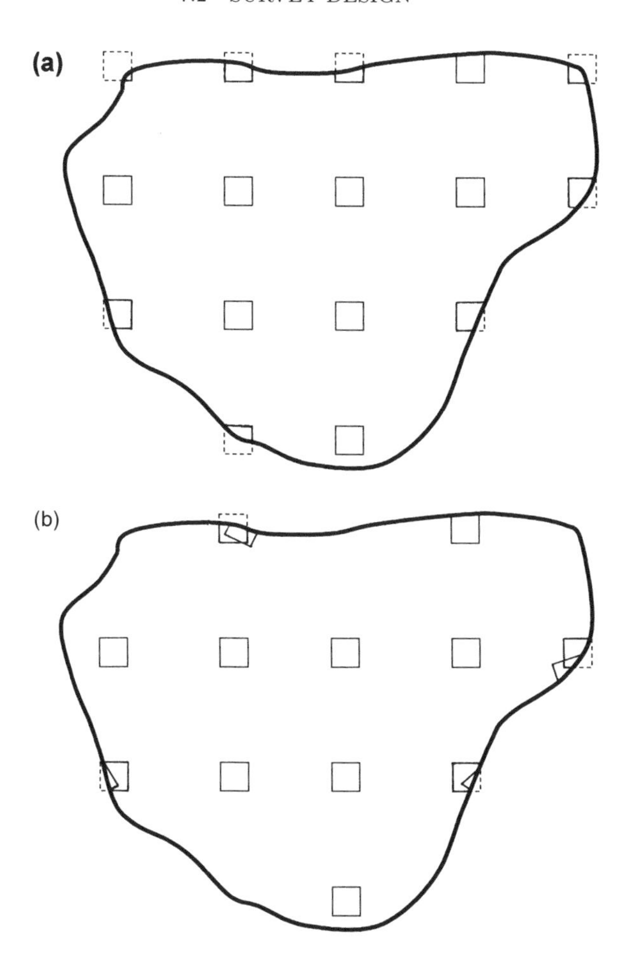

Fig. 7.3. A practical design for line or point transect surveys is to select a sample of closed circuits (rectangles here). The circuit provides the route for a line transect survey, or points are equally spaced along this route. We illustrate two examples of a systematic sample of circuits, in which (a) partial circuits are all sampled and (b) partial circuits that are at least 50% inside the area are sampled, and are extended to compensate for the loss of cover at the edge, due to discarding some of the partial circuits. In this example, a simple reflection rule has been adopted, except where this gives unacceptable overlap within the circuit, in which case the reflected segment is offset slightly.

illustrated in Fig. 7.3b, is to 'reflect' the circuit at the boundary according to some rule.

Both the circuit and the saw-tooth designs share the problem that there is overlap of successive covered strips, caused by the turns in the designs. This is a scale-dependent problem that is not a concern if the effective strip

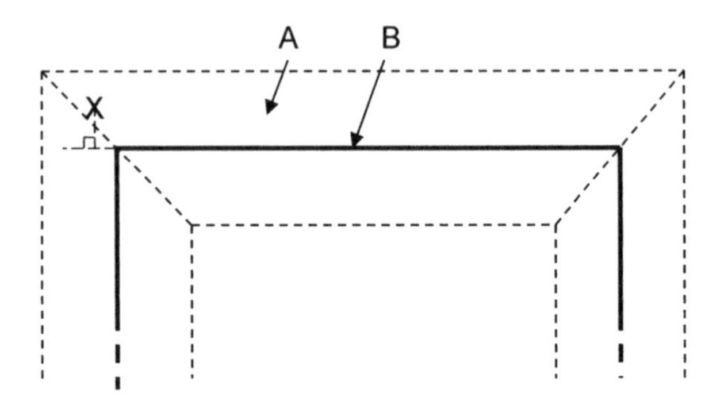

Fig. 7.4. A solution for handling corners in saw-tooth and circuit designs. If an object X is detected in trapezium A, it is recorded as if it were detected from the corresponding line segment B, along with its perpendicular distance from that line segment (extrapolated if necessary).

width is small relative to the perimeter of the rectangle or the length of each saw-tooth. It can be solved by defining the surveyed strips as illustrated in Fig. 7.4.

A more subtle drawback of both methods is the potential dependence of the encounter rate between the line segments which are not independently selected and are spatially connected. The potential for dependence and subsequent underestimation of the variance will depend on the scale of the sampling unit relative to the survey region. In the case of the circuit design, the problem is readily solved by ensuring that the circuit is the sampling unit, not the individual line segments or points that make up the circuit. For saw-tooth designs, if each line segment of the saw-tooth traverses the full width of the study area, then together, these segments should provide good spatial cover, and may be treated as independent sampling units. The proximity and length of the line segments and the scale of the spatial autocorrelation of density will determine the amount of dependence. Dependence in the saw-tooth design can be avoided by replicating the design with different random starting positions; however, achieving a reasonable number of replicates may be difficult. Alternatively, the correlation between adjacent line segments can be examined and incorporated or the variance can be estimated from the residuals of a spatial model (Buckland *et al.* in preparation).

Care must be taken such that the transect direction does not parallel some physical or biological feature, giving an unrepresentative sample. For example, if a series of parallel lines were on or near fence rows, the sample would be clearly unrepresentative (Guthery 1988). Consideration should also be given to possible gradients in density. If a substantial transition in density is thought or known to exist, then the transects should be placed parallel to the direction of the gradient (Fig. 1.3). This would also be true if points were to be placed systematically along lines. For example, if there is a strong density gradient perpendicular to a linear physical feature (e.g. density decreases away from a coastline as bathymetry increases), then a design in which lines are parallel to this gradient and hence perpendicular to the linear feature should be considered. Sampling across the gradient at an angle other than perpendicular (e.g. saw-tooth design) will accomplish the same goal. If the transects were orientated parallel to the linear feature, the variance of the encounter rate estimated from the replicate lines would incorporate unnecessarily the trend in density.

From a sampling theory perspective, a point transect survey is most easily designed as a random sample of points (within any chosen strata). However, if the study region is large, the amount of time to travel from point to point is likely to be excessive, and occasional pairs of points may be quite close together. A systematic grid of points overcomes the latter problem, but not the former. This consideration has led ornithologists, in particular, to place a series of points along a transect line, which is a form of cluster sampling. Thus, there might be 30 lines, each having, say, 10 sampling points. These should not be analysed as if they are 300 independent samples; instead, each line of 10 points is a sampling unit for the purpose of estimating variance.

If a terrestrial survey is to be repeated over time to examine time trends in density, then the lines or points should be placed and marked permanently. Sampling of duck nests at the Monte Vista National Wildlife Refuge has been done annually for 37 years using permanent transect markers set up in 1963 (numbered plywood signs atop 2.5-m metal poles). Repeated sampling should be done at time intervals large enough so that the stochastic errors of successive samples are not highly dependent. Annual surveys are often effective; data from any repeat sampling within a year might be pooled into a single sample as noted above, to avoid strong correlations in the time series. If an area is to be sampled twice within a short time period, one could consider using a system of transects running north-south on the first occasion and another set of transects running east–west on the second occasion. This scheme, although using overlapping transects, might give improved coverage. However, other schemes might be considered if a strong gradient in density was suspected.

Point transects should also be permanently marked if the survey is to be repeated. One must be cautious that neither the objects of interest nor predators are attracted to the transect markers (e.g. poles and signs would not be appropriate for some studies if raptors used these markers for perching and hunting). A good cover map or set of aerial photographs would aid in establishing sample points and in relocating points in future surveys. In addition, a cover map or false colour infrared image might be useful in defining stratum boundaries.

7.2.2 *Sample size*

A basic property of line and point transect sampling theory is that it is the absolute size of the sample that is important when sampling large populations, not the fraction of the population sampled. Thus, if $L = 2400$ m (corresponding to, say, $n = 90$) was sufficient for estimating the density of box turtles on $1\,km^2$ of land, it would also be sufficient for the estimation of density on $25\,km^2$ of land (assuming the sampling was done at random with respect to the turtle population).

The size n of the sample is an important consideration in survey design. If the sample is too small, then little information about density is available and precision is poor. Verner (1985) notes that some surveys have had very small sample sizes ($n \doteq 10$); almost no information about density is contained in so few observations and little can be done regardless of the analysis method used.

As a practical minimum, n should usually be at least 60–80 for reliable estimation of the detection function, or of average density within a (possibly stratified) study area. Even then, the components of variance associated with both n and $\hat{f}(0)$ (line transects) or $\hat{h}(0)$ (point transects) can be large. If the population is clustered, the sample size (i.e. the number of clusters detected) should usually be larger to yield similar precision for the abundance estimate of individuals, substantially so if the variance of cluster size is large. For most populations, the component of variance due to estimating mean cluster size is substantially smaller than those for encounter rate or effective area of the survey, but some populations can show extreme variability in cluster size, with many small clusters and a few very large ones. Such populations are difficult to survey by distance sampling methods. If the large aggregations are known to occur only at certain times of the year, distance sampling surveys should avoid those times.

In some circumstances, cluster size is negatively correlated with cluster density, and in this case, density of individuals may be estimated with higher precision than cluster density (Borchers *et al.* 1998a).

Sample sizes required are often quite feasible in many survey situations. For example, in aerial surveys of pronghorn (*Antilocapra americana*), it is possible to detect hundreds of clusters in 15–20 h of survey time. The long-term surveys of duck nesting at the Monte Vista National Wildlife Refuge have detected as few as 41 nests and as many as 248 nests per year since 1963. Effort involved in walking approximately 360 miles per year on the refuge requires about 47 person days per year. Cetacean surveys may need to be large scale to yield adequate sample sizes; in the eastern tropical Pacific, dolphin surveys carried out by the US National Marine Fisheries Service utilize two ships, each housing a cruise leader and two teams of three observers, together with crew members, for 4–5 months in a survey year. Even with this effort, sample sizes are barely sufficient for estimating trends over eight or more years with adequate precision, even for the main stock of interest.

Sample size in point transects can be misleading. One might detect 60 objects from surveying k points and believe that this sample contains a great deal of information about density. However, the area sampled increases with the square of distance, so that many of the observations are actually in the tail of $g(r)$ where detection probability is low. Detections at some distance from the point may be numerous partially because the area sampled is relatively large. Thus, sample size must be somewhat larger for point transect surveys than line transect surveys. As a rough guideline, the sample size for point transects should be approximately 25% larger than that for line transect surveys to attain the same level of precision. This suggests a minimum sample size of around 75–100 for estimating a detection function, or average density within a study area.

Generally, w should be set large in relation to the expected average distance (either $E(x)$ or $E(r)$). The data can be easily truncated during the analysis, but few (if any) detected objects should be ignored during the actual field survey because they are beyond some preset w, unless distant detections are expensive in terms of resources. For example, dolphin schools may be detected during shipboard surveys at up to 12-km perpendicular distance. These distant sightings add little to estimation and are likely to be truncated before analysis, so that the cost of taking these data is substantial (closing on the school, counting school size, determining species composition) relative to the potential value of the observations. A pilot study would provide a reasonable value for w for planning purposes.

Although we focus discussion here on sample size, the line or point is usually taken as the sampling unit for estimating the variance of encounter rate, and often of other parameter estimates. Thus a sample size of $n = 200$ objects from just one or two lines forces the analyst to make stronger assumptions than a smaller sample from 30 short lines. The dubious strategy of dividing individual lines into contiguous segments, and taking these as the sampling units, can lead to considerable underestimation of variance (Section 3.6.4).

7.2.2.1 *Line transects*

The estimation of line length to be surveyed depends on the precision required from the survey and some knowledge of the encounter rate (n_0/L_0) from a pilot study or from comparable past surveys. Here it is convenient to use the coefficient of variation, $cv(\hat{D}) = \widehat{se}(\hat{D})/\hat{D}$, as a measure of precision. One might want to design a survey whereby the estimated density of objects would have a coefficient of variation of 0.10 or 10%; we will denote this target value by $cv_t(\hat{D})$. Two general approaches for establishing a satisfactory choice for total line length are outlined.

First, assume that a small-scale pilot study can be conducted and suppose n_0 objects were detected over the course of a line (or series of lines) of total length L_0. For this example, let $n_0 = 20$ and $L_0 = 5\,\text{km}$. This information allows a rough estimate of the line length and, thus, sample size required to reach the stated level of precision in the estimator of density. The relevant equation is

$$L = \left(\frac{b}{\{cv_t(\hat{D})\}^2}\right)\left(\frac{L_0}{n_0}\right) \tag{7.1}$$

where

$$b \simeq \left\{\frac{var(n)}{n} + \frac{n \cdot var\{\hat{f}(0)\}}{\{\hat{f}(0)\}^2}\right\} \tag{7.2}$$

While a small pilot survey might be adequate to estimate L_0/n_0 for planning purposes, the estimation of b poses difficulties. However, the value of b appears to be fairly stable and Eberhardt (1978b) provided evidence that b would typically be between 2 and 4, independent of n. Burnham *et al.* (1980: 36) provided a rationale for values of b in the range 1.5–3. They recommended use of a value of 3 for planning purposes, although 2.5 was tenable. They felt that using a value of 1.5 risks underestimating the necessary line length to achieve the required precision. Another consideration is that b will be larger for surveys where the detection function has a narrow shoulder. Here we use $b = 3$ so that

$$L = \left(\frac{3}{(0.1)^2}\right)\left(\frac{5}{20}\right) = 75.0\,\text{km} \tag{7.3}$$

Equating the following ratios

$$\left(\frac{L_0}{n_0}\right) = \left(\frac{L}{n}\right) \tag{7.4}$$

and solving for n gives $n = 300$; we estimate that there will be 300 detections given $L = 75\,\text{km}$, although the actual sample size will be a random variable. Thus, to achieve a coefficient of variation of 10%, one would need to conduct 75 km of transects and expect to detect about 300 objects.

A pilot study to estimate L_0/n_0 can be quite simple. No distances of detected objects from the line need to be measured, and if object density does not vary much, a value as small as 10 for n_0 might be adequate. Thus, random transects of a predetermined length L_0 are surveyed, and the number of detections n_0 recorded. The value of w used in the pilot study should be the same as that to be used in the actual survey. Alternatively, the ratio might be taken from the literature or from one's experience with the species of interest. Of course, the results from the first operational survey should always be used to improve the survey design for future surveys.

Now suppose that a more extensive pilot survey has been conducted, yielding n_0 detections. Then b can be estimated from the data as $\hat{b} \simeq n_0 \cdot \{cv(\hat{D})\}^2$ (Burnham *et al.* 1980: 35). Thus the data from the pilot survey are analysed to obtain the coefficient of variation $cv(\hat{D})$. Substituting $\hat{b}$ into eqn 7.1, the line length required to achieve the target precision is given by

$$L = \frac{L_0 \{cv(\hat{D})\}^2}{\{cv_t(\hat{D})\}^2} \tag{7.5}$$

For this approach to be reliable, n_0 should be in the 60–80 range; it is perhaps most useful when refining the second year of a study, based on the results from the first survey year.

Many surveys are limited by money or labour restrictions such that the maximum line length is prespecified. Thus, it is advisable to estimate the likely coefficient of variation to assess whether the survey is worth doing. That is, if $cv(\hat{D})$ is too large, then perhaps the survey will not provide any useful information and, therefore, should not be conducted. The equation to use is

$$cv(\hat{D}) = \left(\frac{b}{L(n_0/L_0)} \right)^{1/2} \tag{7.6}$$

For the example, if practical limitations allowed only $L = 10$ km,

$$cv(\hat{D}) = \left(\frac{3}{10(20/5)} \right)^{1/2} = 0.274 \text{ or roughly } 27\% \tag{7.7}$$

The investigator must then decide if this level of precision would adequately meet the survey objectives. If for example $\hat{D} = 100$, then a coefficient of variation of 27% yields an approximate 95% log-based confidence interval of [59, 169].

If animals occur in clusters, the above calculations apply to precision of the estimated density of clusters. That is, $\hat{D}$ becomes $\hat{D}_s$, the number of animal clusters per unit area. For clustered populations, a pilot survey yields an estimate of the standard deviation of cluster size,

$$\widehat{sd}(s) = \sqrt{\frac{\sum_{i=1}^{n} (s_i - \bar{s})^2}{n-1}} \tag{7.8}$$

The coefficient of variation of mean cluster size for a survey in which n clusters are detected is then

$$\frac{\widehat{se}(\bar{s})}{\bar{s}} = \frac{\widehat{sd}(s)}{\bar{s} \cdot \sqrt{n}} \tag{7.9}$$

For the case of cluster size independent of detection distance, we have

$$\{cv(\hat{D})\}^2 = \{cv(\hat{D}_s)\}^2 + \left[\frac{\widehat{sd}(s)}{\bar{s}}\right]^2 \cdot \frac{1}{n} \tag{7.10}$$

Now we substitute $n = L \cdot (n_0/L_0)$ and $\{cv(\hat{D}_s)\}^2 = (b/L) \cdot (L_0/n_0)$ to get

$$\{cv(\hat{D})\}^2 = \frac{b}{L} \cdot \frac{L_0}{n_0} + \left[\frac{\widehat{sd}(s)}{\bar{s}}\right]^2 \cdot \frac{1}{L} \cdot \frac{L_0}{n_0} = \frac{1}{L} \cdot \frac{L_0}{n_0} \cdot \left[b + \left\{\frac{\widehat{sd}(s)}{\bar{s}}\right\}^2\right] \tag{7.11}$$

We must select a target precision, say $cv(\hat{D}) = cv_t$. Solving for L gives

$$L = \frac{L_0\left[b + \{\hat{sd}(s)/\bar{s}\}^2\right]}{n_0 \cdot cv_t^2} \tag{7.12}$$

Suppose that a coefficient of variation of 10% is required, so that $cv_t = 0.1$. Suppose further that, as above, $n_0 = 20$, $L_0 = 5$ and $b = 3$, and in addition $\hat{sd}(s)/\bar{s} = 1$. Then

$$L = \frac{5 \cdot (3+1)}{20 \cdot 0.1^2} = 100\,\text{km} \tag{7.13}$$

rather than the 75 km calculated earlier for 10% coefficient of variation on $\hat{D}_s$.

Paradoxically, these formulae yield a more precise estimate of population size for a population of (unknown) size $N = 1000$ animals, for which 50 animals are detected in 50 independent detections of single animals, than for a population of 1000 animals, for which 500 animals are detected in 50 animal clusters, averaging 10 animals each. This is partly because finite population sampling theory is not used here. If it was, variance for the latter case would be smaller, as 50% of the population would have been surveyed, compared with just 5% in the first case. A disadvantage of assuming finite population sampling is that it must be assumed that sampling is without replacement, whereas animals may move from one transect leg to another or may be seen from different legs. Use of finite population corrections is described in Section 3.6.6.

In some studies, animals occur in loose agglomerations. In this circumstance, it may be impossible to treat the population as clustered, due to problems associated with defining the position (relative to the centreline) and size of animal clusters. However, if individual animals are treated as the sightings, the usual analytic variance estimates are invalid, as the assumption of independent sightings is seriously violated. Resampling methods such as the bootstrap (Section 3.6.4) allow an analysis based on individual animals together with valid variance estimation.

7.2.2.2 *Point transects*

The estimation of sample size and number of points for point transect surveys is similar to that for line transects. The encounter rate can be defined as the expected number of detections per point, estimated in the main survey by n/k. Given a rough estimate n_0/k_0 from a pilot survey and the desired coefficient of variation, the required number of sample points can be estimated as

$$k = \left(\frac{b}{\{cv(\hat{D})\}^2} \right) \cdot \left(\frac{k_0}{n_0} \right) \tag{7.14}$$

As for line transect sampling, b may be approximated by n_0 multiplied by the square of the observed coefficient of variation for $\hat{D}$ from the pilot survey. If the pilot survey is too small to yield a reliable coefficient of variation, a value of 3 for b may again be assumed. If the shoulder of the detection function is very wide, this will tend to be conservative, but if detection falls off rapidly with distance from the point, a larger value for b might be advisable. Some advocates of point transects for estimating bird densities argue that detection functions for point transect data are inherently wider than for many line transect data sets, because the observer remains at each point for some minutes, ensuring that all birds within a few metres of the observer are recorded, at least for most species. For line transects, the observer seldom remains still for long, so that probability of detection might fall away more rapidly with distance from the line.

Having estimated the required number of points k, the expected number of objects detected in the main survey should be approximately $k \cdot n_0/k_0$. Suppose a pilot survey of 10 points yields 30 detected objects. Then, if the required coefficient of variation is 10% and b is assumed to be 3, the number of points for the main survey should be $k = (3/0.1^2) \cdot (10/30) = 100$, and roughly 300 objects should be detected.

The above calculations assume that the points are randomly located within the study area, although these procedures are also reasonable if points are regularly spaced on a grid, provided the grid is randomly positioned within the study area. If points are distributed along lines for which separation between neighbouring points on the same line is appreciably smaller than separation between neighbouring lines, precision may prove to be lower than the above equations would suggest, depending on variability in density; if objects are distributed randomly through the study area, precision will be unaffected.

For point transect surveys of clustered populations, no problems arise for estimating density of individuals beyond those encountered by line transect sampling. Equation (7.10) still applies, but the expression $n = k \cdot n_0/k_0$ should be substituted, giving

$$\{cv(\hat{D})\}^2 = \{cv(\hat{D}_s)\}^2 + \left[\frac{\widehat{sd}(s)}{\bar{s}} \right]^2 \cdot \frac{1}{k} \cdot \frac{k_0}{n_0} \tag{7.15}$$

In eqn (7.14), $\{cv(\hat{D})\}^2$ is replaced by $\{cv(\hat{D}_s)\}^2$. Solving for $\{cv(\hat{D}_s)\}^2$ and substituting in the above gives

$$\{cv(\hat{D})\}^2 = \frac{1}{k} \cdot \frac{k_0}{n_0} \cdot \left[b + \left\{ \frac{\widehat{sd}(s)}{\bar{s}} \right\}^2 \right] \tag{7.16}$$

Selecting a target precision $cv(\hat{D}) = cv_t$ and solving for k gives

$$k = \frac{k_0\{b + [\widehat{sd}(s)/\bar{s}]^2\}}{n_0 \cdot cv_t^2} \tag{7.17}$$

Continuing the above example, now with clusters replacing individual objects, the number of points to be surveyed is

$$k = \frac{10 \cdot \{3 + [\widehat{sd}(s)/\bar{s}]^2\}}{30 \cdot 0.1^2} \tag{7.18}$$

If the pilot survey yielded $\widehat{sd}(s)/\bar{s} = 1$ (a plausible value), then

$$k = \frac{10 \cdot (3+1)}{30 \cdot 0.1^2} = 133 \tag{7.19}$$

so that roughly 133 points are needed.

7.2.2.3 *Stratification*

For stratified survey designs, the formulae for sample size determination are more complex. The objective here is to determine the best allocation of sampling among the strata to achieve minimum variance for total abundance. For line transect sampling, if strata are chosen for density or abundance comparisons (e.g. habitat types or trend analysis), sample size should be determined for each stratum using the formulae from Section 7.2.2.1 and the target precision for each stratum should be chosen based on the objective of the comparison. The starting point for a given stratum is the formula

$$var(\hat{D}) = D^2[\{cv(n)\}^2 + \{cv(\hat{f}(0))\}^2] \tag{7.20}$$

Each of the two squared coefficients of variation is proportional to $1/E(n)$, hence

$$var(\hat{D}) = D^2 \left[\frac{b_1}{E(n)} + \frac{b_2}{E(n)} \right] \tag{7.21}$$

where $b_1 = var(n)/E(n)$ and $b_2 = E(n) \cdot var\{\hat{f}(0)\}/\{f(0)\}^2$. Now use

$$E(n) = \frac{2LD}{f(0)} \tag{7.22}$$

to get

$$var(\hat{D}) = D^2 \left[\frac{(b_1 + b_2) f(0)}{2LD} \right] = \left[\frac{D}{L} \right] \left[\frac{(b_1 + b_2) f(0)}{2} \right] \tag{7.23}$$

If, over the different strata, the detection function is the same, then $f(0)$ and b_2 will be the same over strata. This is often a reasonable assumption. It is plausible that b_1 may be constant over strata; this can be checked by estimating $b_1 = var(n)/E(n)$ in each stratum. If these conditions hold, then for stratum v,

$$var(\hat{D}_v) = \left[\frac{D_v}{L_v} \right] K \tag{7.24}$$

for some K, which can be estimated. To allocate total line length effort, $L = \sum L_v$, we want to minimize the sampling variance of the estimated total number of objects in all strata, $\hat{N} = \sum A_v \hat{D}_v$, where A_v is the size of area v and summation is over $v = 1, 2, \ldots, V$. If we pretend that each $\hat{D}_v$ is independently derived (it should not be under these assumptions), then

$$var(\hat{N}) = \sum [A_v]^2 var(\hat{D}_v) = K \sum [A_v]^2 \left[\frac{D_v}{L_v} \right] \tag{7.25}$$

For fixed L, it is easy to minimize eqn (7.25) with respect to the L_v. The answer is expressible as the ratios

$$\frac{L_v}{L} = \frac{A_v \sqrt{D_v}}{\sum A_v \sqrt{D_v}} \tag{7.26}$$

The total effort L comes from

$$[cv(\hat{N})]^2 = \left[\frac{K}{L} \right] \left[\frac{\sum A_v \sqrt{D_v}}{\sum A_v D_v} \right]^2 \tag{7.27}$$

Equation (7.26) suggests that allocation should be proportional to $A_v \sqrt{D_v}$. This result is derived under an inconsistency in that given the assumptions made, $f(0)$ should be based on all distance data pooled and the estimators would look like

$$\hat{D}_v = \frac{n_v \cdot \hat{f}(0)}{2L_v} \tag{7.28}$$

and

$$\hat{N} = \left[\sum_{v=1}^{V} \frac{A_v n_v}{2L_v} \right] \hat{f}(0) \tag{7.29}$$

The first order approximate variance of eqn (7.29) is expressible as

$$var(\hat{N}) = N^2 \left[\frac{f(0)}{2} \right] \left[\frac{b_2}{\sum L_v D_v} + b_1 \sum \frac{(N_v/N)^2}{L_v D_v} \right] \tag{7.30}$$

from which we get an expression for the squared coefficient of variation of $\hat{N}$ in this case of using pooled distances to get one $\hat{f}(0)$:

$$[cv(\hat{N})]^2 = \left[\frac{b_2 f(0)}{2L}\right]\left[\frac{1}{\sum \pi_v D_v} + R\sum \frac{p_v^2}{\pi_v D_v}\right] \tag{7.31}$$

where

$$R = \frac{b_1}{b_2} \tag{7.32}$$

$$p_v = \frac{N_v}{N} = \frac{A_v D_v}{\sum A_v D_v} \tag{7.33}$$

and the relative line lengths by stratum are

$$\pi_v = \frac{L_v}{L} \tag{7.34}$$

Thus, given L, the allocation problem is to minimize eqn (7.31).

We can use the Lagrange multiplier method to derive the equations to be solved for the optimal $\pi_1, \pi_2, \ldots, \pi_V$. Those equations can be written as

$$\frac{D_j}{(\sum \pi_v D_v)^2} + R\frac{p_j^2}{\pi_j^2 D_j} = \left[\frac{1}{\sum \pi_v D_v} + R\sum \frac{p_v^2}{\pi_v D_v}\right], \quad j = 1, \ldots, V \tag{7.35}$$

Fixed point theory can sometimes be used to solve such equations numerically; in this case, it seems to work well. The previous V equations are rewritten below and one must iterate until convergence to compute the π_j. This method is related to the EM algorithm in statistics (Dempster *et al.* 1977; Weir 1990).

$$\pi_j = \sqrt{\frac{R \cdot p_j^2 / D_j}{[1/\sum \pi_v D_v + R\sum(p_v^2/\pi_v D_v)] - D_j/(\sum \pi_v D_v)^2}} \quad j = 1, \ldots, V \tag{7.36}$$

We programmed these in SAS, explored their behaviour, and concluded that a good approximation to the optimal π_j is to use $\pi_j = p_j$, $j = 1, 2, \ldots, V$. Thus, approximately in this case of pooled distance data,

$$\pi_j = \frac{A_j D_j}{\sum A_v D_v} \tag{7.37}$$

Note the relationship between eqns (7.37) and (7.26). Optimal relative line lengths (i.e. $\pi_1, \ldots, \pi_V$) should fall somewhere between the results of eqns (7.26) and (7.37); the exact values of $\pi_1, \ldots, \pi_V$ are not as critical to the precision of $\hat{D}_1, \ldots, \hat{D}_V$ as the total line length L.

7.3 Survey protocol and searching behaviour

Line and point transects are appropriately named because so much that is critical in this class of sampling methods is at or near the line or point. Survey protocol and the observer's search behaviour must try to optimize the detection of objects in the vicinity of the line or point, and search effort or efficiency should decrease smoothly with distance. The aims are to ensure that the detection function has a broad shoulder, the probability of detection at the line or point is unity ($g(0) = 1$), and animals are detected prior to movement in response to the observer.

Ideally, provided $g(0) = 1$, one would like to collect distance data with a very broad shoulder. The choice of an adequate model for $g(y)$ is then relatively unimportant, and D can be estimated with good precision, given adequate sample size. For many studies, proper conduct of the survey can achieve high detection probabilities out to some distance. Many of the methods employed to ensure $g(0) = 1$ also help to widen the shoulder of the detection function.

Survey data for which the detection function drops off quickly with distance from the line or point, with a narrow shoulder and long tail, are far from ideal (Fig. 2.1). Model selection is far more critical and precision is compromised. Occasionally, little can be done at the design stage to avoid spiked data, but usually, such data indicate poor survey design or conduct: perhaps insufficient search effort on the line, poor allocation of search effort near the line or point, poor precision in distance or angle estimation, or failure to detect objects prior to responsive movement towards the observer.

Developing a successful survey protocol to meet these goals will depend on the population being surveyed, the survey platform (e.g. boat, aircraft and on foot), and the observer(s) and their searching behaviour. We discuss particular aspects of survey protocol that depend on the survey platform and population in Sections 7.6–7.9. We focus here on aspects of the survey protocol related to searching behaviour.

Whether an animal is detected depends on its brightness, contrast, size and shape, behaviour, relative motion and its distance from the observer (Koopman 1980). Just as important, how an observer searches determines whether the animal is detected. Most of us have experienced being startled by finding a person nearby in a crowded room after having looked well beyond them. Quite often we have a tendency to search at some distance away from us and rely on peripheral vision at closer distances. Where observers concentrate their search effort clearly affects where they will detect animals.

It is important to remember that observers are human beings. Understanding and making allowances for human nature can be very important to the success of the survey. As humans, we like to be rewarded, we are often competitive, and we get tired and sometimes bored. All of these aspects of our human nature can affect how observers search and how effective that search is. Although there are no guarantees with humans, there are ways to develop survey protocols which are more likely to be successful. We can often reduce the dependence on human nature through use of technology or trained dogs.

Whether we are hunters or bird-watchers, our natural inclination is to maximize the number of animals we detect and we may be further influenced by competition with other observers in a survey. In distance sampling, detecting many animals is not better if, by doing so, animals are missed near the transect line. While an observer's interest and competitive nature are useful to maintain attention throughout the survey, the observer must understand that adhering to the protocol is more important than detecting many animals; quality of data is more important than quantity. They also must understand that the success of the survey largely hinges on satisfying the assumption that $g(0) = 1$. The observer should be trained to search in a fashion which attempts to detect all animals in an inner region around the line or point and either use peripheral vision outside the inner region or to search outside after thoroughly searching the inner region. Specific approaches will be different for line and point transects as described below.

Observers like to be rewarded and their reward is detecting an animal. When observers detect an animal, they want to record the detection. If you have chosen an *a priori* truncation width, the survey protocol should allow the observer to record a detection even if it will not ultimately be used in the analysis. If observers exclude detections during the survey based on an estimated distance, there is a risk that they will unconsciously estimate distance to include the observation that they worked hard to detect. This can be particularly important in low density surveys. This behaviour was noted in the surveys at the Monte Vista National Wildlife Refuge (Section 8.4) where $w = 8.25$ or 12 feet in differing years. For nearly all years, there were more observations in the last distance category than expected. Had the observer been able to record the observations but record that the distance was beyond w, we expect the heaping at w would not have occurred.

While it is not a necessity, there is some benefit in defining a truncation width that encourages observers to focus their effort to ensure that $g(0) = 1$. The problem of unconscious bias can be avoided by using a truncation width in the field which is greater than the width you intend to use in the analysis. Obviously, the observers should be unaware of this. Other approaches for aerial surveys are discussed in Section 7.6.

Surveys that allow for several observers to search simultaneously provide some additional benefits and possible pitfalls. Observers can be used in a team or they can function independently or semi-independently. A team of observers searching different but possibly overlapping areas improves the chances of detecting animals which are only visible at discrete times (e.g. whales which surface or deer blocked by vegetation). The search pattern of the observer team should focus primarily on ensuring that $g(0) = 1$ and secondarily on increasing the effective search width. Observers are naturally competitive in a team approach. A mildly competitive atmosphere can be beneficial to maintain observer attentiveness; however, competition can be a disadvantage if observers focus on detecting many animals rather than ensuring that all animals close to the line or point are detected. Competition can be avoided by isolating observers that are searching separate areas. If observers are searching overlapping areas, double counting by isolated observers can be avoided by using communication with a coordinating observer or by identifying duplicates in the analysis based on spatial and temporal 'closeness'. Isolated, independent observers who search the same area provide data that can be used to estimate $g(0)$, allowing the assumption that $g(0) = 1$ to be removed (Buckland *et al.* in preparation).

Fatigue can compromise accurate data, so consideration should be given to the time spent surveying each day. The quality of the data should not be compromised to extend the survey. Certainly it is unreasonable to believe that an observer can remain at peak searching ability throughout a 7–10-h day. Fatigue may play a larger role in aerial surveys or foot surveys in difficult terrain. Scheduled rest breaks can help extend observer effectiveness and rotation of observers in a multi-observer survey can help maintain interest. The focus should be to collect adequate high quality survey data rather than copious lesser quality data.

7.3.1 *Line transects*

In line transect surveys, the above aims might be enhanced by moving slowly, emphasizing search effort on and near the line, having two or more observers traverse the transects, or using aids to detection such as binoculars. In surveys carried out by foot, the observer is free to use a trained dog, to walk slowly in clumps of heavy cover and faster in low or less suitable cover, or stop frequently to observe. Observers may leave the centreline temporarily, provided they record detection distances from the transect line, not from their current position. In many surveys, it is good practice to look behind occasionally in case an object that was hidden on first approach can be seen. In general, observers should spend the majority of their time searching forward and near the line to ensure $g(0) = 1$ and to detect animals prior to any movement in response to the survey platform. Search effort should be allocated so that it falls off smoothly with distance from the line, to ensure that the detection function also falls off smoothly. If animals are typically moving around, then the average speed of the observer should be at least two or three times faster than that of the animals (Hiby 1986), as otherwise, density is overestimated. If this is not possible, then methods outlined in Section 6.5 should be considered.

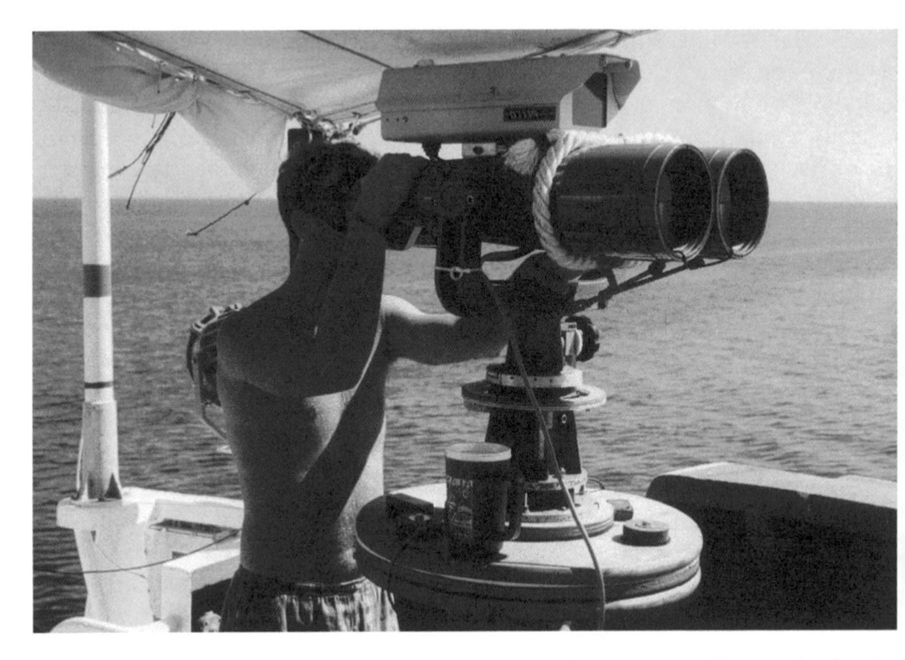

Fig. 7.5. Tripod-mounted 25×150 binoculars used in surveys of large schools of dolphin in the eastern tropical Pacific Ocean.

The survey must be conducted so as to avoid undetected movement away or toward the line in response to the observer. To achieve this, most detection distances should exceed the range over which objects might respond to the observer. If a motorized observation platform is used, the range of response might be reduced by using platforms with quiet motors or by travelling faster. In surveys carried out by foot, the observer can ensure more reliable data by moving quietly and unobtrusively. Detection distances can be improved by use of binoculars. If detection cues are continuous, high power binoculars might be used, for example tripod-mounted $25\times$ binoculars on shipboard surveys of dolphins that typically occur in large schools (Fig. 7.5). If cues are discrete, for example whales surfacing briefly, or songbirds briefly visible amongst foliage, lower magnification is necessary, so that field of view is wider. Indeed, binoculars are often used only to check a possible detection made by naked eye. In some studies, one observer might scan continuously with binoculars while another searches with the naked eye.

Conducting line transect surveys with several observers is one of the best ways to help satisfy the assumptions. Aerial surveys commonly employ two observers, one covering each side of the aircraft. In addition, another observer might guard the centreline through a belly port or bubble window, or a video camera may be used. Shipboard surveys frequently use three or more observers on duty at any one time, with one observer guarding the centreline, and the other two scanning each side and overlapping the centreline. It is also useful to have at least one of the observers looking forward to detect animals prior to movement. For example, in shipboard surveys, one or more observers may use high-powered binoculars focusing far ahead (Fig. 7.5), while other observers search closer with low-powered binoculars or naked eye. In terrestrial surveys, more than one observer is often required for safety reasons, and field protocol can be developed to take advantage of this.

Certain types of double counting can be problematic. If the objects of interest are immobile objects such as nests, then the fact that a particular nest is detected from two different lines is fully allowed under the general theory. Double counting becomes a potential problem only if mobile objects are surveyed such that the observer or the observation platform chases animals from one line to another, or if animals 'roll ahead' of the observer, hence being counted more than once (e.g. 'chain flushes' in surveys of grouse). Movement in response to the observer that leads to double counting should be recognized and avoided in the planning and conduct of a survey.

7.3.2 *Point transects*

For point transect surveys, the longer the observer remains at each point, the more likely is the probability of detection at the point to be unity, and the broader is the shoulder of the detection function. This advantage is offset by movement of objects into the sampled area, and possible double counting of objects, both of which lead to overestimation of density. Optimal time to spend at each point might be assessed from a pilot study. In some cases, it might be useful to observe the point from a short distance away and record distances to objects of interest before any disturbance caused by the approach of the observer. Another option is to wait at the point a short period of time before recording, to allow objects to resume normal behaviour. As for line transects, binoculars may be useful for scanning, for checking possible detections, or for identifying the species. After the recording period, the observer may find it necessary to approach an object detected during that period, to identify it. Detection distances can also be measured before moving to the next point. If distances are assessed by eye, the task is made easier by use of markers at known distances. The ready availability of laser binoculars and similar devices means that, for most land-based surveys, there is little justification to rely solely on distances estimated by eye.

If the radius of each point is fixed at some finite w, one could consider the population 'closed' and use a removal estimator to estimate the population size N (White *et al.* 1982: 101–19). To keep the option of this approach open, the time (measured from the start of the count at the point) at which each object is first detected should be recorded. The count period may then be divided into shorter time intervals, and data for each interval pooled across points. The relevant data would be the number of objects detected in the first time interval, the number of new objects detected in the second time interval, and so on. However, we warn against this approach, as it is extremely sensitive to any kind of object movement, and can generate density estimates that are an order of magnitude too high.

We caution against the use of playback vocalizations to elicit an aural response, as has been done in some bird surveys. Regular use of tapes in the territories of some species can cause unacceptable disturbance. Also, if animals are attracted towards the point and are not detected prior to their movement, large positive bias in density estimates can result. Klavitter (2000) demonstrated that a large positive bias would result from using playback vocalizations in surveys of the endangered Hawaiian hawk. He used radio-tagged birds to show that birds within 1 km of the observer moved up to 800 m towards the observer prior to detection.

7.4 Data measurement and recording

Data should be measured and recorded with a method that is both accurate and easy. Accurate measurement of distance, angle, line length and cluster size is quite important. The observer must work carefully and avoid errors in recording or transcribing data. Data measurement and recording should be easy and reliable and should not detract from searching. Ancillary data, such as sex, species and habitat type, are often taken and they can be useful as covariates to model detection probability or to explain differences in density. However, ancillary data collection should not interfere with searching or degrade the quality of the primary measurements; too often, data are gathered with little thought given to their relevance or value.

Below we describe some methods of data measurement and recording but our description is far from exhaustive. We further describe some platform-specific approaches in Sections 7.6–7.8. In choosing a particular method, accuracy, reliability and ease should be your objectives. In particular, the careful measurement or estimation of distances near the line or point is critical. Every possible effort must be made to ensure that accurate measurements are made, prior to any undetected movement, of all objects on or near the line or point. This cannot be overemphasized.

7.4.1 *Distance measurement*

The basic data to be recorded and analysed are the n distances. For point transects, analyses are based on point-to-object distances r, but for line transects, the widely used methods require that the shortest (perpendicular) distance x between a detected object and the line is estimated or measured. For surveys of immobile objects (e.g. nests), it is typically easiest to estimate or measure perpendicular distance directly from the closest point on the line. However, for many surveys of animals, by the time the observer reaches the closest point on the line, the object may not be visible or may have moved in response to the observer's presence. These problems are minor for aircraft surveys in which the speed of the observation platform is sufficient to render movement of the object between detection and the point of closest approach unimportant. For shipboard surveys of marine mammals, sighting distances are frequently several kilometres, and it may take up to half an hour to arrive at the point of closest approach. Further, for many surveys, it is necessary to turn away from the centreline when an animal cluster is detected, both to identify and to count the animals in the cluster. Hence the natural distance to record is the sighting or radial distance r; by recording the sighting angle θ also, the shortest distance between the animal and the line, i.e. the perpendicular distance x, may be calculated as $x = r\sin(\theta)$ (Fig. 1.4). Here we describe methods of measuring distance, whether it is perpendicular or radial distance.

7.4.1.1 *Ungrouped distances*

Accurate measurements of distance should be one of the primary goals of any distance sampling protocol. Biased measurements of distance will translate to bias in estimated density and abundance. 'Heaping', rounding to convenient values, can result in bias or problems in model fitting depending on the type and amount of heaping. Random measurement error is of lesser importance but can still cause problems if the errors are large relative to the average measurement.

Distance can be measured in a variety of ways from the crude but simple method of pacing to the sophistication of laser range finders. In some circumstances, it may be necessary to estimate by eye rather than measure distance. The nature of the surveyed population and the survey platform (e.g. on foot, boat, aircraft) will largely dictate the methods that can be used reliably and reasonably. We describe four different approaches to distance measurement: direct, range finder, declination angle and visual estimation.

Direct methods are suitable for situations in which it is reasonable to traverse to the location of the detected object (e.g. animal). For land-based surveys, pacing may suffice, but for distances to about 30 m, we recommend the use of a steel tape. For shorter distances, a yard or metre rule could be used for measurements and simultaneously for 'beating' at vegetation (e.g. in nest surveys). For longer range measurements, distance can be computed from the geographic positions on the line and the object's location, which can be accurately obtained from differential Global Positioning System (GPS) corrected for selective availability error. Holt and Powers (1982) used this approach with the less precise Global Navigation System (GNS). This latter method may only prove feasible for aerial surveys with relatively few detections. Unlike the direct methods, the remainder of the methods for distance measurement allow the observer to remain on the line or point.

Range finders measure the line of sight distance from the observer to the object. These include both optical range finders, based on parallax that require manual focusing, and more sophisticated laser range finders that use a visible or infrared laser beam to project a spot of light onto a target. The reflected beam is used to measure the distance. Binoculars with a built-in laser range finder are also available. The range and accuracy of these devices vary and are dependent on the price. When the survey platform height is appreciable (e.g. aerial surveys), distance along the ground $x = \sqrt{d^2 - h^2}$ should be computed from the line of sight distance (d) and platform height h. Some survey lasers calculate this distance internally.

Using trigonometry, distance can also be measured using a known height of the observer's eye above the ground and the angle of declination between the object and either the horizon, a location at a known distance, or horizontal. This is the basis for measuring distance using binoculars with reticle marks, theodolites used by surveyors and clinometers which have been used by foresters to measure tree height (Avery 1967). The angle of declination described here should not be confused with the horizontal angle used to convert a radial distance to a perpendicular distance (Section 7.4.2).

For shipboard surveys, radial distances are frequently estimated using reticles or graticules, which are marks on binocular lenses (Fig. 7.6). The observer records the number of marks down from the horizon to the detected object. This number may be transformed to a distance from the observer using a modification of a method described by Lerczak and Hobbs (1998), as follows (Fig. 7.7).

Let

$R =$ radius of the Earth $\simeq 6370\,\text{km}$

$v =$ vertical height of the binoculars above the sea surface

$\delta =$ angle of declination between successive divisions on the reticle

$\phi =$ angle between two radii of the Earth, one passing through the observer and the other passing through any point on the horizon, as seen by the observer

$= \cos^{-1}\{R/(R+v)\}$

Now suppose that the reticle reading is d divisions below the horizon, so that the angle of declination between the horizon and the sighting is

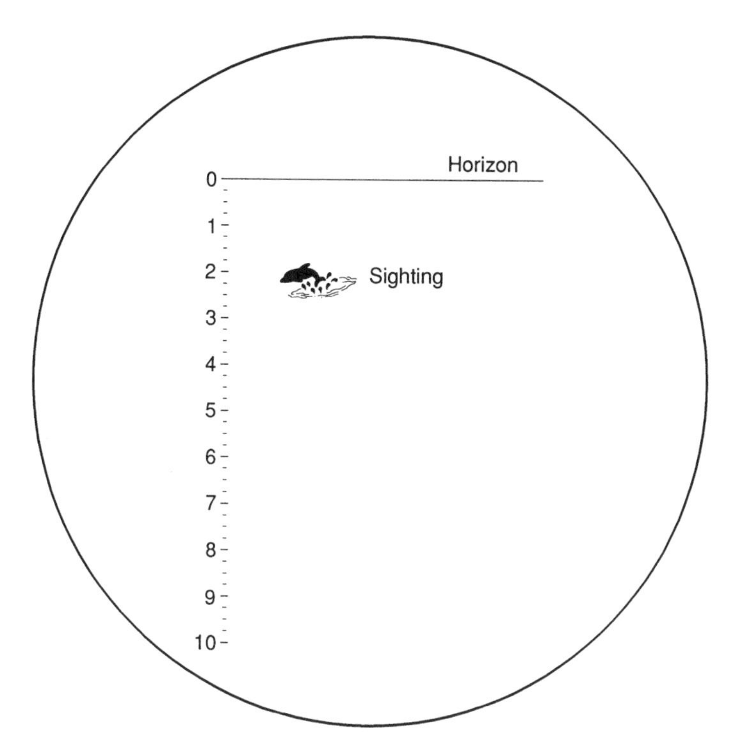

Fig. 7.6. Diagram of reticles used on binoculars on shipboard surveys of marine mammals. Use of these marks allows the computation of sighting distance (Fig. 7.7).

$\psi = d \cdot \delta$. Then the sighting distance is approximately

$$r = \frac{R + v - \sqrt{R^2 - r^2}}{\tan(\phi + \psi)} \tag{7.38}$$

This is a quadratic in r, and the smaller root provides the solution we require:

$$\begin{aligned} r &= \cos(\phi+\psi)\Big[(R+v)\sin(\phi+\psi) \\ &\quad - \sqrt{R^2\sin^2(\phi+\psi) - v(2R+v)\cos^2(\phi+\psi)}\Big] \\ &\simeq \cos(\phi+\psi)\Big[R\sin(\phi+\psi) - \sqrt{R^2\sin^2(\phi+\psi) - 2Rv\cos^2(\phi+\psi)}\Big] \end{aligned} \tag{7.39}$$

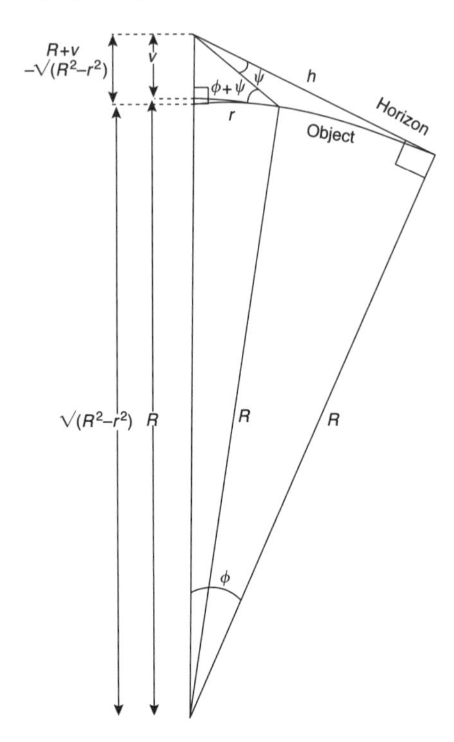

Fig. 7.7. Geometry of the procedure for computing sighting distances from ocular data for shipboard surveys of marine mammals. Reticles provide an estimate of ψ, the angle of declination of the detected object from the horizon, which must be converted into the distance r from the observer to the object.

For example, if the observer's eyes are 10 m or 0.01 km above sea-level, the angle between successive divisions of the reticle is 0.1°, and the reticle reading is 3.6 divisions below the horizon, then

$$\phi = \cos^{-1}\{6370/(6370+0.01)\} = 0.10^\circ \quad \text{and} \quad \psi = 0.36^\circ$$

so that

$$r = \cos(0.46^\circ)\Big[6370\sin(0.46^\circ) - \sqrt{6370^2\sin^2(0.46^\circ) - 2\times 6370\times 0.01\cos^2(0.46^\circ)}\Big] = 1.26\,\text{km}$$

Note that the horizon is at $h = R\tan(\phi) = 11.3$ km (Fig. 7.7). These calculations ignore the effects of light refraction, which are generally small for sightings closer than the horizon. This method can be modified to measure the angle of declination from a location at a known distance (e.g. a shoreline) for situations in which a horizon is not visible (Lerczak and Hobbs 1998).

Clinometers that measure the angle of declination relative to an imaginary horizontal line work on the same principle as binoculars with reticles. In contrast to binoculars which provide a more accurate angular measurement over a narrow field of view (typically 1.5–5°), clinometers can be used over the entire range for the angle of declination (0–90°) but are typically less accurate. Clinometers are used in situations in which the survey platform is a considerable height above the ground as in aerial surveys (Dahlheim *et al.* 2000) or for measuring distances to nearby objects which are outside the field of view of binoculars focused on the horizon (Bengtson *et al.* 1995). The angle of declination from manual clinometers is measured by aligning the clinometer with the object and reading the angle from a scale marked in one-degree increments. Electronic clinometers work on the same principle but have a digital display and some can download to a computer. In using the clinometer during aerial surveys, it is essential that the observer keep both eyes open to avoid missing a nearby sighting while making a measurement; with manual clinometers, the observer may have a tendency to close one eye in trying to read the scale. With some practice, an observer can become adept at keeping both eyes open and reading the scale quickly; however, a clinometer may prove impractical in aerial surveys with high densities. The height of the observer's eye above the ground or sea surface v must be recorded also. Distance x to the observation is computed as $x = v \tan(90° - \phi)$, where ϕ is the angle of declination in degrees. For most distances measured with a clinometer, this simple formula is sufficient. Lerczak and Hobbs (1998) provided a formula that accounts for the curvature of the earth and is more accurate for angles of declination $< 1°$.

We discourage the use of visual observation alone in estimating distances and angles; however, we recognize that it may be necessary in some situations. Unless the observers are unusually well trained, such a procedure invites heaping of measurements (at best) or biased estimates of distance with different biases for different observers (at worst). Scott *et al.* (1981b) found significant variability in precision among observers and avian species, but no bias in the errors. Often, even simple pacing is superior to ocular estimation. Only use visual estimation as a last resort, and to maintain consistency throughout the survey, incorporate a training and testing programme that is used throughout the survey. Where possible, use a technique such as hauling a float behind a boat at known distance to help the observers calibrate continually.

7.4.1.2 *Grouped distances*

It is sometimes difficult to measure distances exactly, in which case they might be recorded only by distance interval. Thus we might note that an object was detected somewhere between, say, 0 and 40 m, and not measure its exact distance. During the course of the survey, a count n_1 will be

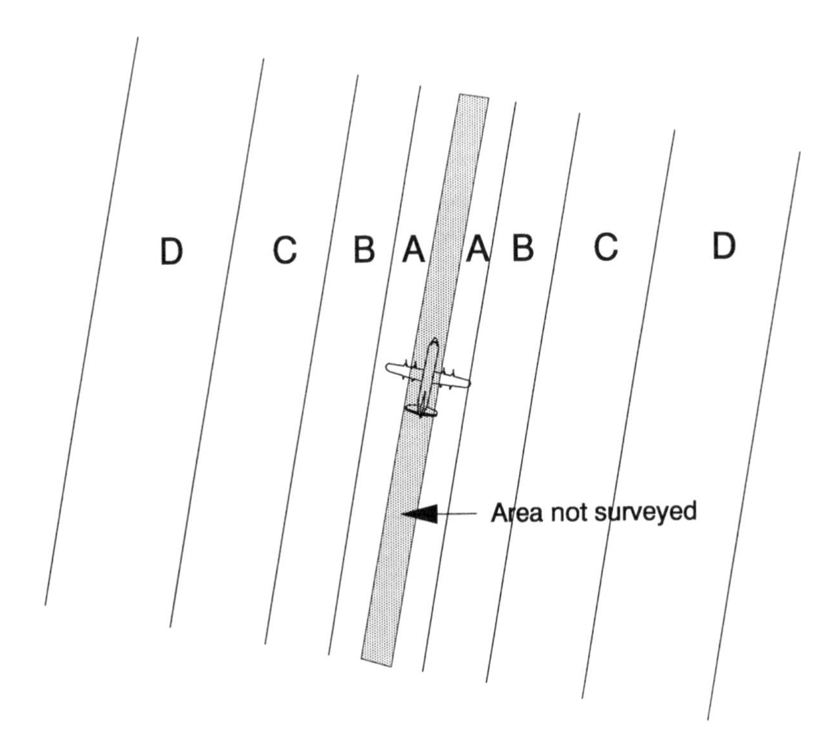

Fig. 7.8. The area below an aircraft can be excluded as shown (shaded area). Here grouped data are recorded in four distance intervals (A–D) of increasing widths with distance. Adapted from Johnson and Lindzey (unpublished).

made of objects in this first distance interval. The survey results will be the frequencies $n_1, n_2, \ldots, n_u$, corresponding to the u distance classes, with total sample size $n = \sum n_i$.

In general, let c_i denote the perpendicular distance from the line or point separating interval i from interval $i+1$. These 'cutpoints' are therefore at $0 = c_0 < c_1 < c_2 \cdots < c_u = w$. In the case of left truncation, $c_0 > 0$. Note, also, that $c_u = w$ can be finite or infinite. The grouped distance data are the frequencies n_i of objects detected in the u intervals defined by the cutpoints; n_i is the number of objects in distance interval i, which spans distances in the range (c_{i-1}, c_i).

Often the selection of the c_i will largely be a natural consequence of the measurement method. For example, nearly all aerial surveys collect grouped data by dividing the field of view based on a defined set of angles of declination. In some cases, $c_0 > 0$ because visibility is impaired below the aircraft (Fig. 7.8). Markers of some type are typically attached to the aircraft to delineate the angles of declination, and hence the distance intervals on the ground for a known height above ground (Fig. 7.9). This

Fig. 7.9. Airplane wing struts can be marked to delineate boundaries of the distance intervals on the ground. Other marks on the side window of the airplane are used to allow observers to align their eyes correctly, to ensure correct classification of detected animals to distance intervals. Compare with Fig. 7.8. From Johnson and Lindzey (unpublished).

is equivalent to the use of a clinometer with fixed angle intervals (see Section 7.6). Thus the cutpoints are fixed by the choice of angles of declination for the markers, from which the corresponding distances are calculated, given the altitude at which the survey is conducted. The accuracy of an angle measurement at 65° from the horizontal should be approximately the same as the accuracy at 10°, but a one-degree error at 10° represents a much larger error in distance. Intervals based on equal angle divisions provide a natural division of distances, with narrower intervals close to the line.

In some situations (e.g. ocular estimation or pacing), there may not be a natural choice for distance intervals based on the measurement method. We offer the following general guidelines in defining distance intervals. If at all possible, there should be at least two intervals in the region of the shoulder of the detection function. In general, the width of the distance intervals should increase with distance from the line or point, because larger distances will tend to be less accurately determined and the information content of the intervals is most important at distances close to the line or point. The width of each distance interval might be set so that the n_i would

Table 7.1. Suggested relative interval cutpoints for line and point transects. An appropriate value for Δ must be selected by the user

Number of intervals, u	Suggested relative interval cutpoints for line transects	Suggested relative interval cutpoints for point transects
4	Δ, 2Δ, 4Δ, ∞	2Δ, 3Δ, 4Δ, ∞
5	Δ, 2Δ, 3.5Δ, 5Δ, ∞	2Δ, 3Δ, 4Δ, 5.5Δ, ∞
6	Δ, 2Δ, 3Δ, 5Δ, 7Δ, ∞	2Δ, 3Δ, 4Δ, 5Δ, 6.5Δ, ∞
7	Δ, 2Δ, 3Δ, 4.5Δ, 6Δ, 8Δ, ∞	2Δ, 3Δ, 4Δ, 5Δ, 6Δ, 7.5Δ, ∞
8	Δ, 2Δ, 3Δ, 4Δ, 5.5Δ, 7Δ, 9.5Δ, ∞	2Δ, 3Δ, 4Δ, 5Δ, 6Δ, 7Δ, 8.5Δ, ∞

be approximately equal. This rough guideline can be implemented if data from a pilot survey are available.

Alternatively, if the underlying detection function is assumed to be approximately half-normal, then Table 7.1 indicates a reasonable choice for group interval widths for various u, where Δ must be selected by the biologist. Thus for a line transect survey of terrestrial animals, if $u = 5$ distance intervals were required, and it was thought that roughly 20% of detections would be beyond 500 m, then $\Delta = 100$ m, and the interval cutpoints are 100 m, 200 m, 350 m, 500 m and ∞. The grouped data would be the frequencies $n_1, n_2, \ldots, n_5$.

As a guideline, u, the number of distance classes in line transect surveys, should not be less than four and five is much better than four. Too many distance intervals tend to defeat the advantages of such grouping, as classification of objects into the correct distance interval can become error-prone and time-consuming. In addition, the use of too many distance intervals distracts attention from the main goal: detections near the line or point. As few as two distance intervals (i.e. 'binomial' models) are sometimes used in point transect surveys (Buckland 1987a) and in line transect surveys (Järvinen and Väisänen 1975; Beasom *et al.* 1981), although only single-parameter models can then be fitted, and the adequacy of their fit cannot be tested. In general, the use of between five and seven distance intervals will be satisfactory in most surveys.

It is commonly thought that all objects in the first distance interval must be detected (i.e. a census of the first band). This is incorrect; the width of this interval might be 40 m and it is not necessary that all objects in the 0–40-m band be detected. Of course, as the shoulder in the data is broadened, there are significant advantages in estimation. As a guideline, we recommend that the probability of detection should not fall appreciably below unity over at least the first two intervals.

An advantage of collecting grouped data in the field is that exact distances are not required. Instead, one merely classifies an object detected into the proper distance class. Collection of grouped data allows a relaxation of the assumption that distances are measured exactly. Instead, the assumption is made only that an object is counted in the correct distance interval. Thus, if an object is somewhere near the centre of the distance class, proper classification may be easy. Problems occur only when the object is detected near the boundary between two distance intervals. If this is of concern, a simple solution is to record which detections are close to a cutpoint. These are then split between the two intervals, so that a frequency of one-half is assigned to each (Gates 1979). Of course, a reduction in the number of distance intervals will result in fewer incorrect classifications.

7.4.2 *Angle measurement*

The accuracy of the sighting angle (θ) measurement is critical for line transect surveys that rely on the computation of perpendicular distance from the measurement of sighting (or radial) distance and angle. The accuracy of angles close to zero is particularly important for larger sighting distances. Observers need to understand the importance of accurate measurement of angles and must be trained to avoid rounding angles to convenient values. Rounding is best avoided by using an objective angle measurement. If binoculars are tripod-mounted, sighting angles can be accurately measured from an angle ring on the stem of the tripod. If binoculars are hand-held, angle boards (Fig. 7.10), perhaps mounted on ship

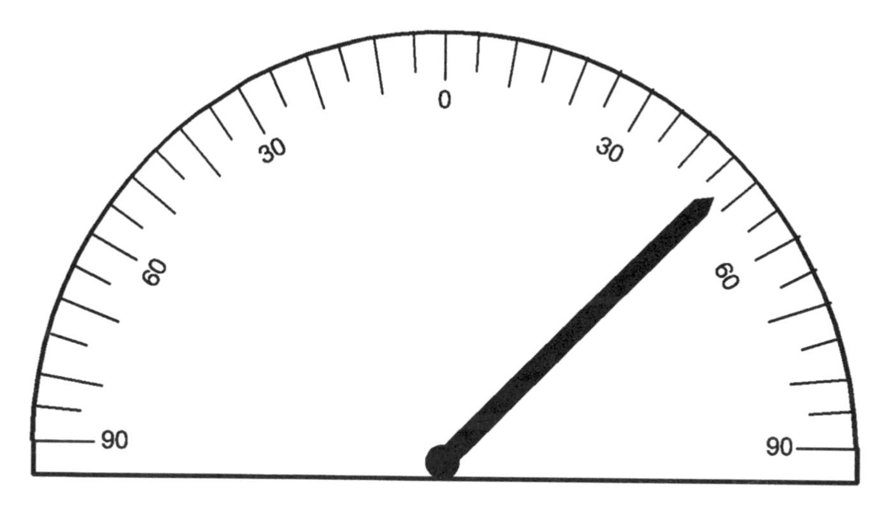

Fig. 7.10. Sighting angles can often be more accurately estimated by the use of an angle board as shown here. Such devices can be hand-made and are useful in many applications of distance sampling.

railings, may be found useful; although accuracy is likely to be poor relative to angle rings on tripods, it should still be appreciably greater than for guessed angles. Compass measurements of angle may be useful in situations where the line is well-marked and the bearing for the line is known. Distance and angle experiments should be carried out if at all possible, both to estimate bias in measurements and to persuade observers that guesswork can be poor!

7.4.3 *Distance measurement error*

The accuracy of distance measurements is critical to the success of distance sampling. Although it is unrealistic to expect that each distance measurement is exact, every possible effort must be made to ensure that the measurements are not biased and potential error is minimized. Calibration, testing and ongoing training for distance measurement should be incorporated into every survey programme.

Symmetric and unbiased errors in distance measurement will create a negative bias in density, but the bias will typically be small in the case of line transect sampling unless the measurement error is large. Chen (1998) showed that the expected bias was a function of the error magnitude and the width of the shoulder of the detection function (i.e. values of x such that $g(x)$ is approximately 1). When the standard deviation of the measurement errors was approximately one-half the width of the detection function shoulder, the expected bias in density was less than -3% (Chen 1998). As a rough guide, the goal should be to achieve consistently measurement errors for distances near the line less than the width of the detection function shoulder.

Unbiased measurement errors in detection distances for point transect sampling are more problematic. Suppose the distances are recorded in intervals of equal width. Because the expected number of objects present in an interval increases with distance from the point, if a constant proportion of detections is misclassified for each interval, with equal numbers being misclassified in each direction, intervals near the point, with few objects present, will record a net gain of objects from further intervals. Thus too many objects are recorded near the point, and density is overestimated. For most point transect surveys, it is possible to avoid this problem by measuring distances accurately, either using laser range finders or by measuring or pacing the distances, although surveys which depend largely on aural cues (Section 7.8.8) will require careful observer training (Section 7.5). However, for cue counting, precise distance estimation, and hence bias in density estimates, is more problematic (Section 6.4). In this case, it is important that the rounding and/or imprecision in distance estimates be taken into account in analyses. Hiby *et al.* (1989) describe a method for doing this, using an independent observer survey design. Independent observer data allow the imprecision in distance estimation to be quantified, by using the recorded distances from the two observers to the same duplicate detection. If reliable estimates of imprecision and/or rounding are available from separate experiments for example, these could be used in analysis instead of independent observer data. Methods for incorporating the uncertainty in distance estimation in this case remain to be developed. A better solution than either of these options is to ensure that distances are estimated precisely, if this is possible.

Biased distance measurements are also problematic. A negative bias in distance measurements will result in a positive bias in density and a positive bias in distance measurements will result in a negative bias in density. Again, bias in density estimates from point transect sampling is greater than for line transect sampling. For example, if distances are consistently overestimated by 10%, then line transect estimates of density will be biased low by 9% (1/1.1), while those from point transect sampling will be biased low by over 17% ($1/1.1^2$). The potential for bias will depend on the method for distance measurement. Distance measurements from clinometers or binoculars with reticles depend on an accurate measurement of the observer's eye height above the observation surface. If the height is incorrectly measured, the bias will translate directly to the distance measurements. Calibration with a known set of distances is typically the easiest way to avoid introducing bias.

Bias can also result if measurements are subject to rounding. A small degree of rounding would not be a problem if it was symmetric. However, observers often have a tendency to round inexactly and to 'heap' measurements to convenient values. Heaping at zero distance is especially problematic, again illustrating the need to know the exact location of the line or point. Well-marked, straight lines are needed for line transect surveys. Upon detection of an object of interest, the surveyor must be able to determine the exact position of the line or point, so that the proper measurement can be taken and recorded. If sighting angles are being measured, a straight line is needed so that the angle is well-defined. Observers may also have a tendency to record objects detected just beyond w as within the surveyed area (Section 7.3). This might be called 'heaping at w' and was noted in the surveys at the Monte Vista National Wildlife Refuge (Section 8.4).

Visual estimation of distances is the least favourable method and most likely to be subject to bias and large errors. Visual estimation should only be used when no other method is practical. When it is the only alternative, it is essential that observers are trained and tested (Section 7.5). Testing should mimic the real situation as close as possible. The test data should be routinely recorded to be incorporated into a measurement error model (Chen 1998; Schweder *et al.* 1997) if the measurements are biased or the errors are too large to be ignored.

Measurement accuracy is also important if the distances are recorded in intervals (grouped). Even though the exact distance does not have to be recorded, the observation must be classified into the correct distance interval. For example, reliability of grouped distances collected from aerial surveys using angle marks on airplane struts depends on the angle measurements for the marks, the altitude measurement, and the observer's ability to align marks to define the angle (distance) interval. In a field study of measurement error in aerial surveys, Chafota (1988) placed 59 bales of wood shavings (22.7 kg each) in short grass prairie habitat in northeastern Colorado to mimic pronghorn. A fixed-wing aircraft (Cessna 185) was flown at 145 km/hr at 91.4 m above the ground to investigate detection and measurement errors. Four line transects were flown, using existing roads to mark the flight path ($L = 83.8$ km). The centreline of the transect was offset 60 m on both sides of the plane because of the lack of visibility below and near the aircraft (Fig. 7.8). Coloured streamers were attached to the wing struts of the aircraft to help the observer in delineating distance intervals (0–25), (25–50), (50–100) and (100–400) m. No marks were put on the window, thus the observer had only a 'front sight'. Neither the pilot nor the observer had experience in line transect surveys, although both had had experience with aerial strip transect sampling, and neither had knowledge of the number or placement of the bales. The observer did not have training in estimating distances. The performance of the observer on two assessments was reported.

In the first assessment, 59 bales were placed in the 0–25 m band to assess the observer's ability to detect objects on or near the centreline (which was offset 60 m). Here the observer detected 58 out of 59 objects in the first band (0–25 m), and the undetected bale was at 22.9 m. However, six of the 58 were recorded as being in the 26–50-m band. Worse, two bales were classed in the 50–100 m band and an additional two bales were classed in the 100–400-m band. Chafota (1988) suggested that possibly the aircraft was flown too low or went off the flight line during part of the survey, thus leading to the large estimation errors.

The second assessment employed 53 bales, including one outside the 400-m distance. The results are shown in Table 7.2. Here, the detection was quite good, as one might expect in the case of relatively large objects placed in short grass prairie habitat. Only one of the 53 bales went undetected (193.9 m). However, the tendency to exaggerate distances is quite clear. Chafota (1988) stressed the need for training in the estimation of distances, an effective pilot study and a carefully designed field protocol. We would concur with these recommendations and add the need for window marking to be used in conjunction with the streamers on the wing struts, an accurate altimeter to maintain the correct altitude, and a navigation system that allows accurate flight lines and positioning (see Johnson and

Table 7.2. Performance of an observer in detecting bales of wood shavings placed at known distances from the centreline in short grass prairie habitat (from Chafota 1988)

Distance interval (m)	Actual frequencies	Observed distance interval (m) 0–25	25–50	50–100	100–400	>400
0–25	21	8	12	1	0	0
25–50	14	1	3	10	0	0
50–100	12	0	1	9	2	0
100–400	5	0	0	1	3	0
>400	1	0	0	0	0	1
Recorded frequencies:		9	16	21	5	1

Lindzey unpublished). Chafota (1988) also offered insight into the effects of measurement errors on $\hat{D}$ from the results of Monte Carlo studies.

It is possible to record sighting distances and sighting angles as grouped; however, this procedure is not recommended except under unusual circumstances. Grouped sighting distance and angle data must be analysed either using a two-dimensional model for the distances and angles (Schweder *et al.* 1999) or by transforming the grouped sighting distances and angles into grouped perpendicular distances ('smearing'). Even when the sighting distances and angles are not grouped, rounding errors in the data cause problems. Angles have often been recorded to the nearest 5°, so that an animal recorded to have a sighting distance $r = 8$ km and sighting angle $\theta = 0°$ will have a calculated perpendicular distance of $x = 0$ km, when the true value might be $x = 350$ m or more. Since estimation of abundance depends crucially on the value of the fitted probability density for perpendicular distances evaluated at zero distance, $\hat{f}(0)$ (Burnham and Anderson 1976), the false zeros in the data may adversely affect estimation. The problem is widespread, and more than 10% of distances are commonly recorded as zero, even for land surveys in which distances and angles are apparently measured accurately (e.g. Robinette *et al.* 1974). Possible solutions, roughly in order of effectiveness, are as follows:

1. Record distances and angles more accurately.
2. Use models for the detection function that always have a shoulder.
3. Group the data before analysis.
4. 'Smear' the data (see below).
5. Use radial distance models.

Only the first of these solutions comes under the topic of this chapter, but we cover the others here for completeness, and to emphasize that solution 1, better survey design, is far more effective than the analytic solutions 2–5.

7.4.3.1 *Accurate measurement of distances and angles*

Improving accuracy in measuring angles and distances is certainly the most effective solution. It may be achieved by improving technology, for example by using binoculars with reticles (graticules) or range finders, and using angle boards or angle plates on tripods (Section 7.4.1). Most important is that observers must be thoroughly trained (Section 7.5), and conscientious in recording data; there is little benefit in using equipment that enables angles to be measured to the nearest degree if observers continue to record to the nearest 5°. Training should include explanation of why accuracy is important, and practice estimates of distances and angles for objects whose exact position is known should be made, under conditions as similar as possible to survey conditions. Where accurate measurements are known to be critical, observers should be tested, and those failing the tests should not be used.

7.4.3.2 *Use of models with a shoulder*

If many perpendicular distances are zero, a histogram of perpendicular distances appears spiked; that is the first bar will be appreciably higher than the rest. If, for example, the exponential power series model is fitted to the data, it will fit the spike in the data, leading to a high value for $\hat{f}(0)$ and hence overestimation of abundance. Models for which $g'(0)$ is always zero (i.e. the slope of the detection function at $x = 0$ is zero) are usually less influenced by the erroneous spike, and are therefore more robust. This does not always follow; if distance data fall away very sharply close to zero, then only very slowly at larger distances, the single-parameter negative exponential model is unable to fit the spike, whereas the more flexible two-parameter hazard-rate model can. If the spike is spurious, the negative exponential model can fortuitously provide the more reliable estimation (Buckland 1987b), although its lack of flexibility and implausible shape at small perpendicular distance rule it out as a useful model. Models that use the half-normal or uniform key tend to be relatively robust to rounding of distances to zero.

7.4.3.3 *Grouping the distance data*

If data are grouped such that all perpendicular distances that are likely to be rounded to zero fall in the first interval, the problem of rounding errors should be reduced. This solution is less successful than might be anticipated. First, interval width may be too great, so that the histogram of perpendicular distances appears spiked; in this circumstance, different line transect models can lead to widely differing estimates of object density (Buckland 1985). Second, the accuracy to which sighting angles are recorded often appears to be quite variable. If a detection is made at a large distance, the observer may be more intent on watching the object than recording data; in cetacean surveys, the animal may no longer be visible when he/she estimates the angle. Thus for a proportion of sightings, the angle might only be recorded to the nearest 10° or 15°, and 0° is a natural value to round to when there is considerable uncertainty. An attempt to impress upon observers that they should not round angles to 0° in minke whale surveys in the Antarctic led to considerable rounding to a sighting angle of 3° on one vessel!

Despite its limitations, prudent grouping of the data can salvage a line transect data set for which distances have been rounded to zero, provided that the histogram of grouped distance data is not unduly spiked.

7.4.3.4 *Smearing*

The concept of 'smearing' the data was introduced by Butterworth (1982b). Although often criticized, for example by Cooke (1985), the technique has become widely used for data from cetacean shipboard surveys. It is an attempt to reduce the effects on the estimates of recording inaccurate locations for detections, through rounding sighting distances and angles to favoured values. When rounding errors occur, the recorded position of an animal may be considered to be at the centre of a sector called the 'smearing sector' (Fig. 7.11); the true position of the animal might be anywhere within the sector. Butterworth and Best (1982) assigned each perpendicular distance a uniform distribution over the interval from the minimum distance between the sector and the centreline to the maximum distance, and selected a distance at random from this distribution to replace the calculated perpendicular distance. Hammond (1984) compared this with assigning a uniform distribution over the sector, selecting a new sighting distance/angle pair at random from the sector and calculating the corresponding perpendicular distance. He also investigated assigning a normal distribution to both the distance and the angle instead of a uniform distribution. He concluded that the degree and method of smearing had relatively little effect on estimation of $f(0)$, but that estimation under either method was improved relative to the case of unsmeared data.

If the calculated perpendicular distance data are grouped before analysis, it is unnecessary to sample at random from the assumed distribution within the smearing sector. For example if smearing is uniform over the smearing sector, the sector can be considered to have an area of unity, and the proportion of the sector within each perpendicular distance interval may be calculated. This is carried out for each observation and the resulting proportions are summed within each interval. They can then be rounded to the nearest integer values and line transect models applied in the normal way for grouped data. Alternatively the methods of Section 3.3.2 for grouped data follow through when the 'frequencies' are not integer, so

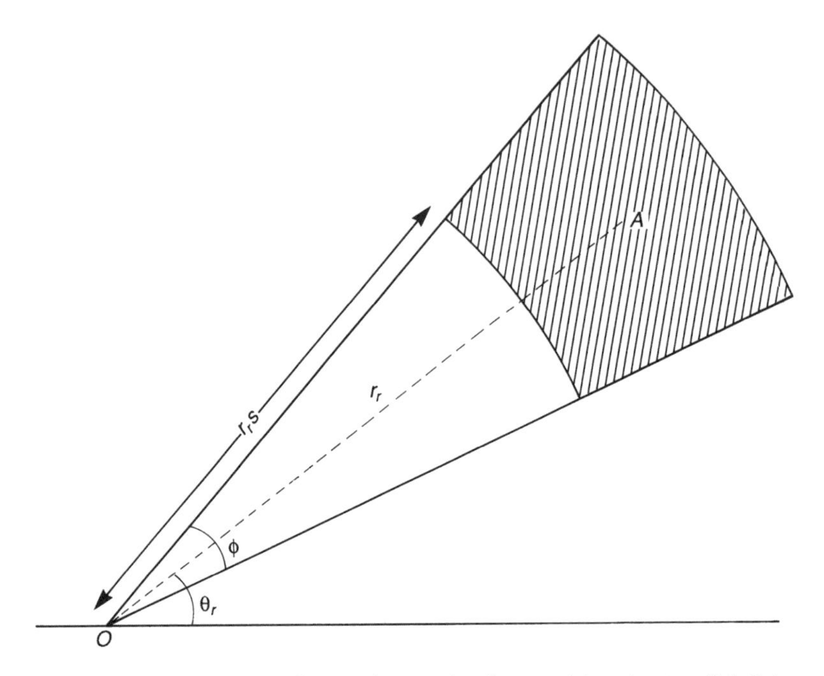

Fig. 7.11. The observer at O records an animal at position A, at radial distance r_r and with sighting angle θ_r. The true position of the animal is considered to be anywhere within the shaded smearing sector. The size of the sector is determined by smearing parameters ϕ and s.

that rounding is not required. This approach is described by Buckland and Anganuzzi (1988a).

Values must be assigned to the smearing parameters to control the level of smearing. Butterworth (1982b) incorporated time and vessel speed in his routine, since distance was calculated as speed multiplied by time taken to close with a whale or whales. The values for the smearing parameters were selected in a semi-arbitrary manner, by examining the apparent accuracy to which data were recorded. Hammond and Laake (1983) chose the level of smearing in a similar way, although the method of smearing was different; the semicircle ahead of the vessel was divided into smearing sectors so that any point within the semicircle fell in exactly one sector. Objects (in this case, dolphin schools) recorded as being in a given sector were smeared over that sector. Butterworth *et al.* (1984) used data from experiments with a buoy that was fitted with a radar reflector to estimate smearing parameters. None of these offer a routine method for smearing, whereas the angle and distance data contain information on the degree of rounding, suggesting that estimation of the smearing parameters from the data to be smeared should be possible. Buckland and Anganuzzi (1988a) suggested an *ad hoc* method for this. Denote the recorded sighting distance and angle by r_r and θ_r, respectively, and the corresponding smearing parameters by s (with $0 < s < 1$) and ϕ (Fig. 7.11), to be estimated. Then the smearing sector is defined between angles $\theta_r - \phi/2$ and $\theta_r + \phi/2$, and between radial distances $r_r \cdot s$ and $r_r \cdot (2 - s)$. Smearing is uniform over the sector, and grouped analysis methods are used, so that Monte Carlo simulation is not required (above). This is the method recommended by Buckland and Anganuzzi, although they also considered two improvements to it. First, rounding error increases with distance from the observer, so that a recorded distance of 1.3 km say is more likely to be rounded down to 1.0 km than 0.7 km is to be rounded up. This may be accounted for by defining the smearing sector between radial distances $r_r \cdot s$ and r_r/s. Second, there are fewer observations at greater perpendicular distances, since the probability of detection falls off. Hence smearing should not be uniform over the smearing sector, but should be weighted by the value of a fitted detection function at each point in the sector. The recommended method therefore has two identifiable sources of bias. One leads to oversmearing, and the other to undersmearing. Buckland and Anganuzzi concluded that the more correct approach did not lead to better performance, apparently because the two sources of bias tend to cancel, and considered that the simpler approach was preferable.

Buckland and Anganuzzi (1988a) estimated the smearing parameters by developing an *ad hoc* measure of the degree of rounding in both the angles and the distances. In common with Butterworth (1982b), they found that errors seemed to be larger in real data than the degree of rounding suggests. They therefore introduced a multiplier to increase the level of smearing and investigated values from 1.0 to 2.5. They noted that undersmearing was potentially more serious than oversmearing, and recommended that the estimated smearing parameters be multiplied by two, which would be correct for example if an angle between 5° and 10° was rounded at random to either endpoint of the interval rather than rounded to the nearest endpoint.

The above methods are all *ad hoc*. Methodological development is needed here to allow the rounding errors to be modelled.

7.4.3.5 *Radial distance models*

Because rounding errors in the angles are the major cause of heaping at perpendicular distance zero when data are recorded by sighting angle and distance, it is tempting to use radial distance models to avoid the difficulty. Such models have been developed by Hayne (1949), Eberhardt (1978a), Gates (1969), Overton and Davis (1969), Burnham (1979) and Burnham and Anderson (1976). However, Burnham *et al.* (1980) recommended that radial distance models should not be used, and Hayes and Buckland (1983) gave further reasons to support this recommendation. First, hazard-rate analysis indicates that r and θ are not independently distributed, whereas the models developed by the above authors all assume that they are. Second, hazard-rate analysis also suggests that if detectability is a function of distance r but not of angle θ, then the expected sighting angle could lie anywhere in the interval 32.7–45°, whereas available radial distance models imply that it should be one or the other of these extremes, or use an *ad hoc* interpolation between the extremes. Third, all models utilize the reciprocal of radial distances, which can lead to unstable behaviour of the estimator and large variances if there are a few very small distances. Fourth, despite claims to the contrary, it has not been demonstrated that any existing radial distance models are model robust.

A model might be developed from the hazard-rate approach (Schweder *et al.* 1999), but it is not clear whether it would be pooling robust, or whether typical data sets would support the number of parameters necessary to model the joint distribution of (r, θ) adequately. We therefore give a strong recommendation to use perpendicular distance models, and improve field methods to achieve accurate sighting distance and angle measurement.

7.4.4 *Cluster size*

Ideally, the size of each cluster observed would be counted accurately, regardless of its distance from the line or point. In practice, one may only be able to estimate the size of the clusters, and such estimates may be biased. There may be a tendency to underestimate the size of clusters at the larger distances, which leads to underestimation of $E(s)$, the mean size of clusters in the population, if $\bar{s}$ is used to estimate $E(s)$. On the other hand, small clusters are less likely to be detected at large distances than larger clusters, which generates bias in the opposite direction, and we cannot assume that the two biases cancel. Care is therefore required both in estimation of cluster sizes and in modelling the mean size of clusters in the population.

Survey design and conduct should attempt to minimize the difficulties in measuring cluster size. More than one observer may be required to make an accurate count of cluster size. Photography has been found useful in many aerial surveys, allowing cluster size to be determined accurately later. It may be possible to leave the centreline to approach the more difficult clusters, and hence obtain a more accurate count. Sometimes it may be reasonable to obtain estimates of average cluster size from the data in only the first few distance bands for which both size-biased and poor estimation of cluster size are less problematic.

If detections beyond distance w are truncated in the field, clusters should be recorded only if the centre of the cluster is within w of the line, but the size of detected clusters should include all individuals within the cluster, even if some of the individuals are beyond w. If the centre of a detected cluster is beyond w, it should not be recorded and no individuals in the cluster should be recorded, even though some individuals might be within the sampled area ($<w$). This is a crucial difference between line transect sampling and strip counts. In the latter, all individuals within the strip and none beyond are counted, irrespective of whether they are in clusters, and of where the cluster centre is. Sometimes, clusters extend over a significant area, and its centre can be difficult to determine. In such cases, it is important to develop clear, if arbitrary, criteria for determining where the cluster centre is. Without such criteria, there will be a tendency to underestimate the distance of the cluster from the line, because the individuals closest to the observer tend to be seen first, and their distance from the line recorded.

Cluster size may vary seasonally. For example, Johnson and Lindzey (unpublished) found that pronghorn populations split into small groups of nearly equal size in the spring, whereas much larger and more variable clusters were found during the autumn and winter months. Distance sampling surveys are more effective when there are many small clusters than when there are relatively few, larger clusters, because the number of detections for a given effort is larger, and the encounter rate, which generally makes the largest contribution to variance of the abundance estimate, is then estimated with better precision.

Calibration is sometimes used to correct recorded cluster sizes. For example, a helicopter may be used to verify the size of a sample of dolphin schools, using photographs. A ratio or regression estimator might then be used to correct the recorded sizes of schools that were not photographed. Such an approach should be used with caution if one of the regression methods of Section 3.5.4 is used to correct for size bias, because that method also corrects for underestimation of the size of clusters far from the line or point. In this case, calibration might need to be restricted only to those clusters detected close to the line or point. In the case of surveys conducted in 'closing mode', the observers (usually on a ship or in an aircraft) approach the detected cluster before estimating cluster size, thus avoiding the tendency to underestimate the size of more distant clusters.

7.4.5 *Line length measurement*

The development of the GPS and the recent removal of introduced error (selective availability) has made measurement of line length a trivial task for most surveys. Line length is easily computed from GPS measured positions of the line end points or by summing the lengths for a sequence of positions along the line. The length (L) in nautical miles between two positions expressed in degrees $[(\text{Lat}_1, \text{Lon}_1), (\text{Lat}_2, \text{Lon}_2)]$ is

$$\begin{aligned} L = 60 \cdot \cos^{-1}[&\sin(\text{Lat}_1) \cdot \sin(\text{Lat}_2) \\ &+ \cos(\text{Lat}_1) \cdot \cos(\text{Lat}_2) \cdot \cos(\text{Lon}_2 - \text{Lon}_1)] \end{aligned} \tag{7.40}$$

Typically, GPS positions are recorded as degrees (ddd), minutes (mm) and tenths (t) of minutes. Decimal degrees can be computed as ddd + mm.t/60. For example, $120°30.5'$ is equivalent to $120.5083°$. The length measurement can be converted to kilometres by multiplying L by 1.852 km/n.m.

Note that the default trigonometric functions of many computer programming languages and application packages assume that angles are measured in radians. Thus care is required to select functions for which angles are measured in degrees. Alternatively, latitude and longitude may be converted from degrees to radians by multiplying by $\pi/180$, and if the inverse cosine function of eqn (7.40) is evaluated in radians, it must be converted to degrees by multiplying by $180/\pi$ before multiplying by 60, to obtain L. As a check on your calculation, two positions at the same longitude but separated by one degree in latitude will be 60 n.m. apart. Likewise, at the equator, longitudes separated by one degree will be 60 n.m. apart.

Care is also required when applying eqn (7.40) for locations near the equator or meridian. The equator is readily handled by defining latitude to be positive say north of the equator and negative to the south. Similarly, the meridian may be handled by defining western longitudes to be negative and eastern longitudes to be positive. This solution works whether the two points are either side of the meridian at 0° or either side of the meridian at 180°. For example if $\text{Lon}_1 = -179°$ (=179°W) and $\text{Lon}_2 = +178°$ (=178°E), then $\cos(\text{Lon}_2 - \text{Lon}_1) = \cos(357°) = \cos(3°)$.

Even with the removal of selective availability, the positions are subject to an error as large as 10 m. Hence positions along the line should not be taken so frequently that measurement error significantly inflates the estimated line length. Line length measurements derived from GPS or map positions treat the observation surface as a horizontal plane. For land surveys of uneven terrain, the surface is not a plane, but a valid estimate of abundance will be obtained if detection distances, line lengths and area are all measured with respect to a two-dimensional projection of the study region (Section 7.9.4). When the study region is large, curvature of the Earth must be allowed for when estimating line length as well as the size of the study region. Distance measurements derived from GPS readings allow for this, but measurements based on flat maps are usually better replaced by ones derived from a GIS.

For surveys in which lines are relatively short, more direct measurements of line length (e.g. theodolite, cable and steel tape) may be more appropriate. Whatever method is used, the accuracy of the measurement is important. Any bias in line length measurement directly biases the density or abundance estimator.

7.4.6 *Ancillary data*

Ancillary data are useful to understand and describe the variables that affect abundance and distribution of the surveyed population and the variables that affect detection probability. The former are typically related to habitat-related features of the environment and can often be obtained from a GIS data source before or after the survey. In many cases, it may be necessary or useful to record the habitat variables during the course of the survey to provide a more accurate measure or to calibrate a measurement from a remote source (e.g. ice concentration and sea surface temperature).

Estimation of detection probability and density sometimes can be improved by incorporating variables through stratification of the data set or inclusion into the detection model (Section 3.7; Buckland *et al.* in preparation). The potential list of variables will depend on the particular survey and population. However, in general, environmental factors that affect visibility (e.g. glare and precipitation) and behavioural or other cues that lead to detections (e.g. blow or breach of a whale) should be considered. If more than one observer is used, the design should allow estimation by individual observer. In line transect surveys, it can be useful to partition and record the detections and cluster size by whether they are to the left or right of the centreline. Examination of these data may allow a deeper understanding of the detection process that might be useful in the final analysis. For example, for a given leg of a marine survey, glare may be worse one side of the line than the other, and such data potentially allow the effect of glare to be quantified.

In point transect surveys, it might be useful to record a sighting angle θ_i (where $0° \leq \theta_i \leq 360°$) for each detected animal. We might define 0° to be directly ahead of the direction of approach to the point by the observer (Fig. 7.12). Analysis of such angles could be used to identify a disturbance effect by the observer approaching the sample points.

We caution against recording too much ancillary data. Do not collect unnecessary data that might be only remotely useful. Ancillary data collection should be streamlined so that it does not hinder searching and collection of the primary measurements for abundance estimation.

7.4.7 *Data recording*

Ease, accuracy and reliability should be the goals of any method devised for data recording. Data recording may be as simple as paper and pencil or as sophisticated as computer data entry. However, sophisticated and highly technical methods should only be used when they are necessary or they improve the quality and quantity of data collected. Highly technical approaches are prone to failure and they are not useful if it means arriving back at the office without some or all of the data. A backup method is a good insurance policy. We offer some general guidelines on data recording

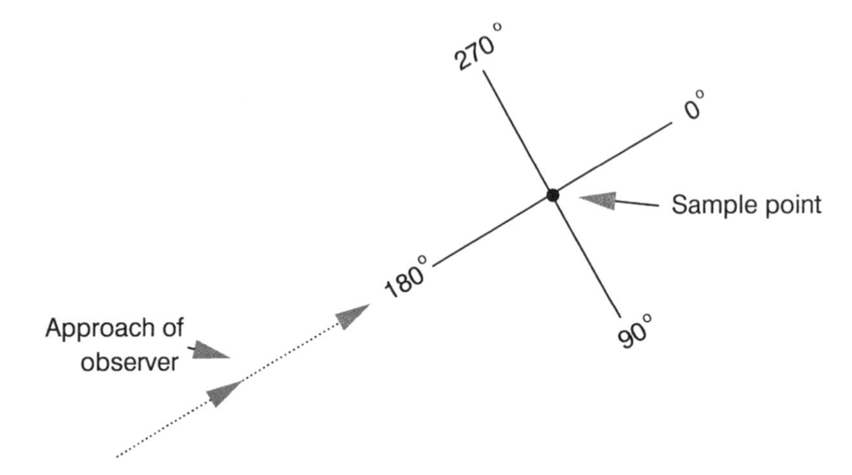

Fig. 7.12. Disturbance by an observer approaching a sample point can often be detected by recording angles θ ($0° \leq \theta \leq 360°$) where 0° is directly ahead of the observer's direction of approach. Thus, an angle is recorded for each object detected. These angles might be recorded by group (e.g. 45–135°, 136–225°, 226–315° and 316–45°).

but do not attempt to recommend any single method. The best method will depend on the particular circumstances of each survey, such as type of survey platform, behaviour of the study population and population density.

For many situations, a paper and pencil is both the easiest and most reliable method for data recording. Use a well-designed and structured form rather than a field notebook. This reduces the chances of forgetting to record key information on a detection, and provides an orderly method for transcribing the data into a computer. Figure 7.13 shows two examples of simple recording forms; another was presented by Burnham *et al.* (1980: 34). The format for such field forms will depend on the specifics of each survey, and a pilot survey is useful for identifying improvements. Note-taking on various aspects of the survey should be encouraged, and these can be recorded on separate sheets or in a field notebook.

The process of data recording should not detract from searching. Sightings are often clustered together, and sightings may be missed if the observer stops searching to record data. This is particularly problematic for aerial surveys. One solution is to use an accompanying person for data recording. An alternative is audio recording, to allow the observer to record data while searching continuously. There are three drawbacks to audio recording of data. Firstly, without a form or some other method to prompt for the data, it is quite easy to forget to announce some of the data, or announce it in an unintelligible way if the observer is interrupted or pressed

Study area ____________ Cloud cover (%) ______ Wind speed ______

Observer name ____________ Date ______

Line number ______ Line length (km) ______ Start time ______ End time ______

Sighting number	Perpendicular distance	Covey size	Number of Males	Number of Females	Number of Unknown
1	____	___	___	___	___
2	____	___	___	___	___
3	____	___	___	___	___
⋮					

Sighting number	Covey size	Perpendicular distance interval (m) 0 – 50	50 – 100	100 – 150	150 – 250	250 – 400
1	___	___	___	___	___	___
2	___	___	___	___	___	___
3	___	___	___	___	___	___
⋮						

Fig. 7.13. Two examples of a hypothetical recording form for a line transect survey of grouse. The example at the top is for taking ungrouped perpendicular distance data for coveys and the sex of covey mates as ancillary information. The example at the bottom allows for recording of covey sizes and grouped perpendicular distance data. Information on each line, such as its length and the proportion of that length in each habitat type, would be recorded just once on a separate form. Most surveys are somewhat unique, requiring specialized forms for use in the field.

for time. This problem can be largely avoided by developing and following a strict protocol on the words used and the order in which the data are announced. The second drawback is the very real possibility of equipment failure. Duplicate recording devices, a regular check of the equipment and a reliable power supply are some ways to minimize potential data loss. The final drawback is the nuisance of transcribing data from the recordings.

Voice-activated tape recorders are useful to avoid spending an equal number of hours back in the office transcribing tapes. It is wise to begin each data notation with an unnecessary word or two (e.g. 'record data' or 'sighting') to initiate recording and avoid losing the first few syllables if the recorder responds too slowly. Voice-activated microphones are not viable in situations with high-volume ambient noise, or if a complete recording is needed to synchronize the observations with another time-based data record.

Transcription time can be reduced without voice-activated microphones by recording the audio tape onto the computer as a WAV file and analysing it with WAV file editing software. The computer recording requires an equal number of hours as the survey but can be done in the background. After recording, the WAV file is displayed graphically and the data can be quickly transcribed by positioning the file at the peaks in the WAV file that represent voice-recorded data. This is quite effective as long as ambient noise is not so high that it hides the peaks. The WAV editing software provides a time stamp relative to the beginning of the recording and this can be synchronized to real time by announcing the time onto the tape in the field.

Transcription of data to a computer should be done promptly, so that missing or unreadable data can be rectified while the day's fieldwork is fresh in the observers' minds. The distance data should also be examined as soon as they have been entered into the computer, or even plotted out by hand while still in the field, so that problems with distance measurement, search patterns or animal movement prior to detection can be detected early and resolved, without compromising the entire survey.

In some cases it might be feasible to use a laptop computer or pen-based hand-held computer (Garrett-Logan and Smith 1997) to record data. As with audio recording, computer failure is a real possibility and a backup or duplicate recording method (e.g. audio recording) is advisable if it is very costly or not possible to repeat a survey. Computer software should be thoroughly tested before and during the survey to make sure that all data are being recorded correctly. A pilot survey is helpful to make certain that the data can be recorded accurately under field conditions. In aerial surveys, it is quite feasible to overload a person entering data onto the computer, and without a backup, data may be lost. While computer data entry avoids wasting time transcribing data in the office, a well-designed paper form may allow the observer to record more reliable data. Computers are particularly useful and essential in recording a continual stream of data directly from another device such as a GPS or altimeter in aerial surveys. Many GPS units have a limited capacity to record position fixes at specified time intervals. This can provide either a primary or backup method of recording positions that can be downloaded later into a computer. Observations can be linked to specific geographic positions if time is recorded with the observation data. An audio recording of observation data can be easily synchronized to a computer stream of GPS position data by announcing a time stamp from the computer clock or the GPS time onto the audio record. Audio or paper recording of the time and positions at the beginning, ending and intermediate times during the survey provides a backup should a computer or GPS fail to record positions. For audio recording of data, it is also useful to announce intermediate and ending time stamps to ameliorate problems in synchronization if the audio tape stretches.

Video cameras can be used to provide a continuous record of the surveyed habitat or as an alternative method of observation which can be reviewed after the survey. A video camera has been mounted in an aircraft to record the area near the line in pronghorn surveys in Wyoming (F. Lindzey, personal communication). The video can be studied after the flight in an effort to verify that no animals were missed on or near the line, or it can be used as the primary method of counting animals in the strip beneath the aircraft that is not visible to the observers. Bergstedt and Anderson (1990) used a video camera mounted on an underwater sled pulled by a research vessel to obtain distance data. Video cameras can also be used simultaneously as a primary or backup audio recording device with a continuously recorded time stamp.

7.5 Training observers

Large-scale surveys usually employ several technicians to gather the data to be used in estimation of density. This section provides some considerations in preparing technical staff members for their task.

Perhaps the first consideration is to interest the staff in the survey and its objectives, and to familiarize them with the study area and its features. A clearly defined survey protocol should be developed and discussed thoroughly by all team members. The protocol should cover all aspects of the survey, including searching methods, species identification, distance measurement, data recording, etc. Again, a small-scale pilot survey will be highly beneficial. People with prior experience are helpful to a team effort. Discussions held at the end of each day of surveying can be used to answer questions and listen to suggestions. A daily review of histograms of the incoming data may reveal problems to be corrected (e.g. heaping).

In some cases, it may be possible to use model animals to train observers in many aspects of distance sampling. Styrofoam models of desert tortoise have been used in Nevada, USA. In that study, estimates of $g(0)$ were made; further details are given by Anderson *et al.* (in press).

Observers must be carefully trained in species identification and become familiar with relevant information about the biology of the species of interest. Particular attention must be given to activity patterns and calls or songs, or other cues of the species. Some time in the field with a good biologist is essential. A training period should be used to validate identifications made by each observer. For bird surveys, Ramsey and Scott (1981a) recommended that observers' hearing ability be tested, and those with poor hearing be eliminated. The same applies to those with poor vision. If most animals are located aurally, then the assumption that they are not counted more than once from the same line or point may be problematic. If for example a bird calls or sings at one location, then moves unseen by the observer to another location and again vocalizes, it is likely to be recorded twice. Training of observers, with warnings about more problematic species, can reduce such double counting.

Training of observers in measuring or estimating distance is essential. It is particularly difficult to estimate distances to purely aural cues; Reynolds *et al.* (1980) used an intensive two-week training period, during which distances to singing or calling birds were first estimated and then checked by pacing them out or by using range finders. This was done for different species and for different cues from a single species. In marine surveys, a radar-reflecting buoy provides a good target for training and testing in distance measurement. Observers should be trained and tested even when measurement devices such as reticle binoculars or clinometers are used. Technicians should have instruction and practice in the use of all instruments to be used in the survey (e.g. range finders, compass, GPS and two-way radios).

Basic safety and first aid procedures should be reviewed in planning the logistics of the survey. In particular, aircraft safety is a critical consideration in aerial survey work (e.g. proper safety helmets, fire-resistant clothing, fire extinguisher, knowledge of emergency and survival procedures). Radio communication, a good flight plan for aerial surveys, and an emergency locator transmitter (ELT) are important for surveys in remote areas or in rugged terrain. Fortunately, many conservation agencies have strict programmes to help ensure aircraft safety for their employees.

7.6 Aerial surveys

Surveying from an aircraft is advantageous in many circumstances. A large area can be surveyed quickly in a short time. In some cases, an aerial view can also provide better visibility through vegetation or under water. The potential problems of animal movement are often avoided because the aircraft speed is typically much greater than any target animal. However, it is these same advantages that provide a unique set of problems and challenges to satisfying the assumptions of distance sampling, in particular that $g(0) = 1$. The flight altitude can often put a non-trivial distance between the object and the observer. In some cases, objects on the line are not visible because flat side windows do not provide downward visibility. Even with downward visibility, vegetation or water surface conditions may obstruct visibility. While reducing the problems with animal movement, flight speed also limits the time for detecting and identifying objects and recording data.

We describe some pitfalls of aerial surveys that we have observed and provide some general recommendations for aircraft characteristics and survey protocols. We realize that each particular aerial survey has its own unique characteristics and we do not attempt to prescribe a single protocol. However, we believe too little attention is paid to the details of applying line transect sampling from aircraft and the resulting analysis is therefore fraught with problems. Quang and Lanctot (1991) have even suggested an alternative approach for analysing line transect data derived from aerial surveys to cope with these specific problems. We believe that many of the problems are preventable with good field technique, and that analytical solutions should be used only as a last resort.

Surveys with locally high densities provide a particular challenge for aerial surveys. Surveys of only one side of the transect have been used to cope with high density or aircraft weight limitations. We discuss those issues in Sections 7.9.2 and 7.9.3, because they are also problems with other survey platforms. Here we focus on survey characteristics, search protocol and distance measurement issues as they pertain to aerial surveys.

7.6.1 *Aircraft and survey characteristics*

Fixed-wing aircraft, helicopters, microlights and even airships (blimps) (Hain *et al.* 1999) have proved useful as aerial survey platforms. However, some types of aircraft are far better for biological surveys than others (Fig. 7.14). The ideal survey aircraft would be able to fly slowly, manoeuver easily, and provide unrestricted forward and downward visibility. Engine noise, range, capacity and cost may also influence the usefulness of an aircraft for a particular application. We will focus here on visibility and how it affects the assumption that all animals are detected on the line.

Visibility from an aircraft varies depending on the size, shape and position of the windows. These factors determine whether downward visibility is restricted, and define the viewing area and hence the length of time an area is in view for a given speed. A view forward of the aircraft is useful to provide ample time to sight and identify objects. A side view is needed to obtain measurements of perpendicular distance. A downward view is needed to enable animals near the line to be detected. A flat side window provides better forward visibility if it is wider and better downward visibility if it is closer to the bottom of the aircraft. Most flat side windows are not optimal for surveying because the bottom of the window is often located just below shoulder level, which provides a better view of the horizon and sky, and restricts the downward viewing angle to a maximum of 55–65° below the horizon. For an aircraft flying at an altitude of H m,

Fig. 7.14. Some aircraft are specifically designed for aerial observation. Note the forward and lateral visibility and the high wing on the aircraft. Helicopters, while more expensive, often provide similar advantages plus the ability to hover or proceed more slowly than a fixed-wing aircraft.

a restricted viewing angle of $\theta°$ will restrict the view to a strip of width $2 * H * tan(90 - \theta)$. For example, if $H = 200$ m and $\theta = 65°$, the obstructed strip will be 186 m (93 m on each side of the line), and 280 m for $\theta = 55°$.

If downward visibility is restricted (e.g. flat side windows), it is possible to exclude the obstructed-view area from the analysis with left-truncation. However, this is not optimal for several reasons. The results of left-truncation can be quite dependent on the model if there is not a shoulder in the detection curve for the data that are observed. Even if there is a shoulder, the problems that plague one-sided transects (Section 7.9.3) can also apply here. The width of an obstructed-view strip will vary depending on the height of the observer. Thus it is absolutely critical that the delineation (e.g. window and strut marker) is clear and accurate. Observers should be allowed to record an observation even if it is within the obstructed-view strip (Section 7.3). Movement in and out of the obstructed-view strip can be problematic if it often occurs while the ground just outside the strip is in view. These problems can be overcome in some situations, but unrestricted downward visibility is preferable. If bubble windows or a belly port is not an option, a down-looking video camera may be useful as a replacement.

We have found that a reliable configuration for aerial surveys of marine mammals is to place observers at side bubble windows and one observer at a belly port. For surveys flown at an altitude of 183 m (600 feet), the observer at the belly port can typically observe 100 m on either side of the line, which means that side-window observers can restrict their downward viewing angle to 65–70° or higher, which relieves some neck strain. A combination of flat side windows and a belly port can also provide complete viewing coverage, although overlap between observers is useful. Using observers only at bubble windows can provide unobstructed viewing. However, regardless of the window configuration, the searching pattern of the observers is crucial, as we discuss below.

Within the capabilities of the aircraft, there is some latitude for choosing both an altitude and a speed for the survey. Any choice involves trade-offs which affect the time available to detect and identify the target animals and record the data, and survey coverage. The optimal values for a survey will depend on the target animal and the aircraft. In selecting an altitude and survey speed, it is important to consider how the choice may influence whether detection near the line is certain, whether animals are disturbed and move prior to being detected, and how well distance can be measured.

An appropriate speed and altitude above the ground can be selected after some preliminary survey work. The choice of speed and altitude balance survey coverage area and detection probability. If it is assumed that $g(0) = 1$, then the choices should be made to satisfy this assumption, with survey coverage as a secondary concern. A higher altitude provides more time to view an area because the distance within view increases in proportion to altitude, but the perceived size of the object is reduced, and detection and discrimination will be more difficult. The optimal altitude will depend on the size of the object, sighting and identification cues, visibility conditions and how easily an object can be discriminated from other like-species or objects. If the altitude is lowered, a concomitant decrease in speed is needed to maintain the same viewing time. In some situations, it may be necessary to increase altitude to prevent movement prior to detection, and in other applications, a lower altitude may be useful if the movement increases the observer's ability to detect or identify (Erickson *et al.* 1993). As an example, an altitude of 120–180 m is used for small schools of dolphins or porpoise, but large whales are often surveyed at an altitude of 300 m.

Flight speed affects the survey range and the amount of available viewing and handling time for an observation, although circling can provide additional recording time in many surveys. Helicopters and airships have the advantage of being able to adjust speed, but for fixed-wing aircraft, the slowest possible flight speed is typically the best choice, and an altitude adjustment is needed to increase the viewing time any further. Flight speed can become important in areas of high animal density, if the observers reach saturation (Section 7.9.2).

Both the animal and its habitat, and the aircraft and its flight characteristics, influence the observer's ability to detect all animals near the line ($g(0) = 1$), yet there are only a few limited options for changing these factors. In some cases, they may determine whether aerial surveys are even possible. However, the observers and their searching protocol (pattern) may be controlled, which will certainly influence whether $g(0) = 1$.

Peel and Bothma (1995) assessed how three different aerial platforms compared for surveys of impala: helicopter (which proved most accurate), microlight and fixed-wing aircraft. Grigg *et al.* (1997) used an ultralight aircraft to survey kangaroos. We anticipate that microlights, ultralights and remote-controlled model aircraft will see increasing use in wildlife surveys. Model aircraft especially, coupled with video equipment, offer the potential for inexpensive surveys, or a second platform for shipboard surveys to confirm cluster sizes or species identity, without putting observers and pilots at risk.

7.6.2 *Search and survey protocol*

Searching protocol in an aerial survey is particularly important in ensuring that $g(0) = 1$ because animals are in view for a relatively short time period. Consider an observer searching forward and directly below the aircraft through a belly port, while surveying at 90 knots (46.3 m/s) and 150 m above the ground. If the angle of view forward along the line was 63° (e.g. achieved with an observer's eye 15 cm above a window 30-cm wide), a target will be in the 300-m field of view for 6.5 s. This is a very short time, but is probably sufficient to detect and identify most objects that are visible. However, if the observer is looking through a side bubble window, they can search from the line to the horizon. With the above parameters, if an observer spends more than 6.5 s searching away from the vicinity of the line, that area will not be searched and may result in missing an observation. Thus, it is essential that observers are constantly searching, and they should not stop searching to record data or for any other reason during an active survey. Aerial survey data should always be recorded by another person or via tape recorder.

Positioning observers at side bubble windows with downward visibility will not guarantee that $g(0) = 1$. Aerial observers positioned at side windows tend to concentrate their search effort away from the line. If they do, they will almost surely miss animals near the line. The aerial survey data described by Quang and Lanctot (1991) demonstrate this problem in which the peak of the distance histogram is away from the line. To improve the searching protocol, it is important to consider why an observer positioned at a side window in an aircraft often concentrates away from the line.

Firstly, it is mentally and physically easier to view from about a 45° downward angle out to the horizon because it reduces neck strain and reduces the relative motion of the ground (e.g. compare focusing on fence-posts along a road out the side-window of a car relative to the horizon). Second, an observer is rewarded by more detections at further distance. This may seem like a contradiction to the concept that targets are more difficult to detect at distance, but it is not. Within a short period of time, an observer can only effectively search a field of view of some angle, θ. For example, assume that $\theta = 10°$. At an altitude of 150 m, an observer who searches at downward angles between 40° and 50° from the horizon will view a linear distance of 53 m, while an observer searching between 30° and 40° would view 81 m and would see 50% more sightings if detection probabilities were equal.

The solution is obvious but not necessarily easy to accomplish. If there is only a single observer on each side, the observers must spend most of the time searching in the vicinity of the line. They should search forward and continually scan, so as to avoid being hypnotized and staring but not searching. The observers need to understand that seeing everything close to the line is more important than seeing many objects. However, observers are human, and more direct measures may be more successful. One approach is to darken the top portion of the window, which would block the observer's field of view to some chosen width. An alternative is to use an additional observer positioned at a belly port, or a down-looking video camera. Data from the different platforms might simply be pooled, or treated as double-platform data to allow estimation of $g(0)$ (Section 7.9.5). Laake *et al.* (1997) found that observer experience influenced the resulting value of $g(0)$ for harbour porpoise. They suggested that, if a third observer is positioned at a belly port, it should be the most experienced observer, to ensure that $g(0)$ is as close to one as possible. Less experienced observers can gain experience observing from the side windows.

The idea proposed by Quang and Lanctot (1991), for which certain detection is assumed to occur at some distance greater than zero, and left-truncation both have merit in some cases because the detection curve may not decrease monotonically, even with an adequate search protocol. If the objects are not continuously available to be seen (e.g. underwater or blocked by vegetation), and the observer's forward field of view is more limited close to the line, the chances for detection (i.e. time available to be seen) will increase with distance, while the probability of detecting available objects at a given moment decreases with distance. The resulting data may be non-monotone, as demonstrated by Hain *et al.* (1999) for right whale (*Eubalaena glacialis*) observations. In these situations, it is unlikely that $g(0) = 1$, and some adjustment will be necessary for the objects that are unavailable to be detected (Sections 6.3 and 7.9.5; Buckland *et al.* in preparation).

7.6.3 *Distance measurement*

In aerial surveys, perpendicular distance can be measured by one of three methods: angle of declination, geographical position and visual estimation. We are unaware of any plausible situation in which visual estimation would be preferred or required, and recommend against its use in aerial surveys. Use of GPS to measure the position of a sighting relative to the line requires flying over the observation to obtain its position. This is not precise for short distances, and will only be effective if the density is very low; even then, it wastes valuable flight time. However, it may be necessary in very uneven terrain. Measuring the angle of declination and computing perpendicular distance from a known altitude (Section 7.4) is a more direct and efficient method, but is not without its problems. Survey lasers may provide a more effective option for some surveys, especially in hilly terrain.

The angle of declination can be measured with a clinometer, to derive an ungrouped perpendicular distance (Section 7.4.1.1), or distances can be measured in intervals (grouped) using specific angle intervals (Section 7.4.1.2). The latter can be accomplished by aligning two sets of marks that define the angle. The marks can be on the window and an airplane strut (Guenzel 1997). For aircraft without struts (e.g. helicopters), the marks could be placed on a three-dimensional plexiglass sighting triangle attached to the inside of a window (Bengtson *et al.* 1995), or on a bubble window and a plexiglass strip mounted on the inner window frame. Two sets of markers are required (like the front and rear sight on a rifle) to define the declination angle reliably. It is insufficient to use a single set of marks, which requires the assumption that the observer holds their head at a particular position to define the correct angle, because small shifts in head position can yield large errors in distances. Observers should be cautioned not to assign objects to distance intervals until they are nearly perpendicular to the aircraft. If such assignment is attempted while the object is still far ahead of the aircraft, there is a tendency to assign the object to a nearer distance interval. (This problem is related to parallax.)

To compute the distance or distance interval, both approaches require that the altitude is known. They also assume that the aircraft is level, and the heading is parallel and directly above the imaginary line on a flat surface. Aircraft are not always perfectly level and parallel during flight. Deviations are termed pitch (forward/backward tilt), yaw (side-to-side tilt) and crabbing (turned to the left or right of the flight direction). If there is any pitch, it is typically only a few degrees, and will not influence measurement of declination angle. A consistent yaw will decrease distance measurements on one side of the aircraft and increase measurements on the opposite side. Pilots can typically minimize yaw, but turbulence may cause bouncing and inaccurate readings. Theoretically, clinometers will be not be biased by yaw, but bouncing and gravitational forces will make readings less reliable. Measurements of any kind will not be accurate and should not be attempted during flat or banked turns. Pilots will use crabbing to maintain a heading in a cross-wind. If observers are positioned at side windows and crabbing is towards the right, the right-side observer will be at a disadvantage, because observations close to the centreline are in the field of view for less time, appearing suddenly from underneath the aircraft. The left-side observer will have a better field of view, and should warn the right-side observer of objects that may appear on the right-side. Observers can minimize the impact of crabbing by making distance measurements when objects are perpendicular, and by communicating with each other to avoid double counting. Measurement error of perpendicular distance only becomes substantial ($>$15%) when the crabbing angle exceeds 30°, at which time flying is not particularly safe.

An accurate altitude measurement is essential in computing perpendicular distance from the declination angle. Altitude can vary for measurements from a clinometer as long as altitude is constantly measured, because each angle is translated to an ungrouped distance. Some altimeters have the capability to download a constant stream of altitude measurements into a computer. However, altitude must remain constant or nearly so to use a single set of distance intervals from data collected as angle intervals (e.g. strut markers). It is theoretically possible to analyse grouped distances using different sets of intervals, but the current version of program Distance assumes a single set of intervals.

Hilly or mountainous terrain (Section 7.9.4) can be problematic even if altitude is constantly monitored. Objects that are at a higher elevation than the line will be measured at a smaller declination angle and greater distance than the equivalent flat surface or horizontal separation distance. Likewise distance will be underestimated for objects at a lower elevation than the line. In very uneven terrain, declination angles will not provide reliable measurements and an alternative approach should be used, such as a survey laser that automatically calculates the horizontal distance to a detected object. Trenkel *et al.* (1997) found that such a survey laser was very effective for estimating the distances of detected red deer (*Cervus elaphus*) from the line during helicopter surveys in hilly terrain.

In general, a method which measures an angle interval is preferable to using a clinometer, with the exception of surveys with low density and variable altitude. It is much easier to observe and record an angle interval, especially when several objects are detected nearly simultaneously. A nearby object is likely to be missed when measuring the angle with a typical clinometer used in forestry applications. Classifying locally high densities of objects into angle intervals is much easier without any interruption in searching.

7.7 Marine shipboard surveys

Ships and boats are often used for surveys of marine mammals and seabirds. Ships can be used to survey large areas of the ocean that are often inaccessible to land-based aircraft, and they travel more slowly, which provides more time to detect objects. Marine mammals are intermittently below the water surface; hence, they are more likely to be seen at the surface within the observer's field of view when the observer is travelling slowly. Also, the distance between the observer and the object can be less from a ship than from an aircraft, which enhances the observer's ability to detect and identify small objects such as seabirds.

These advantages of a shipboard survey may also create difficulties. In particular, both marine mammals and seabirds may move in response to an approaching ship. Responsive movement and in particular attractive movement such as bow-riding by Dall's porpoise can create considerable positive bias (Buckland and Turnock 1992). Even non-responsive (random) movement biases density unless the objects travel slowly relative to the speed of the ship (Yapp 1956; Hiby 1986; Hiby and Hammond 1989). Random movement is particularly important in the case of seabirds, which can travel at speeds much greater than typical vessel speeds. Section 6.5 outlines methods for coping with fast-moving objects such as seabirds.

Marine mammals are only visible when they are at the surface to breathe, and for many species, it is unlikely that $g(0) = 1$. Diving seabirds present a similar problem of lesser magnitude. Methods for coping with uncertain detection on the trackline and movement are described in Buckland *et al.* (in preparation). Objects that are not continuously available to be seen present difficulties in estimating $g(0)$ as described in Section 6.3. Here we describe some potential problems with ship surveys and provide some general recommendations for vessel characteristics and survey protocols.

7.7.1 *Vessel and survey characteristics*

As a survey platform, an ideal vessel would provide a stable viewing platform at a sufficient height above the surface of the water. Stability will improve viewing conditions and precision of distance and angle measurements. Platform height will affect visibility, stability and measurements of distance based on declination angles from the horizon. The optimal platform height will depend on the particular survey conditions. Surveys of Antarctic seals in the pack-ice conducted from a 32-m tower on an icebreaker provided good viewing conditions to detect seals that are often hidden by hummocks in the ice (Bengtson *et al.* 1995). A 32-m platform was stable for surveys through ice, but for most ships, a lower height is necessary for marine surveys unless the sea is calm. The platform height should be as high as possible without unnecessarily compromising stability.

As platform height decreases, the effective viewing range will decrease as birds or marine mammals are hidden by waves. For lower platform heights, errors in distance measurements from binoculars with reticles will become more prevalent because of changes in height due to ship movement through the waves, and imprecision in measurement of small declination angles.

Size, speed and engine noise may also influence the suitability of a ship as a survey platform. Speed and engine noise may affect the amount of responsive movement and the distance at which marine mammals and birds detect the vessel. The vessel must be of sufficient size to carry the complement of observers and data recorders. Small rigid or inflatable boats are used effectively in surveys of coastal seabirds that are typically detected within 200 m of the boat and distances can be reliably estimated by eye. The crew typically consists of one or two observers and a data recorder. However, depending on its size and layout, a small boat may constrain options for conducting independent observer surveys to estimate $g(0)$ and possible effects of movement (Section 7.9.5). Reliable distance estimation by eye is also notoriously difficult to achieve, and requires considerable training.

7.7.2 *Search and survey protocol*

Detection of objects before they respond to the vessel should be the focus of search protocols (Section 7.3) for shipboard surveys. High-powered binoculars may be required if marine mammals or birds respond at long distances from the vessel. Marine mammal surveys typically use two observers, with each observer primarily covering one side of the line from 45–60° to 5° or 10° on the opposite side of the line. A data recorder is often used as an additional observer (with or without hand-held binoculars) to search closer to the ship. In this case, care must be taken to ensure that search effort falls smoothly with distance from the line; otherwise, if a significant proportion of objects are missed by the main observers but detected by the data recorder, a detection function similar to Fig. 2.1 might result, and subsequent analysis will be unreliable.

The focus of search should be forward, but this does not mean that objects seen at angles greater than 45–60° should be excluded. If objects are not responding to the vessel, it is valid to include objects that are first detected after they have passed abeam. However, these sightings should be the exception, because in most instances marine mammals or birds will have responded and moved from their original position. In particular, several species of dolphins and porpoise will approach the vessel to swim in front of the ship (bow-riding). Sightings of bow-riding dolphins should not be included unless they were seen prior to their approach towards the ship.

Surveys of marine mammals such as whales and dolphins are often complicated by the need to estimate cluster size (Section 7.4.4) and identify species for detections at long distances. Ship surveys are typically conducted using 'closing mode' (Section 7.8.7) because species identification and cluster size estimates are improved (Barlow 1997). Seabird surveys are conducted in 'passing mode' because detection distances tend to be smaller, seabirds stay in view longer, and densities may make closing mode impractical. Seabird surveys sometimes encounter problems in areas of high density, particularly near coastal colonies (Section 7.9.2). One-sided transects have been used in some surveys to cope with high densities and multiple species (Spear *et al.* 1992), but we recommend other approaches (Section 7.9.3).

The effectiveness of a marine ship survey depends in large part on the environmental conditions encountered during the survey. Wave and swell height and the presence of white-capped waves can severely reduce visibility both laterally and near the line. Visibility is often indexed by the Beaufort scale, which is a measure of wind speed based on features of the sea state such as wave height and the prevalence of 'white caps'. Swell, glare and fog can also reduce visibility. Reduction in lateral visibility is accommodated by the detection function, and visibility measures can be used as covariates in the detection function (Beavers and Ramsey 1998; Marques and Buckland in preparation), but in many cases, these conditions may reduce $g(0)$ well below one. Surveys with independent observers can estimate $g(0)$ and can accommodate covariates such as sea state (Buckland *et al.* in preparation). However, $g(0)$ cannot be adequately estimated under poor visibility conditions if too few detections are made, and suspending the survey may be the best choice.

7.7.3 *Distance measurement*

In most ship surveys, perpendicular distance must be computed from a measured radial distance and angle, and this has often been problematic. Accuracy of distance and angle measurements should be a primary focus in a ship survey. Particularly important are cases where the detected object is some distance ahead of the observer but at a small sighting angle; every effort should be made to measure the angle accurately. Sections 7.4.1–7.4.3 discuss the relevant issues and some suggested measurement techniques. Calibration and testing are important components that should not be overlooked even when distance and angle are measured with devices such as binoculars with reticles. By making measurements of known distances, the platform height can be verified. The platform height is the distance from the water surface to the observer's eye, so for vessels with low platforms, it may be advisable to take account of each observer's height. Also, by calibrating with known distances, the observer's use of the devices can be tested. Testing and calibration are essential when distances are estimated visually. Angles should always be measured at least with an angle board. There is no reason to estimate angles by eye. GPS and radar-reflecting buoys are useful devices to make measurements of known distances for use in distance and angle estimation experiments.

7.8 Land-based surveys

Land-based surveys are very diverse in character, so that it is difficult to give many general guidelines. However, a number of issues can arise that are specific to land-based surveys, and we discuss some of them here.

7.8.1 *Surveys of small objects*

Line transect surveys are being increasingly used for surveys of small objects, such as dung piles (Section 6.2) or flowering plants, that may only be visible within a metre or two of the line. A common problem with such surveys is that a large percentage of detections (commonly up to 40%) is often recorded as on the line. Reliable modelling of the distance data, with so many zeros, is not possible. There are two main reasons for this phenomenon. The first is that the line is not sufficiently well-defined, so that when an observer detects an object within a few centimetres of the line, he or she cannot accurately measure its distance, and so records it as on the line. The second reason is that the objects themselves have size; an elephant dung pile or a pellet group deposited by a moving deer may extend over 40–50 cm. When most detections are within a metre of the line, a substantial proportion of detected objects will actually intersect the line. If these are recorded as distance zero from the line, the data are unanalysable. Solutions to both problems are straightforward, yet even well-trained observers frequently fall foul of them.

7.8.1.1 *Locating the line*

When lines are short, as they typically are for small objects, they might be readily marked with string or rope. The process of marking the line and measuring it can be made faster if the observer pulls a rope or cable of known length. Marques *et al.* (in press) adopted this approach for surveys of sika deer (*Cervus nippon*) pellet groups. The survey design comprised a series of parallel lines through the survey region. Along each line, 50-m surveyed sections alternated with gaps. The length of the gaps equalled the space between successive parallel lines, so that the 50-m sections provided a systematic grid of transects throughout the study region. This enabled the 50-m transects to be treated as sampling units, to give reliable bootstrap estimates of variance. The size of the gaps was adjusted to achieve the required amount of effort in each forest block. A cable of length 50 m was pulled, both to mark the 50-m transects and to measure the gaps between successive transects. On the surveyed sections, the distance from the centre of any detected pellet group to the cable was accurately measured.

If it is not practical to mark the line, but two observers are available, the following strategy may prove effective. The two observers walk in tandem, with one a few metres behind the other, and too far back to see any object detected by the front observer. The rear observer is responsible for directing the front observer, for example by following a compass bearing, or by sighting on a distant object. The front observer carries a measuring pole. When he or she detects an object, the pole is held vertically, and the rear observer directs movement of the pole until it is in line with the direction of travel. In this way, the location of the line has been precisely defined (by where the pole touches the ground) without the rear observer knowing the location of the object. This avoids bias, and the distance of the object from the line can now be measured simply by rotating the measuring pole to the horizontal, while keeping the lower end's location fixed, and reading off the required distance.

Another possibility for marking a line is to have a helper, or in suitable habitat a vehicle travelling at walking pace, move along the line, towing a rope. The observer walks behind, and measures the distance of any detected object from the (moving) rope. In this case, it is crucial that the line is located independently of the positions of objects of interest.

7.8.1.2 *Measuring distances from the line*

If objects are of sufficient size that a significant proportion of detected objects intersect the line, as occurs for example with elephant dung piles, then it is important that an objective criterion is devised to define the centre of any object encountered. The criterion need not give a good approximation to the centre of gravity of the object, but it does need to be simple and unambiguous to apply. For example, for elephant dung piles, the distance of both the nearest edge and the farthest edge of the dung pile might be measured, and the average of these two taken as the distance of the dung pile from the line. If the dung pile intersects the line, one of these distances should be negative (say the edge to the left of the line) and the other positive (to the right). The two distances are averaged, and the sign of the average ignored. (This ensures that a dung pile whose centre really is on or very near the line can be recorded as such.) With pellet groups, a similar strategy can be adopted by measuring to the outermost and innermost pellet – or the two outermost pellets, if the pellet group straddles the line. The criterion used should be written down, and observers should be trained in its application, and in its importance for subsequent analysis of the data.

7.8.2 *Stratification by habitat*

In many surveys, both density and detectability of animals are functions of habitat. If habitat can be accurately recorded before the survey, a stratified design may be used. However, in land-based surveys, it is common to record habitat variables at the same time as the survey is conducted. These variables are often of greater relevance to density and detectability than is available prior to the survey. The simple option is then to design the survey unstratified, collect the data, and subsequently post-stratify both the effort and the detections, to estimate habitat-specific detection functions and densities. This approach also allows modelling of the detection function and density as functions of covariates (Buckland *et al.* in preparation).

It is possible to stratify the design, to allow greater effort in areas of high density, and then to gather better habitat data, to produce refined strata at the analysis stage. However, the analysis must respect the stratification built into the design, and quickly becomes very complex (Marques *et al.* in press).

7.8.3 *Permanent transects and repeat transects*

When the primary purpose of a survey is to monitor changes in abundance over time, greater precision is obtained by covering the same sample of transects (whether lines or points) at each visit (typically once a year). This is because variation in density by spatial location is removed from successive estimates, and so there is greater statistical power for detecting trends.

Transects should be permanently marked, to ensure that they are accurately relocated on each visit. One method of doing this is to mark points by driving pegs or metal stakes into the ground that have a short piece of brightly coloured rope or material attached. For lines, pegs can be used at equally spaced points along each line.

If the full range of a colony or population is being monitored, then it is important to have a design which spans a wider area than the population currently occupies, so that expansion of the range is identified. Without such a strategy, the rate of increase for an expanding population will be underestimated. Once the population expands into this wider buffer zone, the zone may need to be widened further.

There is merit in a survey scheme that retains a proportion of transects on each survey occasion but replaces others by new transects. This provides some continuity over time, for assessing trend, but also gradually widens the spatial cover of survey effort, which allows better estimation of average abundance through the survey period, and increases the information for fitting spatial and spatio-temporal models of density.

In circumstances where permanent transect lines must be cut (Section 7.8.4), it is important that the creation of the line does not affect the study population, as otherwise trends in abundance will be biased. For example, the animals may use the cut transect as a path, or it may allow access to previously inaccessible resources. It may also give access to hunters or others that might disturb or otherwise impact the study population, and hence bias the trend estimates. Marking of permanent lines should also be done with care if it might contribute to additional disturbance or hunting pressure. In relatively open habitat, a limited amount of clearing can provide a route for the observers, while having minimal impact on the study population. In dense habitats, the effects of establishing permanent cut transects are potentially much greater, and should be carefully assessed.

In many line transect studies, especially ones in which transects must be cut (Section 7.8.4), 'semi-permanent' transects are established. These are then surveyed several times within a single field season. Similarly, point transects are often marked, and visited several times in a season. This can be an effective way of increasing sample size for scarce species, so that reliable estimates of the detection function and of abundance can be obtained. Provided the population is thought to be stable within the time period, the distance data from the repeat visits can be pooled. However, repeat visits to the same line or point are not independent, and should not be analysed as if they are visits to different lines or points. This is handled within software Distance by pooling all data from repeat visits to a single line or point, and recording the associated effort as line length multiplied by the number of visits (line transect sampling) or as the number of visits (point transect sampling).

Often in multi-species surveys (Section 7.9.1), repeat visits are made because certain dates are better for certain species. In songbird surveys in northern latitudes, an early date may be required for resident species, before young fledge, while a later date may be necessary for summer migrants. In this case, it is better to decide on which visit or visits will provide appropriate data for which species, from knowledge of their biology, rather than select say the visit that revealed most detections, which leads to overestimation of abundance.

7.8.4 *Cut transects*

In habitats such as rain forests, the vegetation is so thick that transects must be cut, to allow the observer to travel along them. In this circumstance, every effort should be made to minimize the effect of the cutting on the study population. Thus the minimum of cutting should be carried out, and where possible, the transect might be cut one day and surveyed the next, so that the effects of disturbance are likely to be minimal.

If the animals of the study population themselves preferentially use the cut transect, then too many animals will be recorded on the line, and estimation may become unreliable. Similarly, if the vegetation is so thick that the observer's only clear line of view is along the cut transect, he or she will tend to detect animals only as they cross the cut transect, and again, too many animals will be recorded on the line, and estimation will be unreliable.

Cut transect lines have proved successful for a range of surveys, including surveys of monkeys, ape nests (Plumptre and Reynolds 1994, 1996) and elephant dung piles (Walsh and White 1999). In general, the cut lines should not be obvious paths, and should not be perfect straight lines, so that they are not an obvious feature of the landscape. Where poaching or other human access may be problematic, care should be taken to ensure that the line is not visible where it intersects a path.

Cut transect lines represent a substantial investment of effort. For surveys of mobile animals, it may be more cost-effective to walk the cut lines several times during the field season, and pool the distance data. For estimating trends in abundance, it can be useful to establish permanent transects along the cut routes. These issues are discussed further in Section 7.8.3.

7.8.5 *Roads, tracks and paths as transects*

In many studies, it is easier to conduct line transect surveys along roads, tracks or paths (which we will call 'routes') instead of along random lines. Similarly, point transect surveys are sometimes conducted from points equally spaced along routes. If the aim of the study is to estimate abundance within the survey region, then it is essential to check whether such surveys yield biased estimates of density. Without such a check, all that can be said is that the surveys provide estimates of animal density in the vicinity of routes, which may be unrepresentative of the survey region. Nor is it acceptable to argue that these densities provide estimates of relative abundance for the survey region, to allow trends in abundance to be monitored. Trends in abundance along routes may be very different from elsewhere, because of the ease of access for hunters or poachers, or because of greater disturbance, or because of different trends in habitat along routes (such as greater forestry activity), etc.

Few studies that use route transects have attempted to verify whether the resulting density estimates are unbiased for the wider study area. Of those, we are aware of only one that makes a compelling case for route transects. Gill, Thomas and Stocker (1997) found that they were able to estimate deer densities in forest using a thermal imager from a vehicle. In their study area, many forest tracks crossed the site on a regular grid, and densities estimated from their track transects were comparable with those estimated from surveys on foot. Generally, densities along routes tend not to be representative. For example, Jenkins *et al.* (1999) showed that chameleon densities were higher along trail transects than along straight transects away from trails, and recommended against the use of trail transects.

Hiby and Krishna (in press) noted that line transect surveys of dense forest require substantially fewer resources if they use game trails and paths instead of random lines. They discussed the potential biases, but argued that some species at least will be relatively unaffected by the presence of a trail. Survey design can be improved if the network of trails can be mapped, and random segments of trail selected for the design. The random sampling can also be stratified by habitat, to ensure that habitats are represented roughly in proportion to their occurrence in the study region.

Hiby and Krishna (in press) also examined the implications of using curved transects instead of the conventional straight transects. They showed that, with moderate curvature, standard analysis methods could be used provided the usual measurement of the perpendicular distance of a detected animal from the line was replaced by the shortest distance between the animal and the curving transect. Trails with a minimum radius of curvature that is smaller than the width of the shoulder in the detection function should be avoided, as standard analysis methods can then give biased estimates.

Walsh and White (1999) advocated a composite scheme for estimating density of elephant dung piles, in which 'recces' are conducted along existing trails, and at intervals along each recce, a transect is cut perpendicular to the recce and surveyed. Only counts of dung piles are made along the recces, while distance data are gathered from the shorter transect sections. The effective strip width is estimated from the transect distance data, and encounter rate is estimated using both the transect and the recce data, with calibration of the latter if necessary. Because far more ground can be covered in a given time for recces relative to cut transects, this approach potentially gives higher precision for a given amount of effort than one based on cut transects alone.

7.8.6 *Spotlight and thermal imager surveys*

Spotlight surveys can be very useful for animals that may be hidden during the day but are active at night, such as deer (Kie and Boroski 1995; Pierce 2000), red foxes (Heydon *et al.* 2000), badgers, rabbits or kangaroos. The animals are often particularly visible by spotlight because of reflection from the eyes.

Both line and point transect surveys can be conducted using spotlights, but both have their problems. The surveys might be conducted by foot (with batteries in a backpack) or by vehicle. For foot surveys, line transect sampling is likely to be preferred because a higher percentage of time in the field can be spent surveying, leading to larger sample sizes. Another problem with point transects is that, because the observer is stationary, obstructions cause sections of the plot to be in deep shadow. By contrast, when moving along a line, an area that is in shadow from one point may be visible from elsewhere on the line. For either method, there will be some areas within the surveyed strips or circles that cannot be seen. If these areas are potentially sufficient to generate substantial bias, then the methods of Section 6.7 should be considered. Note however that they cannot correct for bias if animal density in hidden areas is atypical.

If spotlight surveys are conducted by vehicle, there are two reasons why point transects might be preferred. The first is that it is easier to drive the vehicle to a random point, by whichever route is passable, than to drive along a random line, which in most environments, would be impossible. The second is that a search conducted from a stationary vehicle is likely to be both safer and more thorough than one conducted from a moving vehicle.

In practice, spotlight surveys are sometimes only feasible when conducted from a vehicle that is restricted to roads and tracks. In this circumstance, the limitations noted in Section 7.8.5 apply. The design should seek to achieve as random coverage of the area as is feasible from roads, with a large number of short routes, selected from the network of quieter roads and tracks in the area, in preference to a small number of long routes. Surveys can be conducted either as line transect surveys, with observers surveying as the vehicle is moving, or as point transect surveys, with the vehicle stopping at random points along the route or, more usually, equally spaced points with a random start.

There are often obstructions such as hedges, walls or buildings alongside a road or track. In this circumstance, in a line transect survey from a vehicle, there is a tendency to stop at a gap, search, then move on to the next gap. This leaves triangular blind-spots behind the obstructions, while further from the obstructions, successive searches overlap. The result tends to be that animals close to the trackline are missed. If possible, field methods should be devised to avoid this problem, perhaps by using a higher vehicle or observation platform, or, when necessary, by an observer moving away from the vehicle to a point where the blind-spot can be searched. At the analysis stage, a possible option might be to left-truncate the distance data (Section 4.3.2), to remove the biased data close to the transect.

Surveys using thermal imagers have similar considerations to those that use spotlights. Again, either a vehicle or a backpack is required. Just as spotlights are very effective for detection because the light reflects in the eyes of animals, thermal imagers highlight any part of an animal that is visible, because of the temperature contrast. Thus, if only an ear remains unhidden, a thermal imager reveals this immediately, whereas the naked eye is unlikely to detect it. Another advantage of thermal imagers is that they can be used at day or night. They can also be used from the air, widening their usefulness further. For an example of successful use of a thermal imager to survey deer, see Gill *et al.* (1997).

7.8.7 *Objects detected away from the line*

When objects are detected at some distance from the line, it can be difficult to record accurate data. For example, if objects occur in clusters, it may be necessary to approach the cluster to estimate the cluster size reliably, although regression methods for correcting for size bias in the detection of clusters (Section 3.5.4) can also correct for bias in estimating the size of clusters away from the line. Similarly, the observer may need to approach a detected object for reliable species identification. Distances of objects from the line may need to be measured or paced out, although survey lasers and laser binoculars are very effective for land-based surveys for distances from around 20 m to several hundred metres.

There is no intrinsic reason why the observer should stay on the line. If more accurate data can be gathered by moving off the line, then field methods should allow this. Similarly, if a search pattern can be employed, involving search away from the line (for example using a trained dog or additional observers), that widens the shoulder of the detection function, then estimation is likely to be improved provided that detection on the line is certain, and probability of detection falls off smoothly with increasing distance from the line.

If the observer does move off the line to verify information on a detected object, then there is a potential for bias due to the detection of additional objects in the vicinity of the original object. This has the effect of widening the effective strip width in areas of high object density, as additional 'secondary' detections are more likely to be made in such areas. If a constant effective strip width is assumed, this leads to overestimation of that width, and hence underestimation of object density. In marine surveys, this problem has long been recognized, and surveys are conducted either in passing mode, in which case the observers stay on the line, and estimate distance and cluster size as best they can, or in closing mode, in which case the observers stop search effort while recording the required data, then resume search effort later. Secondary sightings made while off-effort are excluded from the analysis. If secondary detections are frequent in a land-based survey, there may be a need to implement similar strategies.

7.8.8 *Bird surveys*

Bibby *et al.* (2000) give a comprehensive account of survey techniques for bird populations. Land-based surveys of birds tend to differ from most other land-based surveys for two major reasons: the number of species surveyed is often large, so that field protocols and analysis methods that can be applied to a wide range of species with varying behaviours are preferred; and the proportion of detections that are entirely aural is often high, perhaps over 90% for many species in forested habitats. Both of these factors contribute to the popularity of point transect sampling amongst the ornithological community, despite their lower efficiency relative to line transects. In an environment with a diverse bird community, and possibly difficult terrain, it is easier for an observer to concentrate on a wide range of songs and calls, identify the species and estimate their distance if he or she is standing at a point, rather than walking along a line whose position may be unmarked. Perhaps for these reasons, field comparisons between line and point transect sampling have usually been conducted on a diverse community of songbirds (Section 7.10).

Point counts, in which all detected birds (sometimes within a fixed radius) are counted, are also popular amongst the ornithological community. These do not yield estimates of abundance, because the surveyed area about the point is unknown and unestimated. It is often claimed that such counts are indices of relative abundance. However, because detectability is neither controlled for nor estimated, they can generate strongly biased estimates of trend. Thus trends in time may be compromised by changes of observers with varying abilities to detect different species, by observers who learn from experience to detect more calls and songs, by habitat succession, by trends in traffic noise, and so on. Trends in space are compromised because detectabilities tend to be much higher in open habitat (where for example birds of prey are visible at large distances) than in closed habitat (where birds of prey are seldom visible unless they pass overhead). Comparisons between species are also compromised, because some species are more detectable than others. We therefore advise against the use of point counts. Fixed-radius point counts may prove useful, provided all, or nearly all, birds within that radius are detected, as the count divided by the surveyed area then gives an estimate of absolute density. For multi-species surveys, however, it can be difficult to select a radius that is suitable for all species, and many detections will be wasted because they are beyond the chosen radius. See Rosenstock *et al.* (in preparation) for more discussion of these issues.

For diverse multi-species communities, a composite design, in which line transect sampling is conducted, with point transects at equally spaced intervals along the line, is worth considering. Thus line transect data might be used to estimate abundance of larger species that flee or flush at the approach of the observer, or for smaller species that may escape detection by a stationary observer. Other species might be estimated by both methods. If the two estimates for a species are consistent, they might be combined by taking the average, weighted by the inverse variances of the estimates. If the two estimates differ significantly, then the assumptions of the respective methods should be reviewed, to identify which estimate is likely to be the more reliable. If an estimate of density is by chance on the low side, the variance estimate also tends to be small. Hence in practice, it may be better to weight the two estimates by the inverse squared coefficient of variation, rather than the inverse variance, to avoid downward bias in the abundance estimate.

When many of the detections are purely aural, observer training to estimate detection distances is especially important (Section 7.5). Laser binoculars are now almost an essential piece of equipment for most land-based surveys, even for those in which detections are largely aural. It is the nearer detections for which accurate distance estimation is most critical, and for these, it is usually possible to determine which tree, bush or vegetation patch a singing or calling bird is in. Laser binoculars can then be used to measure the distance to an object (a tree, branch, bush, patch of ground, etc.) that is at about the same distance as the bird. Note that it is much better to measure accurately to a guessed location than to guess the distance to a guessed location. Not only does it remove a source of bias and variability but it also removes rounding error from distance estimates, making the task of modelling the data more straightforward. For distances less than about 20 m, for which laser binoculars do not work, distances might be physically measured or paced. Another option is to use a small laser measure, designed for shorter distances, and commonly used by builders and architects.

In songbird surveys in forest, the assumption that $g(0) = 1$ may be compromised for some species because a bird 'at the point' may in fact be 40 m or more straight up. Field protocol should seek to address this problem. For example, the surveys may be restricted to singing males only, which will be more detectable. For some species, it may be necessary to carry out a subsidiary study, in an attempt to estimate the proportion of individuals that remains undetectable at the point. Note that, for surveys in such habitats, the required detection distances are the horizontal distances of the birds from the point. That is, distances should be measured to a point on the ground vertically below the bird, not to the bird itself.

7.8.9 *Surveys in riparian habitats*

Riparian habitats commonly have two features which make survey design problematic: much of the site might be relatively inaccessible; and the site will usually be long and narrow. Many bird surveys are thus conducted as point transects, located at equally spaced points along the river, or along a road parallelling the river. This has several difficulties. First, the estimated density applies to the strip alongside the river or road, and may be very atypical of densities elsewhere in the site, giving biased estimates of abundance. Second, the design gives too much weight to narrow sections of riparian habitat, and under-represents wide sections. If densities differ between wide and narrow sections, as seems likely, abundance estimation will again be biased.

If accessibility is not a problem, a possible design is to have line transects at equally spaced intervals along the river. Each transect should be perpendicular to the river, and traverse the riparian habitat from one side to the other. If it is problematic to cross the river, the lines can extend from the edge of riparian habitat to the river bank. Essentially separate surveys can then be conducted of each bank. This design samples narrow and wide habitat in correct proportion, with shorter lines in the narrow sections. It also samples the full range of riparian habitat. It may also generate a large number of very short lines, so that efficiency may be compromised.

If line transects perpendicular to the river are not feasible or practical, another option is to have a regular grid of point transects, randomly superimposed on the study area. The points of the design should then be located, and if necessary, made accessible for surveyors. This may entail creating a path or a route for a boat. Data quality might be enhanced by making a raised observation platform at the point.

Because riparian habitats are often long and narrow, point transect surveys may have quite severe bias due to edge effects. This can be circumvented using the methods of Section 6.7. Another solution is to add a buffer zone around the riparian habitat, extending a distance w beyond the riparian habitat, where w is the maximum detection distance to be retained in the analysis. The grid of points is then superimposed on the wider survey region. Some points will fall outside the riparian habitat, but birds in riparian habitat will still potentially be detectable from them. A rapid fall-off of detections from points within the riparian habitat, caused by the presence of non-riparian habitat with low or zero bird density close to those points, is then balanced by detections of riparian birds at larger distances from the non-riparian points. Because distance sampling does not require the birds to be uniformly distributed through the site, the resulting abundance estimate is an estimate of total abundance in the wider survey region, which will equal abundance in the riparian habitat for species that do not occur in the non-riparian habitat.

7.9 Special circumstances

7.9.1 *Multi-species surveys*

Multi-species surveys pose several difficulties that may not always occur in surveys of a single species. These problems are not insurmountable but they do require careful consideration in the planning and conduct of the survey to ensure that distance sampling assumptions are satisfied.

In a multi-species survey, there are likely to be errors in species identification (Bart and Schoultz 1984). These can be minimized but probably not eliminated by proper training and experience. If two or more species are easily confused, they can be grouped in the data collection or they can be recorded separately when the observer feels confident about the identification and recorded as a group when the observer is less certain. Observers should be allowed to record a detection as an unknown species, and they should be trained to focus their effort on identifying the species for detections close to the line or point. Unidentified observations away from the line can be excluded from the analysis, effectively treating them as if they had been undetected.

Some species are more easily detected than others in a multi-species survey. In theory, this is not a concern, but it does raise a number of problems in practice. If densities are desired for each species, an adequate sample size will be needed for each species. If several species are of similar size and provide similar visual or aural cues, then their detections could be pooled to estimate a common detection probability. However, any comparison of densities within the species group hinges on the assumption of equal detection probability. A difficult problem is how to achieve a proper balance in the search effort for different species, to ensure that $g(0) = 1$ for all species. Cryptic species may require much effort searching the vicinity near the line or point, whereas large, highly visible but scarce species may require more search effort at larger distances. Acquiring an adequate sample size for each species, while ensuring that the detection function has a shoulder and $g(0) = 1$, may not be practical for a broad collection of species with widely varying visibilities. Several observers might be used, with each observer focusing on one or more species. The use of several observers may also help avoid 'swamping' observers when surveying high-density populations (Section 7.9.2). Alternatively, a single observer might make repeat visits to each line or point, concentrating on different species at each visit.

We note two further issues for multi-species surveys. Species that are only observed opportunistically, and were not listed as target species at the outset, should not usually be recorded. Such sightings might be recorded when nothing else is happening, but ignored when target species have been detected. There is then the danger that these data may be analysed by another investigator at a later date and misinterpreted. Likewise, it may be worthwhile to examine whether the detection probability of one species is affected by whether another species is detected. If it is, and the effect is ignored, then again there is potential to misinterpret the data, and perhaps to infer a non-existent ecological relationship.

7.9.2 *Surveys of animals that occur at high densities*

Much of the development of distance sampling was motivated by the need to survey sparsely distributed populations, in which we wish to retain the data from all, or nearly all, detected objects. In this circumstance, it may be the only practical approach to estimating abundance. Often, the methods are applied to populations that occur at high densities. In this case, strip counts, or complete counts within a fixed radius of points, should be given more serious consideration. If these are ruled out in favour of distance sampling, then the following issues may require consideration.

For surveys of mobile animals, as density increases, the observer may become 'swamped' with too many cues and sightings. Swamping may occur with a single species, but it is more likely in a multi-species survey. Accurate data recording might be compromised by the number of sightings, calls and other cues experienced during a short time interval (Bibby *et al.* 1985). There are various solutions that involve either reducing the rate or amount of data collected by an observer, or streamlining data collection.

When it is possible, the obvious solution is to increase the observation time by reducing the rate of travel, as long as animals are not moving substantially. Reducing the rate of travel is possible with most survey platforms except for fixed-wing aircraft, which are typically flown close to their slowest speed during surveys. In fixed-wing aircraft and possibly with other survey platforms, it may be useful to retrace a section of the line to re-observe a high density region if the observers were unable to record all of the required data. With GPS capabilities available on most aircraft, re-tracing the line exactly is relatively easy.

The amount of data collected by each observer can be reduced by using more observers to search simultaneously or by reducing the area that is searched. Multiple observers are useful to reduce swamping if the search region can be partitioned (e.g. one observer for each side of the line), the observations can be partitioned (e.g. by species) or the observers can work effectively in a team to divide observations. The latter approach may lead to double counting or failure to record observations near the line if communication is poor. The search area can be reduced by searching only one side of the line or only a portion of the point transect, but we do not recommend this approach (Section 7.9.3). A preferred alternative is to reduce the transect width. If the transect width is only reduced in high density sections, then the analysis must be post-stratified to account for this dual mode of searching, to estimate separate detection probabilities and densities.

In some situations, a high density of observations can be accommodated if the data collection and recording are streamlined. For surveys with high densities, the amount of ancillary data should be minimized. The data that are needed should be collected in the most efficient manner without sacrificing accuracy. Minimizing the effort to collect distances is one way that can speed data collection. The binomial method of Järvinen and Väisänen (1975; line transects) or Buckland (1987a; point transects), in which distances are assigned to one of just two distance intervals, might be considered, especially in multi-species surveys if only estimates of relative abundance are required. In general, the collection of grouped distance data is quicker than for ungrouped distances. For example, in aerial surveys it is much quicker to count animals that fall into particular distance intervals than to attempt to use a clinometer for each observation. In some cases, swamping can be avoided by using audio recording or a second person as data recorder. A concise protocol for announcing data can help to streamline data recording and ensure accuracy (Section 7.4.7). For surveys with multiple observers, separate recorders (persons or audio) may be needed to allow several observers to record simultaneously. Alternatively, separate stereo channels may be used on the same audio recorder.

7.9.3 *One-sided transects*

Although the analysis theory allows the observer to search on only one side of the line (i.e. $\hat{D} = n \cdot \hat{f}(0)/L$), we caution against this practice unless the animal's position relative to the line can be determined reliably and animal movement is relatively minor. If there is a tendency to include animals from the non-surveyed side of the line, then counts near the line will be exaggerated. This is a form of heaping, which causes us to fit a detection function that falls too steeply as distance from the line increases. Hence density will be overestimated, partly because the encounter rate is inflated, but primarily because average detection probability is underestimated. Animals that move from one side of the line to the other while in view of the observer may bias the estimators further. A somewhat analogous but less severe problem can occur with point transects if only a fraction of the circular area (e.g. semi-circle) is surveyed and animals outside of the surveyed area are incorrectly included. However, in that situation, only the encounter rate is improperly inflated and detection probability will be unaffected.

Typically, one-sided transects are considered for aerial surveys to overcome space or weight limitations. When a single observer searches through a flat side window of an aircraft, the line is typically offset a little, due to the lack of visibility below the aircraft. In this case, the problem is also offset; animals closer than the offset line may tend to be recorded as if they are on, or just beyond, it. If the aircraft has forward visibility, such as a helicopter, or downward visibility with side bubble windows, there may be a tendency to include animals from the opposite non-surveyed side of the line.

Several alternatives exist for aerial surveys where forward visibility is good (e.g. helicopter) but only one observer is available. Two options involve searching both sides of the line. First, the observer could search a more narrow transect (smaller w) on both sides of the line. This procedure would concentrate most of the searching effort close to the line and this would help ensure that $g(0) = 1$. Second, and perhaps less satisfactory, the width of the transect could be larger on one side of the line than the other side. This would result in an asymmetric detection function and could be more difficult to model. Theory allows asymmetry in $g(x)$, and, if modelling proved too problematic, one could always truncate the distance data and alleviate the problem. Alternatively, the right-hand data might be analysed independently from the left-hand data. Whatever approach is adopted, measuring perpendicular distance on the pilot's side will be a challenge and may prove unreliable.

Two other options in which the observer searches one side of the line only, and for which positions of detected animals can be more reliably fixed, are the following. An aerial video camera focused either downward or forward could be used to assess an animal's position, and subsequent viewing of the tapes can help ensure that $g(0) = 1$. If animals are likely to move, the camera should be focused forward to detect the animal prior to movement. However, a forward view may prove problematic in surveys with heavy vegetation. Proper alignment of the camera is essential for accurate measurement. If a camera is not possible, a similar but less desirable alternative may be possible if the helicopter has a glass bubble window or viewing port below. The animal's position (left or right of the line) may be more reliably determined by looking downward as the helicopter passes over the animal, or over its original position, if the animal has moved.

If a single observer uses a bubble window on the side of an aircraft, typically the observer's field of view will include a small area on the opposite side of the line. In this case, the only alternative is to measure carefully the animal's position relative to the line either manually (see Sections 7.4.1 and 7.6) or with a video camera as described above. If a clinometer is used, the angle of declination will exceed 90° for an animal on the opposite side. Each detection should be recorded and detections on the opposite side can be excluded.

We have focused our discussion on aerial surveys; however, the same precautions must be considered for one-sided transects in any type of survey. One-sided transects should never be used to avoid observer saturation in high-density surveys (see Section 7.9.2). A survey design issue such as searching only one side of the line illustrates the importance of carefully considering the assumptions of the theory in deciding how best to conduct a survey. Surveying only one side of the line makes the assumptions about movement and measurement error crucial because they will more directly affect the data near the transect centreline. Errors in assigning the detection of an animal to the left or right side of the line are irrelevant if both sides of the line are surveyed, but they are critical if only one side is surveyed. Data near the centreline are most important in obtaining valid estimates of density. If practical considerations lead to the use of one-sided transects, then it is crucial that observers are asked to record detections from the wrong side of the line, in a field that clearly indicates that they were the wrong side. Although such detections will be ignored by the analyst, this allows an observer to record the detection in a way that does not bias subsequent analysis. Note, though, that it may be very difficult to judge whether the animal is on the right or wrong side of the line. In the one-sided helicopter survey of deer reported by Trenkel *et al.* (1997), in which exact measurements of distances were attempted using a survey laser, 10 of the 77 detected animal groups were recorded as on the line. Five (i.e. one-half) of these groups were excluded from the analysis, on the assumption that 50% of such detections would in fact have been on the wrong side.

7.9.4 *Uneven terrain and contour transects*

Surveys are often conducted in uneven terrain. This poses safety problems for both ground-based and aerial observers. For ground-based surveys, point transect sampling may be preferred over the otherwise more efficient line transect sampling, because it is easier to gain access to a random point than to walk along a random line. If the species to be monitored does not lend itself to point transect sampling, perhaps because it flushes or moves away in response to the approach of the observer, then the ideal of randomly located transect lines may have to be compromised. The curved transect methodology of Hiby and Krishna (in press; see Section 7.8) may then be relevant, although the proviso stated there applies: the density estimates along the transects may not be representative of densities throughout the study area.

For aerial surveys, equipment that allows the aircraft to maintain roughly constant altitude above the ground may improve both safety and accuracy of estimated distances from the line. Note that the distance required is the horizontal separation between the animal and the line. For aerial surveys in uneven terrain, estimation of distance from angle of declination will not be reliable unless allowance is made for the fact that the animal may be at a different altitude than the ground vertically beneath the aircraft. Laser range finders with the capability to measure the required horizontal distance are very effective.

In hilly terrain, helicopters may prove more useful than fixed-wing aircraft, as they can more readily manoeuvre to maintain constant height above the ground. Estimation of horizontal, perpendicular distances of detected animals from the line must be made with greater care however, as the platform may not be level or aligned with the transect line, especially in windy conditions.

Quang and Becker (1999) developed methods for contour transects, to estimate bear abundance in mountainous terrain by aerial survey. Using their design, a random point is selected within the survey region, and a fixed-length transect is laid down, with the selected point at its midpoint. The transect follows the contour through that point. The aircraft flies at a fixed altitude above the transect, and the observers search for animals on the uphill side of the aircraft only. Because the searched 'strip' can fold back on itself, the covered area must be calculated with care. It is no longer just the strip width times its length; rather, it is defined by a horizontal projection of the ground area swept out by a horizontal line of length w, perpendicular to the contour, and moving along the transect with one end on the transect. In the bear surveys, there were two observers independently recording bears, and those observed by both observers were identified. This allows logistic modelling of the distances of detected bears from the line, together with any other relevant covariates, such as cluster size and slope, without having to assume that all bears on or near the line are detected. The Horvitz–Thompson estimator, with coverage probability estimated by logistic regression, is used to estimate the number of bears in the covered area, and this estimate is divided by the size of the covered area to estimate density in the wider survey region. This approach to analysis is described in detail in Buckland *et al.* (in preparation).

Another issue in uneven terrain is the need for consistency in the estimation of study area size, detection distances and, for line transect sampling, line lengths. Generally, these should all be projections of the actual measures onto a horizontal plane. Thus the length of a line should be measured from a map of the survey design, or from a comparable horizontal projection; it should not be the distance measured over the ground, which would generate downward bias in the abundance estimate, if area of the survey region is calculated as a horizontal projection (as it almost always is). Fortunately, this bias tends to be small, unless the terrain is particularly steep (in which case contour transects can help to avoid this problem). Marques *et al.* (in press) calculated that for deer pellet group surveys in a hilly area of southern Scotland, an adjustment of just 0.23% was required.

It might be argued that, in hilly areas, horizontal projections underestimate the habitat available to the animals, and therefore overestimate animal density. While this is true in an ecological sense, if the aim is merely to estimate the total number of animals within a defined region, then it is important that line lengths, distances from the line and areas are all measured as horizontal projections. In principle, it would be possible to measure all quantities with respect to the (sloping) ground, but this complicates analysis appreciably, partly because the surveyed strips are no longer simple rectangles with easily calculated areas, but also because the length of the line, or the surface area of a hillside, is a function of the scale it is measured at. The more precisely it is measured, the longer or larger it gets.

7.9.5 *Uncertain detection on the trackline*

Detection probability on the trackline, $g(0)$, can be estimated by a variety of analysis methods that are addressed by Buckland and Turnock (1992), Alpizar-Jara and Pollock (1996), Manly *et al.* (1996), Quang and Becker (1997), Borchers *et al.* (1998a, 1998b), Hiby and Lovell (1998), Laake

(1999), Schweder *et al.* (1999) and Skaug and Schweder (1999). These and other methods for estimating $g(0)$ are described in Buckland *et al.* (in preparation). Here we discuss some general guidelines for field methods and data collection to enable $g(0)$ estimation.

There are numerous ways to address $g(0)$ estimation that are specific to the survey platform and the population of interest and how they can be detected. Some methods use double counts of two or more observers on the same platform (Manly *et al.* 1996; Quang and Becker 1997) and others use double counts from observers on separate platforms (Buckland and Turnock 1992; Hiby and Lovell 1998; Caretta *et al.* 1998), to increase robustness of estimation when the objects (e.g. porpoise) only provide discrete cues (Section 6.3.2) or move in response to the survey platform. Some approaches use different methods of detection such as visual and acoustic (Borchers 1999) or land-based and aerial detection (Laake *et al.* 1997) and others use a sample of marked or radio-tagged animals (Alpizar-Jara and Pollock 1999; Borchers 1999). All of these methods share common features because they are all based on the concept of capture–recapture, or what in this case might be better called sight-resight or mark-resight.

The field methods must provide the necessary data to determine or estimate which of the detections made by the independent or semi-independent observers are of the same object (i.e. duplicates), or which marked/tagged objects have been detected. In most cases, duplicates are assessed in the analysis by proximity, which requires the collection of precise data on location and times of detected cues. Duplicate assessment can be improved by using a computerized data collection system that records accurate times or by audio-recording of independent observers onto separate stereo channels. Observers should be given clear instructions on when they should announce or otherwise denote a sighting, to avoid unnecessary imprecision in timing. For example, in a double-count aerial survey, each observer should be instructed to announce the sighting when it is abeam. Timing is even more important when the objects only provide discrete cues like a whale blow. Matching duplicate cues would be quite difficult if some observers announced the sighting when they saw the cue and others announced the sighting after they collected all relevant data. Precise instructions and well-trained observers are critical.

In some situations, duplicates can be determined in real time, eliminating the potential error and difficulty associated with analytical matching. An intermediary recorder in verbal contact with two independent observers should be able to assess duplicate cues easily, although duplicate animals can be more difficult to identify if observers see different cues from the same animal. The recorder's assessment may be improved if he/she can also see the observers and the detection. Duplicate assessment is not as difficult when the observers are not completely independent. In some methods, a secondary observer detects and tracks objects prior to detection by the primary observer (Buckland and Turnock 1992; Borchers *et al.* 1998a). The secondary observer can be aware of detections by the primary observer and can decide whether the primary observer detected the object he/she was tracking. However, the primary observer must not be aware of detections made by the secondary observer. In this scheme, the best observers should be used as the secondary observers to detect objects well ahead of the primary observer's field of view. Also, the secondary observer should search a wider area than the primary observer, so that a proportion of objects that are initially some distance from the line but that can move into the area searched by the primary observer, through either responsive or non-responsive movement, will be tracked by the secondary observer.

To achieve independence between observers, it is important that observers do not get any visual or auditory cues from each other. Achieving independence is quite easy if the observers are on separate platforms or when the second observer is a video camera or acoustic device. However, it is often difficult to achieve independence between observers on a small plane, boat or vehicle. In some cases, it is too difficult to achieve two-way independence between both observers, but 'one-way' independence can be achieved by placing one observer in front and in view of the other observer. In this case, the tracker method could be used to present trials to a primary observer, or the secondary observer could function in the reverse role to detect objects missed by the primary observer. Analysis for the latter approach is based on a removal estimator (Cook and Jacobsen 1979) which requires some fairly strong assumptions.

Achieving independence between observers does not mean that their detection probabilities for an object will be independent in the statistical sense. Features of the object, the detection cue, or the conditions for detection can create dependence in the detection probability between the observers. Typically this is a positive dependence associated with heterogeneity in detection probabilities. The highly visible objects tend to be detected by both observers, whereas the cryptic objects tend to be missed by both observers. A positive dependence will introduce a positive bias in $g(0)$ and a subsequent negative bias in density. Occasionally, but less frequently, a negative dependence can be introduced with a subsequent negative bias in $g(0)$ and positive bias in density. A negative dependence could occur in a visual–acoustic survey of dolphins if they only vocalized when not in view and they did not vocalize when they were visible. In designing a survey when $g(0) < 1$, the potential variables that could affect detection probability (e.g. visibility, behavioural cues, etc.) should be carefully evaluated for inclusion into the data collection.

In some studies, it is possible to establish some objects at known locations along the transect lines. The proportion of these subsequently detected by the observers provides a simple estimate of $g(0)$. The observers might be offered some bonus for finding these objects, as an incentive to search vigilantly close to the line. The critical assumption is that these objects have the same probability of detection as any of the objects of interest that lie on the line. Anderson *et al.* (in press) used styrofoam models of desert tortoise for this purpose.

7.9.6 *Trapping webs*

Trapping web design issues can be explored using the Windows-based program WebSim, available from http://www.colostate.edu/depts/coopunit/software2.html. WebSim allows the user to explore the effects of varying numbers of traps, trap spacing, animal movement rate, animal density, home range type, and the number of trapping occasions. The user specifies values for these variables. Using simulated populations, the program then estimates a predicted mean squared error and bias. If values are not entered for all variables, predictions are made over a range of values for the missing variable(s). WebSim also generates simulated trapping web data, and links to Distance to estimate density for each simulated data set.

7.10 Field comparisons between line transects, point transects and mapping censuses

Several researchers have attempted to evaluate the relative merits of point transect sampling, line transect sampling and mapping censuses through the use of field surveys. We summarize their conclusions here.

7.10.1 *Breeding birds in Californian coastal scrub*

DeSante (1981) examined densities of eight species of breeding bird in 36 ha of Californian coastal scrub habitat. True densities were established by an intensive programme of colour banding, spot-mapping and nest monitoring. Point transect data were collected by four observers who were ignorant of the true densities. Points were chosen on a grid with roughly 180 m separation between neighbouring points. This gave 13 points, three of which were close to the edge of the study area; only one-half of each of these three plots was covered, so that in effect 11.5 points were monitored. The recording time at each point was 8 min. Each point was covered four times by each of the four observers. Detection distances were grouped into bands 9.14 m (30 feet) wide out to 182.9 m (600 feet), and into bands twice that width at greater distances. The 'basal radius', within which all birds are assumed to be detected, was estimated as the internal radius of the first band that had a density significantly less than the density of all previous bands. Significance was determined by likelihood ratio testing with a critical value of four (Ramsey and Scott 1979). The density of territorial males was estimated using counts of singing males only, unless twice that number was less than the total count for that species, in which case the density of territorial males was estimated from half the total count. This follows the procedure of Franzreb (1976) and Reynolds *et al.* (1980). Only experienced observers were used, and they were given four days of intensive training. One day was spent verifying observers' identifications from calls and songs, one day estimating and verifying distances to both visual and aural detections, and two days carrying out simultaneous counts at points. DeSante found that the point transect data yielded underestimates of density, by about 18% when estimates for all eight species are summed. Individual species were underestimated by from 2% (white-crowned sparrow, *Zonotrichia leucophrys*) to 70% (scrub jay, *Aphelocoma coerulescens*). Correlation between actual density and estimated density across the eight species was good ($r = 0.982$). Variation in bias between observers was small. The use of the method of Ramsey and Scott (1979) undoubtedly contributed to underestimation of density in DeSante's study; the method assumes that all birds within the basal radius are detected, and the basal radius is estimated here from a small number of points, almost certainly giving rise to estimates of basal radii that are too large. An analysis of the original data by more recent methods might prove worthwhile.

7.10.2 *Breeding birds in Sierran subalpine forest*

DeSante (1986) carried out a second assessment of the point transect method, in a Sierran subalpine forest habitat. On this occasion a 48-ha study plot was identified in the Inyo National Forest, California. Methods were similar to the above study, with actual densities estimated by intensive spot-mapping and nest monitoring. Twelve points were established with a minimum separation of 200 m, and count time at each point was eight minutes, preceded by one minute to allow bird activity to return to normal after arrival at the point. Counts were carried out on four days in late June and a further four days in the second week of July. Statistical methodology was the same as for the above study.

Amongst species for which there were at least 25 detections in the point transect survey, bias ranged widely from nearly −70% for white-crowned sparrow to nearly +100% for American robin (*Turdus migratorius*) and Clark's nutcracker (*Nucifraga columbiana*). DeSante discussed the reasons why these results were less encouraging than those obtained in his earlier study. These include a higher proportion of birds missed close to the observer, due to the tall canopy, leading to underestimation, and more double counting of individuals through greater mobility, leading to overestimation. Greater mobility relative to the scrub habitat of his earlier survey occurred because densities were lower and there were more large species, both contributing to larger territories. Further, birds flying over the plot were counted; this is poor practice, leading to overestimation, unless a snapshot approach is adopted (Section 6.5.2). The relatively poor performance of the point transect method may be partially attributable to the fact that just 12 points were covered. DeSante considers that an alternative scheme of relocating points each day would not have significantly increased accuracy. Although this may be true for many species, for those species which tend to sing from favoured song posts, four counts from each of 12 points is appreciably less informative than one count from each of 48 points, even when, as here, the study area is too small to accommodate 48 non-overlapping plots.

7.10.3 *Bobolink surveys in New York state*

Bollinger *et al.* (1988) compared line and point transect estimates with known densities of bobolinks (*Dolichonyx oryzivorus*) in one 17.6-ha meadow and one 12.6-ha hayfield in New York state. Intensive banding and colour marking established population sizes, and whether each individual was present on a given day. Twelve and ten line transects, respectively, of between 200- and 500-m length were established at the two sites, together with 18 and 14 point transects. One or two line transects were covered per day, each transect taking 3–7 min. The observer waited 4 min after arriving at the start of a transect to allow birds to return to normal behaviour, and the lines were covered in both morning and afternoon. A 4-min waiting period was also used for point transects, and a 4-min counting period was found to be adequate. Two points were covered each morning of the study and two each afternoon. Thus, as for DeSante's studies, adequate sample sizes of detections were obtained by repeatedly sampling the same small number of transects. The Fourier series model was applied to both line and point transect data. For point transects, it was applied to squared detection distances, which can lead to poor performance, as noted earlier.

Bollinger *et al.* found that their point transects took longer to survey than line transects on average, but appreciably less time was spent counting. The number of males counted during point transects was slightly greater on average than during line transects, but substantially fewer females, which are more secretive, were counted. Thus, density estimates were obtained for both males and females from the line transect data, but for males only from the point transect data. The Fourier series was unable to fit adequate non-increasing detection functions to the morning point transect counts, which the authors suggest may be indicative of movement of bobolinks away from the observer. Both methods overestimated male abundance, with the point transect method showing the greater bias (mean relative bias of 140%, compared with 76% for line transects). Bias was found to be lower in general for the afternoon count data. Line transect estimates of female densities were approximately unbiased, although there was a suggestion of underestimation during incubation, and overestimation when the young were large nestlings or had fledged. About 25% of bias in male density estimates was attributed to avoidance of field edges by the birds; transects were deliberately positioned so that field edges were not surveyed. Survey design to eliminate or reduce this source of bias is discussed in Section 7.2. Additional bias was considered to be possibly due to 'random' movement of birds, with detection biased towards when the birds were relatively close to the observer, or to attraction to the observer. Movement is very likely to have contributed to the upward bias, as noted in Chapter 5, but attraction towards the observer is difficult to reconcile with the suggestion that poor fits for point transect data might be due to observer avoidance. It may be that the Fourier series model was inappropriate for the squared distance data, as was found by Buckland (1987a) for the point transect data of Knopf *et al.* (1988), rather than that birds avoided the observer.

7.10.4 *Breeding birds in Californian oak–pine woodlands*

Verner and Ritter (1985) compared line and point transect counts in Californian oak–pine woodlands. They also considered counts within fixed areas (strip transects and circular plots) and unbounded counts from both lines and points as measures of abundance. They defined four scales for measures of abundance: a 'nominal scale', which requires information only about occurrence; an 'ordinal scale', which requires sufficient information to rank species in order of abundance; a 'ratio scale', which requires relative abundance estimates – bias should be either small or consistent across species and habitats; and an 'absolute scale', which requires unbiased (absolute) estimates of abundance. They assessed the performance of different survey methods in relation to these scales. True bird densities were unknown. Although the area comprised 1875 ha of oak and oak–pine woodlands in the western foothills of the Sierra Nevada, the study plots were just two 19.8-ha plots of comparable relief and canopy cover, one grazed and the other ungrazed. This study, in common with most others of its type, therefore suffers from repeated sampling of the same small area, and hence non-independent detections.

Sampling took place over eight-day periods, with two line transects and 10 counts covered per day. The transects were 660-m long and were positioned randomly at least 60-m apart each day. The points were located at intervals of 150 m along the transects. The design was randomized and balanced for start time, starting point and count method. All counts were done by a single observer. Four methods of analysis were considered: bounded counts (strip transects of width 60 m and circular plots of radius 60 m); Emlen's (1977) *ad hoc* estimator; Ramsey and Scott's (1978) method; and the exponential polynomial model (Burnham *et al.* 1980). Note that we do not now recommend any of these estimators for songbird data. The Fourier series model was found to perform less well than the exponential polynomial model, so results for it were not quoted. Interval estimates were computed for the exponential polynomial model only. Without more rigorous analysis of the data and with no information on true densities, comparisons between the methods of analysis and between line and point transects are severely constrained. However, the authors concluded that line and point transects showed similar efficiency for determining species lists (for point separation of 150 m and 8 min per point); point transects yielded lower counts per unit time, but would be comparable if point separation was 100 m and counting time was 6 min per point; Ramsey and Scott's (1978) method gave widely differing estimates from line transect data relative to those from point transect data; more consistent comparisons between models were obtained from line transects than from point transects; most species showed evidence of movement away from the observer; the exponential polynomial model was thought to be the most promising of the four methods.

7.10.5 *Breeding birds along the Colorado River*

Anderson and Ohmart (1981) compared line and point transect sampling of bird populations in riparian vegetation along the lower Colorado River. All observers were experienced, and each carried out replicate surveys under both sampling methods in each month from March to June 1980. The distance walked was identical for each sampling method. The line transect data were analysed using the method of Emlen (1971), and the point transect data using method M1 of Ramsey and Scott (1979). Thus, models that can perform poorly were again used, and this may have compromised some of the authors' conclusions. For example, they sometimes obtained inflated density estimates from the point transect data when detection distances less than 30 m were divided into two or more groups, whereas the method performed well when all observations within 30 m were amalgamated into a single group. A more robust method would be less sensitive to the choice of grouping. Anderson and Ohmart concluded that the point transect surveys took longer to complete when the time spent at each point was 8 min, but that times were comparable for recording times of 6 or 7 min. More area was covered and more birds detected using the line transect method, because of the dead time between points for the point transect method. The authors tabulated estimated average densities of 10 of the more common species, which appear to show relatively little difference between line and point transect estimates, or, for most species, between those estimates and estimates from territory mapping, although overall, line transect estimates were significantly lower than mapping estimates. Neither method generated average estimates significantly different from the point transect estimates. However, the authors noted that day-to-day variation in point transect estimates was greater than for line transect estimates, and suggested that at least three repeat visits to point transects are necessary, whereas two are sufficient for line transects. They concluded that the line transect method is the more feasible, provided stands of vegetation are large enough to establish transects of 700–800 m in length, and provided that the topography allows ambulation. They indicated that these transects should be adequately cleared and marked. In areas where vegetation occurs in small stands, or where transects cannot be cleared, they suggested that point transects might be preferable.

7.10.6 *Birds of Miller Sands Island, Oregon*

Edwards *et al.* (1981) compared three survey methods, two of which were line and point transect sampling. They described the third as a sample plot census, in which an observer records all birds that can be detected within a given radius of a point. Distances were not recorded, so corrections for undetected birds cannot be made. The study was carried out on Miller Sands Island in the Columbia River, Oregon. Four habitats were surveyed: beach, marsh, upland and tree-shrub. The method of Emlen (1971) was used for the line transect data and the method of Reynolds *et al.* (1980) for the point transect data. The authors found that significantly more species were detected using point transects than either line transects or sample plots. However, the truncation point for line transects had been 50 m, and for point transects 150 m, and the sample plots were circles of radius 56.4 m, so the difference is unsurprising. Density estimates were found to be similar for all three methods, although the point transect estimate was significantly higher than the line transect estimate in a handful of cases. The methods were not standardized for observer effort or for time spent in the field, making valid comparison difficult.

7.10.7 *Concluding remarks*

More field studies, carefully standardized for effort, would be useful. A large study area, too large for all territories to be mapped, is required for a fair comparison of line and point transect sampling and of mapping methods. Within such an area, line and point transect sampling grids could be set up using the recommendations of this chapter, so that both methods require roughly the same time in the field on the part of the observer. In addition, territory mapping should be carried out on a random sample of plots within the area, again so that time in the field is comparable with each of the other methods. The analyses of the line and point transect data should be comparable, for example using the standard methods of Distance. More than one model should be tried. The precision of each method should then be compared, and an assessment should be made of whether at least one of the methods consistently over- or underestimates relative to the others. If different researchers could agree on a common design, this could be repeated in a variety of habitats, to attempt to establish the conditions and the types of species for which point transect sampling is preferable to line transect sampling or vice versa.

The studies described here tend to favour line transects over point transects. This may partly reflect that line transect methodology has had longer to evolve than point transect methodology. It is important to realize that point transect sampling is essentially passive, whereas line transect sampling is active. For birds that are unlikely to be detected unless they are flushed or disturbed, such as many gamebirds or secretive female songbirds, line transect sampling should be preferred. Very mobile species are also likely to be better surveyed by line transects, provided the guidelines for such species of Section 6.5 are adhered to. For birds that occupy relatively small territories, and which are easily detected at close range, such as male songbirds of many species during the breeding season, point transects may be preferable, especially in patchy or dense habitat. Attempts to estimate abundance of all common species in a community by either method alone are likely to perform poorly for at least some species. If only relative abundance is required, to monitor change in abundance over time, either technique might prove useful. However, bias may differ between species, so great care should be taken if cross-species comparisons are made. Equally, bias may differ between habitats, although well-designed line or point transect studies yield substantially more reliable comparisons across both species and habitats than do straight counts of birds without distance data or other corrections for detectability.

Several other authors compare line transect sampling with census mapping. Franzreb (1976, 1981) gives detailed discussion of the merits of each, concluding that census mapping is substantially more labour intensive, but for some species at least, provides better density estimates. Choice of method must take account of the species of interest, whether density estimates for non-breeding birds are required, the habitat of the study area, resources available and the aims of the study. O'Meara's (1981) comparison of line transects and mapping censuses includes an assessment of the binomial models of Järvinen and Väisänen (1975). These models were found to be more efficient, both in terms of time to record detections into one of just two distance intervals and in terms of variance of the density estimate, than Emlen's (1971, 1977) method, which requires detection distances to be recorded so that they can be assigned to successive bands at increasing distances from the line. Line transect estimates were found to be lower than census mapping estimates, apparently due to imperfect detection of birds at or near the line (Emlen 1971; Järvinen and Väisänen 1975), but estimates could be obtained for twice as many species from the line transect data. Redmond *et al.* (1981) also compared census mapping with the line transect methods of Emlen (1971) and Järvinen and Väisänen (1975), for assessing densities of long-billed curlews (*Numenius americanus*). They also found that the method of Järvinen and Väisänen was easier to apply than that of Emlen, because it requires just two distance intervals, and was far more efficient than census mapping in terms of resources in the case of territorial male curlews. Female curlews were not reliably surveyed using line transects, nor were males during brood rearing.

Several field evaluations have been made of distance sampling theory in which a population of known size or density is sampled and estimates of density made (Laake 1978; Parmenter *et al.* 1989; White *et al.* 1989; Bergstedt and Anderson 1990; Otto and Pollock 1990; Southwell 1994; Anderson and Southwell 1995; Anderson *et al.* in press). Strictly speaking, these are not evaluations of the distance sampling methods, but rather an assessment of the degree to which the critical assumptions have been met under certain field conditions. We encourage more studies of this type as such results often provide insights into various issues. We recommend that the person performing the data analysis should not know the value of the parameter being estimated.

7.11 Exercises

1. In line and point transect sampling, not all objects within the surveyed area $a = 2wL$ (line transects) or $a = k\pi w^2$ (point transects) are detected. Summarize the assumptions under which the density of objects *within the surveyed area* can be reliably estimated by standard distance sampling methods. What further assumption is required, to allow estimation of abundance in the wider study area of size A? Discuss the implications of this assumption for survey design.

2. A line transect survey of a pine forest is to be conducted, to estimate red deer density. It is not feasible to detect and locate most deer before they move away from the observer, due to poor visibility within the forest. For that reason, their dung is to be surveyed. Deer dung occurs in well-defined pellet groups. Below are extracts from an analysis conducted on data from a pilot survey, which comprised lines of total length 450 m. The 'clusters' are the pellet groups, and mean cluster size is the mean number of pellets per group. Hence it is the density of clusters, labelled 'DS' in the Distance output, that is of interest here.

(a) Interpret the output from Distance. You should assess whether the data are adequately modelled, and summarize the key results.

(b) For the main survey, a target coefficient of variation for estimating pellet group density is set at 8%. Estimate the amount of survey effort required to achieve this in each of the following cases.

(i) The estimate of $f(0)$ from the pilot survey, and hence the estimate of pellet group density (DS), are considered to be inadequately estimated from the small pilot sample.

(ii) The estimate of pellet group density is considered to be sufficiently reliable to be used to estimate required effort in the main survey.

(c) If the estimated number of pellet groups deposited per animal per day is 24, with coefficient of variation 5%, and the mean number of days for a pellet group to disappear is 155, with coefficient of variation 12%, obtain an estimate of red deer density in the forest, and the corresponding coefficient of variation, using the pilot survey estimate of pellet group density of 57871 groups/km^2 ($cv = 22.57\%$).

```
Effort         :    0.4500001
# samples      :   18
Width          :  125.0000
# observations:   34


Model  1
   Half-normal key, k(y)=Exp(-y**2/(2*A(1)**2))

       Iter   LN(likelihood)     Parameter Values
      ---------------------------------------------------------------------
         1   -155.004            46.5451
         2    ....

        10   -154.967            48.3636
      ---------------------------------------------------------------------
      Results:
      Convergence was achieved with   10 function evaluations.
      Final Ln(likelihood) value =  -154.96691
      Akaike information criterion =   311.93381
      Bayesian information criterion =   313.46017


Model  2
   Half-normal key, k(y)=Exp(-y**2/(2*A(1)**2))
   Cosine adjustments of order(s) :  2

       Iter   LN(likelihood)     Parameter Values
      ---------------------------------------------------------------------
         1   -154.967            48.3636          0.00000
         2    ....

        11   -154.951            47.7612        -0.509868E-01
      ---------------------------------------------------------------------
      Results:
      Convergence was achieved with   11 function evaluations.
      Final Ln(likelihood) value =  -154.95122
```

```
      Akaike information criterion =   313.90244
      Bayesian information criterion =   316.95517

   Likelihood ratio test between models  1 and  2
      Likelihood ratio test value    =     0.0314
      Probability of a greater value =   0.859392
 *** Model  1 selected over model  2 based on minimum AIC
```

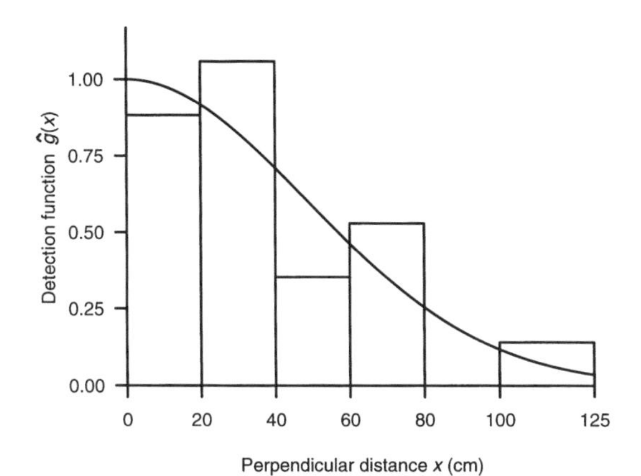

Goodness of Fit Testing with some Pooling

Cell i	Cut Points		Observed Values	Expected Values	Chi-square Values
1	0.00	20.0	10.0	11.01	0.093
2	20.0	40.0	12.0	9.31	0.780
3	40.0	60.0	4.0	6.64	1.051
4	60.0	80.0	6.0	4.01	0.994
5	80.0	125.	2.0	3.03	0.352

```
 Total Chi-square value =      3.2701  Degrees of Freedom =   3
Probability of a greater chi-square value, P = 0.35183
 Combined Estimates:
```

		Estimate	%CV	df	95% Confidence Interval	
Half-normal/Cosine						
	m	1.0000				
	AIC	311.93				
	Chi-p	0.35183				
	f(0)	0.16660E-01	12.83	33	0.12846E-01	0.21607E-01
	p	0.48019	12.83	33	0.37025	0.62278
	ESW	60.024	12.83	33	46.281	77.847

Combined Estimates:

	Estimate	%CV	df	95% Confidence Interval	
DS	57871.	22.57	11	35428.	94530.
D	0.81700E+06	22.96	11	0.49613E+06	0.13454E+07

```
           N       0.10512E+09   22.96      11  0.63837E+08  0.17311E+09


Measurement Units
-------------------------------
 Density: Numbers/Sq. kilometer
     ESW: centimeters
```

3. Table 1 shows encounter rates (n/L), $\hat{f}(0)$, and mean school size estimates $(\hat{E}[s])$, together with their estimated coefficients of variation, for minke whales from a shipboard line transect survey in the Antarctic. Note that there are four strata and that $f(0)$ is estimated separately in the two southern strata, but that the northern strata are pooled to provide a single $f(0)$ estimate for the north. Further, the two southern strata are pooled to provide a single estimate $\hat{E}[s]$ for the south, and the two northern strata are pooled to provide a single estimate $\hat{E}[s]$ for the north.
   (a) Estimate individual abundance in each stratum, together with its *cv*.

   (b) Estimate individual abundance in the north, and individual abundance in the south, together with their *cv*s.

   (c) Estimate total individual abundance in all strata, together with its *cv*.

   (d) Use the results from this survey to decide how best to apportion survey effort between the strata for a second survey of the same area with the same total survey effort.
4. Tables 2 and 3 show estimates from Distance outputs of a line transect survey of a desert tortoise population in two geographic strata

Table 1. Line transect parameter estimates by stratum. Numbers in brackets are *cvs*

<table>
<tr><th>Stratum</th><th>Sample size, n</th><th>Effort $L(nm)$</th><th>Encounter rate, n/L</th><th>$\hat{f}(0)$</th><th>Mean school size, $\hat{E}[s]$</th><th>Area A (nm^2)</th></tr>
<tr><td>Southwest</td><td>122</td><td>457</td><td>0.267 (0.45)</td><td>1.01 (0.10)</td><td rowspan="2">0.80 (0.14)</td><td>58,643</td></tr>
<tr><td>Southeast</td><td>76</td><td>688</td><td>0.111 (0.42)</td><td>1.53 (0.16)</td><td>82,004</td></tr>
<tr><td>Northwest</td><td>10</td><td>310</td><td>0.032 (0.69)</td><td rowspan="2">1.53 (0.16)</td><td rowspan="2">2.67 (0.10)</td><td>137,738</td></tr>
<tr><td>Northeast</td><td>115</td><td>574</td><td>0.200 (0.22)</td><td>165,419</td></tr>
</table>

Table 2. Estimates of $f(0)$ (in km^{-1}) and related data and results. AIC is Akaike's Information Criterion. Estimation was performed using interval data (five intervals in all cases). $P(>\chi^2)$ is the reported probability of obtaining a worse goodness of fit statistic, with degrees of freedom given by df, given that the true model has been fitted

Stratum	$\hat{f}(0)$	$cv[\hat{f}(0)]$	n	AIC	$P(>\chi^2)$	df
All	0.319	0.10	125	396.41	0.91	2
South	0.296	0.11	78	251.11	0.86	2
North	0.415	0.30	47	147.91	0.50	2

Table 3. Effort (in km), encounter rates (n/L in animals per km of transect) and stratum areas

Stratum	L	n/L	cv$[n/L]$	Surface area (km^2)
All	209	0.60	0.12	1680
South	132	0.59	0.13	1040
North	77	0.61	0.24	640

('North' and 'South'). Estimates labelled 'All' are from data pooled over the strata. The animals occur singly. Observers walked along marked lines and used laser range finders to measure distances to tortoises. Desert tortoises move slowly and spend a large proportion of their lives underground and hence invisible.

(a) Decide whether to estimate $f(0)$ separately in each stratum or from the pooled data. Give reasons.

(b) Irrespective of what you decided in part (a), estimate animal abundance and its *cv* using stratified encounter rates and the $\hat{f}(0)$ from pooled data.

(c) From these results, calculate the approximate total effort and effort in each stratum that is required in order to estimate abundance (in the same way) with no more than a 10% *cv*.

(d) Abundance estimates obtained in this way will be negatively biased because animals spend a large proportion of their time underground and hence undetectable. Suggest how you might go about obtaining an unbiased abundance estimate using line transect methods.

5. The figure shows the planned transects for an aerial survey of dolphins in the waters around an island. The island is in the centre of the figure. The dotted lines are depth contours. More than one species

of dolphin occur in the survey area. One species in particular prefers shallow water, within the first depth contour. If the survey is analysed unstratified, the design will bias the abundance estimate of this species upwards. Why? Suggest a better design.

6. The figure shows the perpendicular distance distribution obtained from an aerial survey of seals, using an aircraft with flat windows. The aircraft speed is such that the paucity of sightings close to the trackline cannot be due to animals avoiding the aircraft.

   (a) Suggest at least two plausible reasons for the paucity of sightings in the first two perpendicular distance intervals.

   (b) Suggest how you could change the survey to avoid this problem on future surveys.

   (c) Suggest how you might estimate abundance from these data.

   (d) Only a fraction of the population of seals is 'hauled out' (i.e. not in the water) at any one time. How could you use data from a sample of radio-tagged seals, from which you know the times that these seals were hauled out over periods of some days, to convert the line transect estimates to estimates of absolute abundance?

7. A line transect survey of a population that does not occur in clusters produces $n = 97$ detections from 31 lines of roughly equal length, totalling $L = 85.8\,\text{km}$. Analyses give $\widehat{var}(n) = 184$ and $cv[\hat{f}(0)] = 0.0965$.

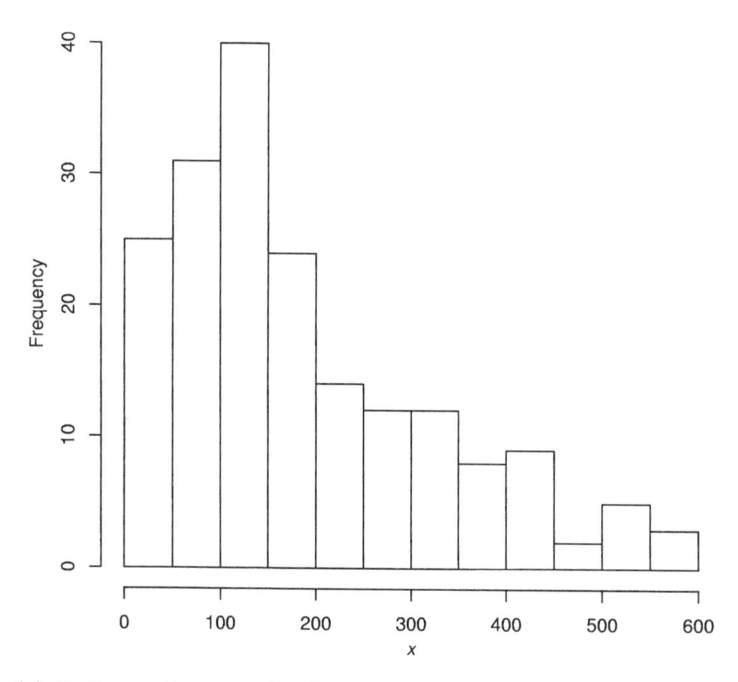

(a) Estimate the quantity b.

(b) A 10% *cv* is required for $\hat{D}$. Estimate the total effort needed to achieve this, and the sample size that can be expected with that amount of effort.

8. A line transect survey yields the following grouped perpendicular distance data.

Distance interval (m)	0-10	10-20	20-30	30-40	40-50	50-70	70-100
Frequency	33	19	20	16	12	11	13

Why are these data problematic to model? What design and field procedures might need attention to avoid this problem in future surveys?

9. In a pilot line transect survey, 90% of the variance in $\hat{D}$ was found to be attributable to the variance of n, 7% to $\hat{f}(0)$ and 3% to $\hat{E}(s)$. Where should efforts to refine the design be concentrated? What design improvements should be considered?

8
Illustrative examples

8.1 Introduction

Several analyses of real data are presented here to illustrate line and point transect analysis, and use of program Distance. Units of measurement of the original study are adhered to, to avoid quoting cutpoints as 'every 0.305 m' instead of 'every foot'. Four of the examples are line transect surveys and three are point transect studies. The first two examples are case studies in which the true density is known. The first of these is a set of line transects to estimate the number of bricks that had been placed on the bed of a region of Lake Huron, as part of a programme to assess the viability of using an underwater video system to monitor numbers of dead lake trout. This is followed by a reanalysis of the wooden stake surveys carried out as part of a student project in sagebrush–grass near Logan, Utah (Laake 1978). The third example is a comprehensive analysis, including a trend analysis, of the data from annual surveys of duck nest density that have been carried out since 1964 at the Monte Vista National Wildlife Refuge in Colorado. An analysis of fin whale data from ship sightings surveys in the North Atlantic illustrates use of stratification.

The first point transect example is an illustration of model selection and stratification on house wren data collected during surveys of riparian vegetation in South Platte River bottomland in Colorado. The second example is an analysis of data on the six species most commonly recorded in songbird surveys of the Arapaho National Wildlife Refuge, also in Colorado. In the final example, an approach for assessing effects of habitat on density using point transect sampling is described, and an analysis of binomial point transect data is carried out in which bird density and detectability in restocked forest plantation in Wales are related to habitat succession.

8.2 Lake Huron brick data

Bergstedt and Anderson (1990) presented the results of 13 independent, line transect surveys of known populations of bricks placed approximately at random on the floor of portions of Lake Huron. The bricks were to simulate dead lake trout, and the purpose of the survey was to explore the feasibility of large-scale surveys of dead lake trout on the lake bottom. The distance data were taken remotely via a video camera mounted on an underwater sled and pulled along the lake bottom by a surface vessel. The data were taken in five distance categories (0, 0.333, 0.667, 1, 1.333, 1.667 m), and these cutpoints were marked on a cathode ray tube on board the vessel. The relationship between the cutpoints on the screen and the perpendicular distance on the lake bottom was defined by calibration studies before the first survey. The true density in all surveys was 89.8 bricks/ha, and this allows an evaluation of the utility of line transect sampling theory. Full details of this survey are given by Bergstedt and Anderson (1990). Our purpose here is to reanalyse the data from the 13 surveys using the methods presented in earlier chapters.

Table 8.1 provides a summary of the estimates of density for each of the 13 surveys under each of five models: half-normal + cosine, uniform + cosine, uniform + simple polynomial, half-normal + Hermite polynomial and the hazard-rate model + simple polynomial. Information related to model selection for the pooled data is presented in Table 8.2. Akaike's Information Criterion (AIC) selects the hazard-rate model with no adjustment terms, and this model fits the grouped distance data well ($p = 0.70$). The remaining four models (half-normal + cosine, uniform + cosine, uniform + polynomial and half-normal + Hermite polynomial) have similar AIC values, but fit the data less well ($0.09 \leq p \leq 0.29$).

Table 8.1. Summary of density estimates and coefficients of variation for the Lake Huron brick data under five models of the detection function (data from Bergstedt and Anderson 1990). True density = 89.8/ha. The final means are weighted by line length

Survey	HN + cos		Uni + cos		Uni + poly		HN + Herm		Haz + poly	
1	138.5	19.5	144.4	15.7	137.1	19.2	126.8	16.6	104.0	16.2
2	96.4	18.7	92.6	32.5	100.0	15.6	93.1	27.1	88.2	15.6
3	88.8	20.4	91.4	19.2	86.2	17.0	88.8	20.4	70.0	17.7
4	84.0	19.7	85.5	15.9	83.2	19.6	77.1	16.0	63.0	16.2
5	78.3	22.0	81.6	20.0	72.9	18.4	78.3	22.0	61.3	18.9
6	84.5	33.8	93.6	14.6	90.4	17.7	84.4	27.5	75.9	16.4
7	80.7	38.2	76.9	26.3	92.4	19.3	85.4	16.6	71.1	16.0
8	122.4	19.1	107.0	15.7	104.1	19.2	122.4	19.1	86.2	16.7
9	96.2	19.5	99.6	15.7	95.7	19.2	96.2	19.5	73.6	16.0
10	72.8	20.2	70.2	26.7	67.7	16.9	67.6	30.0	61.5	16.0
11	89.4	20.9	93.4	19.2	86.1	17.3	89.4	20.9	70.6	18.0
12	92.7	18.7	83.9	15.3	81.6	18.0	81.9	29.2	76.5	18.0
13	56.8	20.7	58.2	20.0	55.1	17.8	56.8	20.7	55.5	27.2
Wt. Ave.	88.7	6.2	88.3	6.0	86.5	6.0	86.2	6.8	72.4	4.7

Table 8.2. Model selection statistics for the pooled Lake Huron brick data. ΔAIC shows AIC−AIC_{min}, where $\text{AIC}_{\text{min}} = 1667.9$ is the AIC value corresponding to the best fitting model

Key function	Adjustment	Total no. of parameters	ΔAIC	p-value*
Half-normal	Cosine	4	3.3	0.09
Uniform	Cosine	3	2.4	0.29
Uniform	Polynomial	3	2.4	0.29
Half-normal	Hermite polynomial	3	4.8	0.06
Hazard-rate	Polynomial	2	0.0	0.70

*Goodness of fit test.

Surprisingly, the hazard-rate model provides the poorest estimates of mean density based on the weighted mean of the estimates (Table 8.1). The hazard-rate model fitted a very flat shoulder to the pooled data and produced estimates that were low in 12 of the 13 surveys. Performance of the other four estimators is quite good (Table 8.1). All were slightly low due, in part, to the results of survey 13. Estimated confidence intervals for mean density covered the true value for all but the hazard-rate model.

The weighted mean of the estimates was very close to the true parameter, except for the hazard-rate model (88.7, 88.3, 86.5, 86.2 and 72.4, respectively). Any of the first four models performs well, especially when one considers that the average sample size per survey was only 45. Despite the poor performance of the hazard-rate model, it was selected as the best of the five models by AIC in 9 of the 13 data sets (surveys).

The reader is encouraged to compare these results with the original paper by Bergstedt and Anderson (1990), which includes discussion of various points relating to possible measurement errors, missing bricks in the first distance category, and potential problems with survey 13. Bergstedt and Anderson (1990) note that some bricks were missed near the centreline, and that the cutpoints drawn on the cathode ray tube, although accurate in the initial calibration, were perhaps compromised by the uneven lake bottom.

8.3 Wooden stake data

Laake (1978) set out 150 unpainted wooden stakes ($2.5 \times 5 \times 46$ cm) within a rectangular area of sagebrush–grass near Logan, Utah, in the spring of 1978 to examine the performance of the line transect method. The stakes were placed in a restricted random spatial pattern such that the number of stakes was distributed uniformly as a function of distance from the line. In fact, each 2-m distance category had 15 stakes present. Stakes were driven in the ground until about 37 cm remained above ground. One stake was placed about every 7 m of transect and alternated between left and right sides of the line. Exact placement was generated randomly within the 7 m section. True density was 37.5 stakes per hectare.

A single, well-marked line ($L = 1000\,\text{m}$, $w = 20\,\text{m}$) was traversed by 11 different, independent observers. The observers were carefully instructed and supervised and fatigue was probably a minor factor as each survey could be completed by the observer in approximately 2 h. Observers traversed the line at different times, thus the data for each of the 11 surveys are independent. The number of stakes detected (n) varied from 43 to 74, corresponding to $\hat{P}_a$ ranging from 0.29 to 0.49. Histograms and estimated detection functions ($\hat{g}(x)$) differed greatly among observers. In field studies, the detection function would also be affected by habitat type and species being surveyed. These factors affect n and make it, alone, unreliable as an index of density.

Two strategies were used for the analysis of these data. First, the data were pooled over the 11 surveys ($n = 642$) and AIC was computed for five models:

Model	AIC	ΔAIC
Half-normal + cosine	2412.9	0.0
Uniform + cosine	2415.2	2.3
Uniform + polynomial	2417.4	4.5
Half-normal + Hermite	2450.8	37.9
Hazard-rate + cosine	2416.4	3.5

Thus, the half-normal + cosine is selected as the best model. In fact, all models seem fairly satisfactory for data analysis except the half-normal + Hermite polynomial model where the likelihood ratio test indicated that the first adjustment term (for kurtosis) was not required ($p = 0.617$). There are options under 'adjustment terms' in the 'detection function' tab of the model definition section of Distance, such as selection method 'forward' (which yields AIC = 2415.1 for this model) or 'look-ahead' set equal to two (also giving AIC = 2415.1), that allow the user to avoid such poor model fits (Section 8.6). Estimates of density are shown in Table 8.3 under the two best models, as suggested by AIC. The estimates of density are quite similar between these two models.

While n varied widely, estimates of density varied from only 26.38 to 38.31, using two models for the detection function (half-normal and uniform key functions with cosine adjustment terms). Confidence interval coverage

Table 8.3. Summary of stake data taken in 1978 in a sagebrush–grass field near Logan, Utah (Laake 1978). Density in each of the 11 surveys was 37.5 stakes/ha. Cosine adjustments were added as required in modelling $f(x)$. For each survey, $L = 1000$ m and $w = 20$ m

Survey no.	Key function	Sample size	Density estimate	cv (%)	Log-based 95% confidence interval
1	Half-normal	72	37.11	19.3	(25.51, 53.99)
	Uniform		30.00	14.5	(22.63, 39.77)
2	Half-normal	48	35.18	19.9	(23.90, 51.78)
	Uniform		36.01	20.1	(24.36, 53.23)
3	Half-normal	74	28.76	15.8	(21.16, 39.10)
	Uniform		29.26	14.8	(21.92, 39.07)
4	Half-normal	59	38.31	19.1	(26.42, 55.52)
	Uniform		33.30	17.5	(23.68, 46.81)
5	Half-normal	59	34.41	19.9	(23.37, 50.66)
	Uniform		29.58	19.0	(20.47, 42.76)
6	Half-normal	72	26.38	16.2	(19.24, 36.17)
	Uniform		27.08	15.6	(19.98, 36.69)
7	Half-normal	55	34.48	19.9	(23.44, 50.72)
	Uniform		34.69	21.1	(23.06, 52.18)
8	Half-normal	61	33.31	20.2	(22.51, 49.30)
	Uniform		34.48	21.3	(22.82, 52.09)
9	Half-normal	46	28.32	21.9	(18.51, 43.31)
	Uniform		23.52	21.2	(15.60, 35.48)
10	Half-normal	43	34.16	20.1	(23.15, 50.42)
	Uniform		32.69	21.1	(21.71, 49.22)
11	Half-normal	53	29.80	17.4	(21.25, 41.80)
	Uniform		31.45	17.8	(22.23, 44.50)
Mean	Half-normal	642	32.75	3.6	(30.54, 35.11)
	Uniform		31.10	3.6	(28.99, 33.36)
Pooled	Half-normal	642	34.37	7.2	(29.86, 39.55)
	Uniform		34.60	7.5	(29.86, 40.08)

cannot be accurately judged from only 11 replicates, but examination of the intervals in Table 8.3 shows no particular indication of poor coverage. Estimates of density are low in all cases, except for survey 4 for the half-normal estimator. Averaging the density estimates over the 11 surveys indicates a negative bias of approximately 13–17%. Pooling the data over the 11 surveys provides an approximate estimate of bias of about −7% to −8%. The main reason for the negative bias seemed to be some lack of model fit near the line. Examination of the histograms and estimated detection functions seems to indicate that models commonly fit poorly near zero distance. Some of the negative bias is due to measurement error for stakes near the line. The exact location of each stake was known, and errors in measurement could be assessed. For example, the information for three stakes (stake numbers 45, 93 and 103) is shown below:

Stake no.	103	45	93
True distance (m)	0.92	5.03	14.96
Ave. distance ($\bar{x}$)	0.73	4.77	14.63
$\widehat{sd}(x)$	0.14	0.15	0.28
$100(\widehat{sd}(x)/\bar{x})$	18.9	3.2	1.9

This suggests that the measurement error is largely due to improper determination of the centreline. Finally, the negative bias is partially the result of observers missing about 4% of the stakes in the first metre and 13% of the stakes in the first two metres. However, this is offset by the tendency to underestimate distances near the line. One observer was seen actually tripping over a stake on the centreline, but still the stake was not detected. Stakes do not move and do not respond to the observer; for field surveys of animals, the relative importance of the different assumptions may be very different. If this survey was to be repeated, we would enlarge the study area and lengthen the line such that $E(n) \simeq 80$. Also, observers would be shown the evidence that stakes near the centreline were occasionally missed, and that measurements were often in error, in the hope that these problems could be lessened.

A second strategy will be illustrated that is less mechanical, requires a deeper level of understanding of the theory, and is somewhat more subjective. The pooled data are displayed as a histogram using 1-m cutpoints (Fig. 8.1). This information suggests several aspects that should influence the analysis. First, the distance data have a long tail, a small mode around 15 m, and considerable heaping at 10 m at the expense of 9 and 11 m. Some heaping at 0, 5 and 15 m is also seen. A histogram based on 0.5-m cutpoints indicated that detection declined rapidly near the line. Thus, the data have some of the characteristics illustrated in Fig. 2.1. The modelling of $g(x)$ will require additional terms to model the extra mode and long tail. Figure 8.1 shows a model with three cosine terms, and the data suggest that this still might underestimate near the line. Thus, a reasonable approach might be to truncate the data at $w = 10$ m in the hope of obtaining a better fit near the line and alleviating problems in the tail of the distribution.

At this point, one could choose a robust model for $g(x)$ and proceed; here we will first use the uniform key with cosine adjustments (the Fourier series model). Fitting these truncated data still required three adjustment

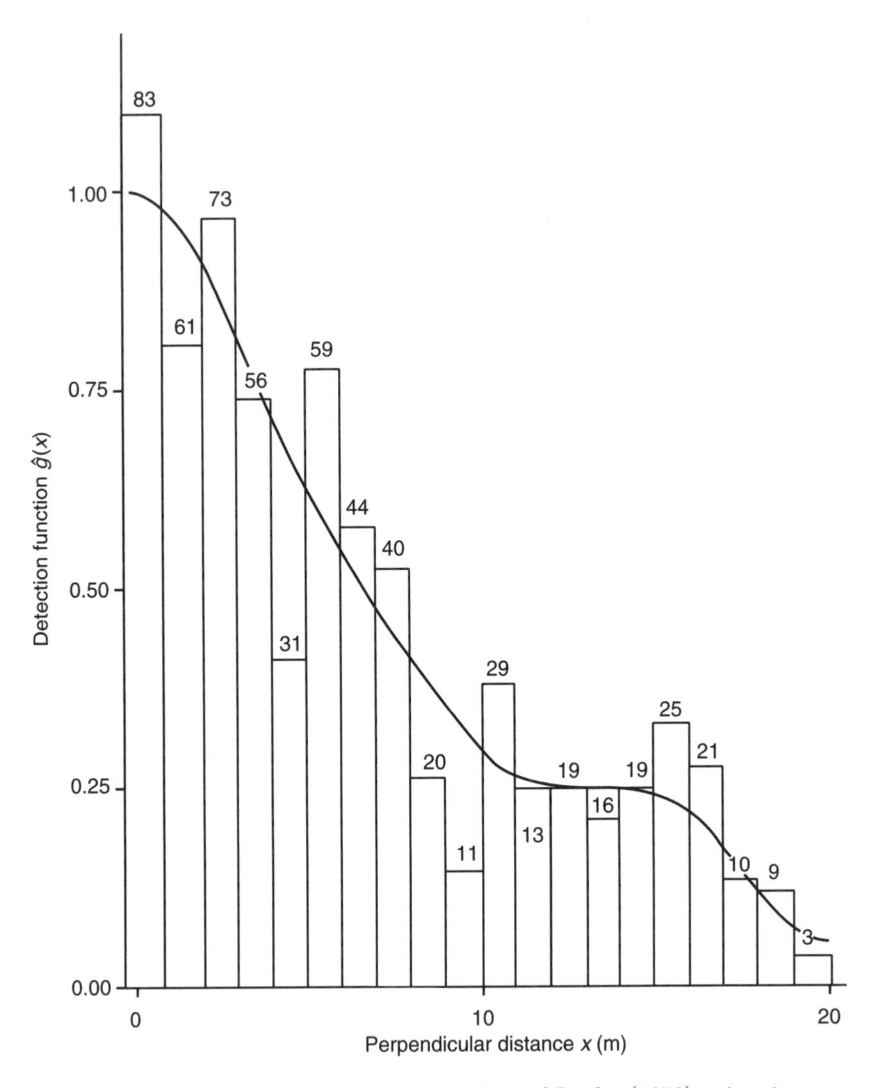

Fig. 8.1. Histogram of the wooden stake data of Laake (1978) using 1 m cut-points. Also shown is a three-term Fourier series fitted to these grouped data. Note the heaping at 0, 5 and 10 m, the relatively long tail in the distribution, and the additional mode near 15–16 m.

terms, but provided a good fit near the line (Fig. 8.2). Some heaping at 5 m now shows more clearly. This model of the pooled data yielded $\hat{D} = 37.71$ stakes per hectare ($\widehat{se}(\hat{D}) = 2.82$), with a log-based confidence interval of (32.58, 43.64). These calculations assume $var(n) = 0$, as the stakes were placed uniformly. In reality, the value of $var(n)$ in this study would be small but non-zero; zero was used only for illustration. Now, the data from each

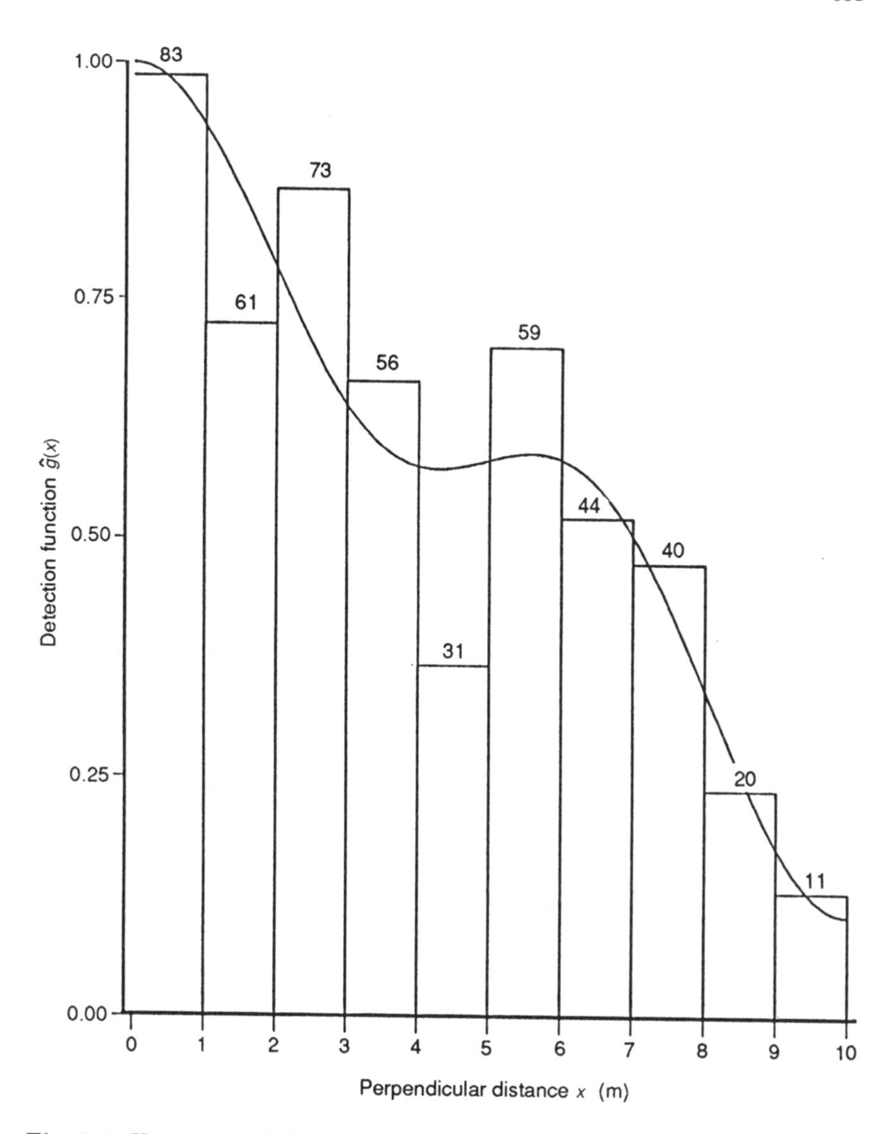

Fig. 8.2. Histogram of the wooden stake data of Laake (1978), truncated at 10 m. The detection function is modelled using a uniform key function and three cosine adjustment terms.

separate survey (i.e. individual person) can be modelled using a three-term cosine series. The average estimate of density from the 11 surveys was 36.99 ($\widehat{sd}(\hat{D}) = 5.04$), again close to the true density of 37.5 stakes per hectare.

Second, having selected a truncation point ($w = 10$ m), one could select a model using AIC, based either on the individual data sets or on the pooled data sets. Here, we will examine only the pooled data. The results are summarized below:

Model	AIC	ΔAIC
Half-normal + cosine	2118.7	6.3
Uniform + cosine	2112.4	0.0
Uniform + polynomial	2114.7	2.3
Half-normal + Hermite	2118.7	6.3
Hazard-rate + cosine	2126.8	14.4

The AIC selects the uniform + cosine model (the Fourier series) as the best model and, as shown above, provides an excellent estimate of density. The other models provided estimates that were inferior (half-normal + cosine = 33.60, uniform + polynomial = 31.47, half-normal + Hermite = 33.59 and hazard-rate + cosine = 31.90), and this could have been judged from the poor model fits near the line (Fig. 8.3). Likelihood ratio tests chose models with one parameter, except the hazard-rate model (two parameters) and the uniform + cosine (three parameters). Only the uniform + cosine model had confidence intervals that covered the true parameter.

Because the authors knew the value of D during these analyses, we cannot claim total objectivity in this example, which happened to provide an excellent estimate. However, the point is that careful review of the distance data can suggest anomalies (e.g. spiked data, long tails, heaping and a second mode), and these can suggest analysis approaches that should be considered equally with the more mechanical approach of using AIC to select the model. We advocate some truncation, especially in cases where there are clear outliers in the distance data. Poor fit of the model to the data near the line should always be of concern. Generally the guidelines outlined in Section 2.5 will serve the analyst well in planning the analysis of distance data.

8.4 Studies of nest density

Studies of duck nest density have been conducted annually since 1964 at the Monte Vista National Wildlife Refuge in Colorado. During the 27 years of these studies, 10 041.6 miles of transect were walked and 4156 duck nests were found. Here we will examine the data for individual species and years to illustrate various points, approaches and difficulties. No attempt is made here to provide a final, comprehensive analysis of these data. Further details are found in Gilbert *et al.* (1996).

Strip transects were originally established on the refuge in a systematic design, running north–south, with 300 feet between transects and $w = 8.25$ feet, giving a 5.5% sample of the entire 18.4 square mile refuge. Each transect centreline was marked by a series of numbered plywood signs attached to a 2.5-m pole (Burnham *et al.* 1980: 32). Only one-half of the original

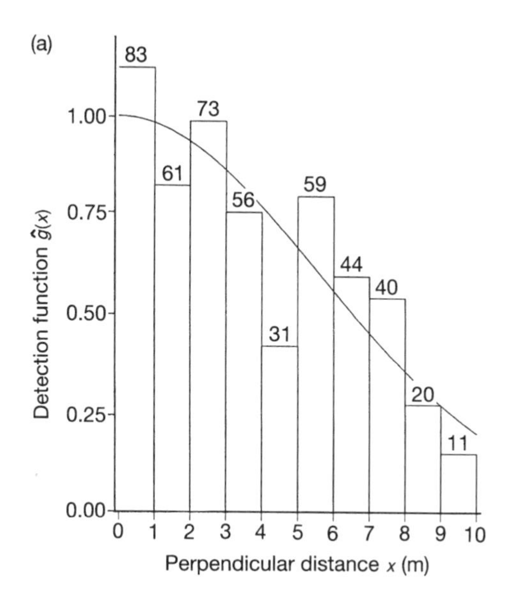

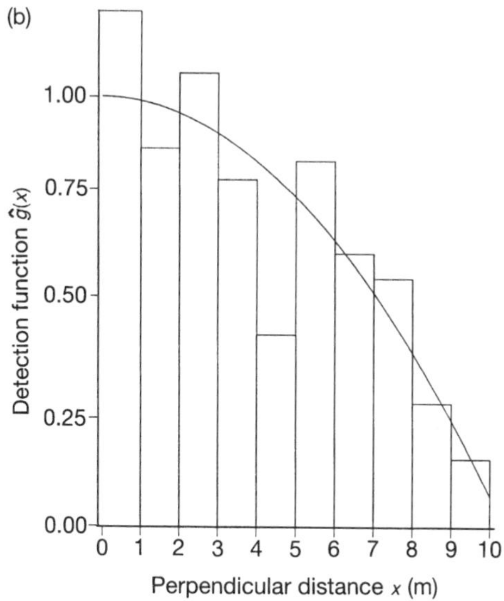

Fig. 8.3. *Continued*

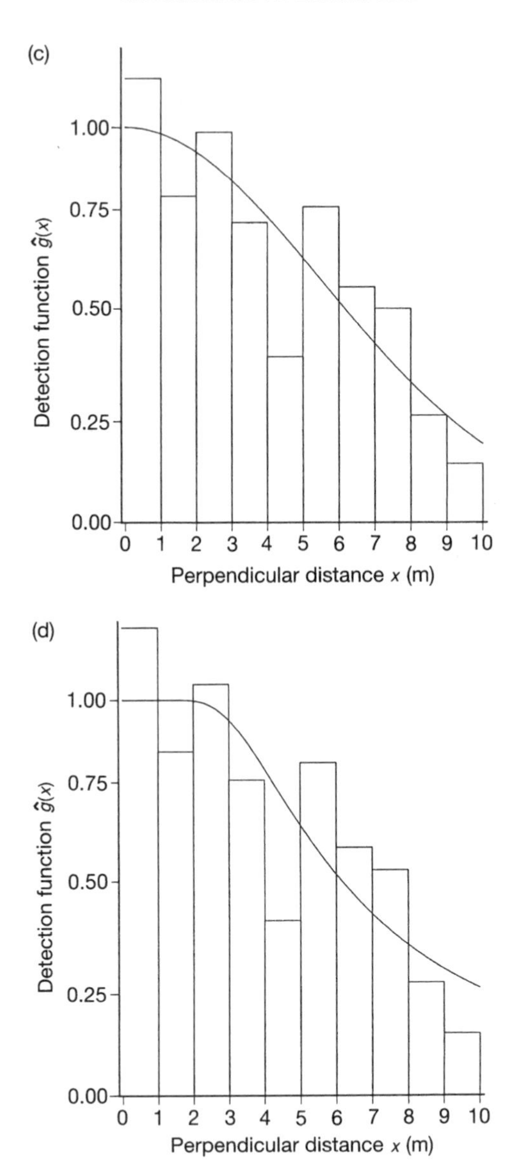

Fig. 8.3. Histograms of the wooden stake data (Laake 1978) and model fits for (a) the half-normal + cosine, (b) uniform + polynomial, (c) half-normal + Hermite and (d) hazard-rate + cosine model.

transects were surveyed during the 1969–90 period, except in 1971 when only one-quarter of the original transects were run and in 1977 when no survey was conducted. During 1969–79, w was increased to 12 feet, but this was changed back to 8.25 feet during the 1980–85 period. Strip width was increased again to 12 feet in 1986–87 and finally changed back to 8.25 feet during 1988–90. Distances to detected nests were not recorded during 1964–66 and 1975–79. These erratic changes were often due to personnel or budget limitations. Transects were searched twice each year to monitor nest density of both early and later nesting species. A third search was made in a few years, but few nests were found and these data are not included in any of the examples given here. The mallard (*Anas platyrhynchos*) was the most common nesting duck, but substantial numbers of northern pintail (*Anas acuta*), gadwall (*Anas strepera*), northern shoveler (*Anas clypeata*) and teal (*Anas cyanoptera*, *A. discors*, and *A. carolinensis*) nests were also found. Species identity could not be determined for many nests; these were classed as 'unknown'.

The refuge, at an elevation of 7500 feet, is characterized by level terrain and high desert vegetation in relatively simple communities. The drier, alkaline sites contain greasewood (*Sarcobatus vermiculatus*) and rabbit-brush (*Chrysothamnus* spp.) on the higher sites, while saltgrass (*Distichlis stricta*) dominates the lower sites. The wetter sites are dominated by baltic rush (*Juncus balticus*), but other species include cattail (*Typha latifolia*), spikerush (*Eleocharis macrosachya*), bullrush (*Scirpus validus*), and sedges (*Carex* spp.). Water is managed from pumped and artesian wells and irrigation sources and a system of dikes and borrow pits allow open water to be interspersed with vegetation cover to create good waterfowl nesting habitat. This area has one of the highest duck nest densities of any in North America.

The original design used strip transects, and it was assumed that all duck nests within the strip were detected. Perpendicular distance data were collected in 1967 and 1968, and it was clear by 1970 that some nests near w remained undetected, even with the narrow width of 8.25 feet (Anderson and Pospahala 1970). Perpendicular distances were recorded in 1969–74 and 1986–90, primarily as a means to relocate nests so that the nest fate could be determined. In years when $w = 12$ feet, it was likely that more nests remained undetected. Thus, distance sampling theory is appropriate to obtain estimates of density, account for different sampling intensities, resolve differences in transect width w, and provide a basis for correction for undetected nests in years when no distance measurements were taken.

8.4.1 *Spatial distribution of duck nests*

On biological grounds it seemed likely that nests were somewhat randomly distributed on the refuge. Thus, one might expect that the variation among the number of nests (n_i) detected by transect line (l_i) would be approximately Poisson, so that the variance in total sample size (n) might be roughly Poisson (i.e. $\widehat{var}(n) \simeq n$). The variance of n was computed empirically for mallard and non-mallard nests for each of 26 years using

$$\widehat{var}(n) = \frac{L \sum_{i=1}^{k} l_i (n_i/l_i - n/L)^2}{k-1} \tag{8.1}$$

where $L = \sum_{i=1}^{k} l_i$ and k = number of replicate lines.

Modelling of $var(n)$ as a function of n is common in statistical application (Carroll and Ruppert 1988). The ratio $\hat{b} = \widehat{var}(n)/n$ was computed for each year for mallard and non-mallard nests (Table 8.4); b is often called a variance inflation factor. A random spatial distribution of nests yields $\hat{b} \simeq 1.0$. The relationship between n and $\widehat{var}(n)$ can be estimated by a weighted linear regression through the origin, where the weight is the

Table 8.4. Sample size (n), estimates of the empirical variance of n ($\widehat{var}(n)$), and their ratio ($\hat{b}$) for mallard and non-mallard duck nests

	Mallards			Non-mallards		
Year	n	$\widehat{var}(n)$	$\hat{b}$	n	$\widehat{var}(n)$	$\hat{b}$
1964	142	210.20	1.48	119	201.27	1.69
1965	140	188.79	1.34	72	94.41	1.31
1966	151	238.01	1.58	109	146.69	1.35
1967	150	255.58	1.70	94	156.29	1.66
1968	176	281.84	1.60	139	240.67	1.73
1969	112	168.79	1.51	118	166.61	1.41
1970	103	166.44	1.62	126	225.92	1.79
1971	66	179.61	2.72	48	139.98	2.92
1972	63	99.10	1.57	64	102.08	1.60
1973	102	160.18	1.57	137	450.11	3.28
1974	69	157.18	2.28	40	54.43	1.36
1975	37	45.49	1.23	28	57.86	2.07
1976	49	61.89	1.26	46	72.48	1.58
1978	20	27.61	1.38	35	49.33	1.41
1979	13	9.73	0.74	40	96.06	2.40
1980	34	23.46	0.69	55	82.93	1.51
1981	50	105.63	2.11	54	79.27	1.47
1982	57	125.91	2.21	53	88.53	1.67
1983	79	123.10	1.56	35	48.69	1.39
1984	112	180.20	1.61	38	50.16	1.32
1985	70	112.88	1.61	82	105.49	1.29
1986	114	205.24	1.80	83	105.90	1.27
1987	130	164.00	1.26	132	195.82	1.48
1988	93	182.94	1.97	57	65.66	1.15
1989	74	103.67	1.40	56	58.63	1.05
1990	56	117.97	2.11	32	37.63	1.18
Wt. Ave.			1.630			1.677

sample size. The point estimate under this approach is a ratio estimator. The estimated value of b for mallards was 1.630 ($\widehat{se} = 0.068$) and 1.677 ($\widehat{se} = 0.111$) for non-mallards. These estimates are not significantly different ($z = 0.36$) and, thus, a pooled estimate of b for all species was computed as $\hat{b} = 1.651$ ($\widehat{se} = 0.063$). This weighted regression has an adjusted correlation of $r^2 = 0.93$ (Fig. 8.4), and provides $\widehat{var}(n) \simeq 1.7n$ which will be used

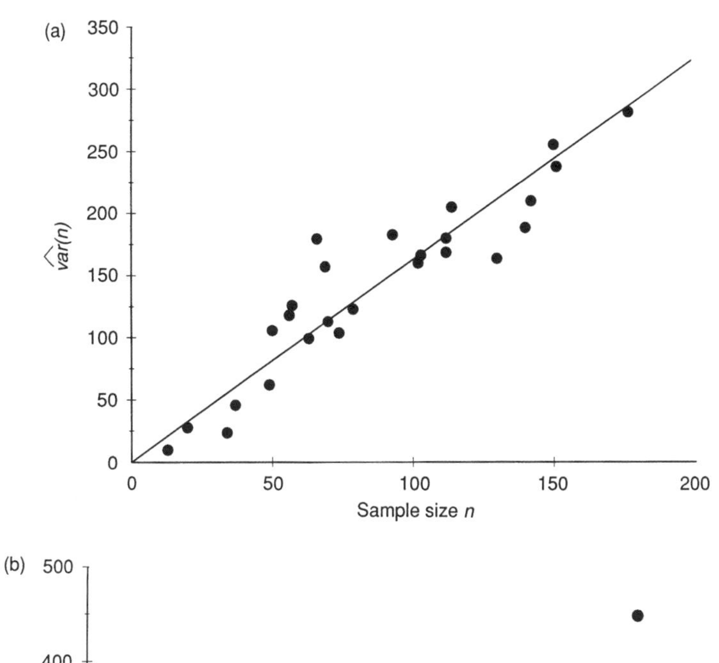

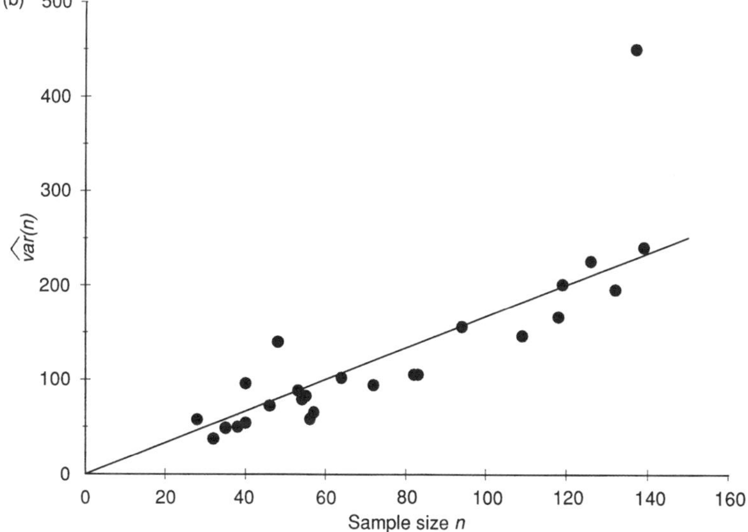

Fig. 8.4. Relationship between the empirical variance in n (i.e. $\widehat{var}(n)$) and sample size (n) for (a) mallard nests and (b) non-mallard nests.

in the remaining material for this example. This reflects some contagion in the distribution of duck nests, related, no doubt, to the variable distribution and quality of nesting habitat on the refuge. Minor species such as the redhead (*Aythya americana*), which nest in specialized habitat types, probably had a very non-random spatial distribution of nests (i.e. $b \gg 1.0$). Note that $var(n)$ and b can be estimated for all years, even those where perpendicular distances were not recorded.

8.4.2 *Estimation of density*

This material focuses primarily on the grouped distance data collected during 1969–74 and 1986–87, years when $w = 12$ feet. Histograms of the distance data are shown in Fig. 8.5 for the mallard, pintail, gadwall, teal, shoveler and unknown species. The first interval includes nests from 0 to 11 inches, the second interval includes nests detected from 12 to 23 inches, etc. Relatively few pintail nests were recorded within 2–3 feet of the centreline, whereas for other species, there appears to have been preferential heaping of nests close to the transect centreline. In each case, except the pintail, many nests found in the second interval (12–23 inches) were probably heaped into the first interval and perhaps into the third interval. Heaping at zero distance is common, and has been reported several times in distance sampling literature (e.g. Robinette *et al.* 1974), especially for data taken as sighting distance (r_i) and angle (θ_i) and then the perpendicular distances (x_i) computed as $x_i = r_i \sin(\theta_i)$.

Estimated detection functions $\hat{g}(x)$ are also plotted in Fig. 8.5, assuming the half-normal key function with Hermite polynomial adjustments, if required. The half-normal key function seems quite reasonable for modelling these data. While there was substantial variation and some obvious heaping in the counts (n_i), the fit appeared fairly good, with the clear exception of that for the pintail (discussed further in Section 8.4.3). Nests of the mallard seem to be most easily detected, as shown by the estimated unconditional probability of detection in the surveyed strip of area $a = 2wL$ ($\hat{P}_a = 0.80$, $\widehat{se} = 0.03$; Table 8.5). Gadwall nests were also easy to detect ($\hat{P}_a = 0.76$, $\widehat{se} = 0.09$), whereas teal, shoveler and unknown species nests were less detectable ($\hat{P}_a = 0.64$, $\widehat{se} = 0.04$; $\hat{P}_a = 0.60$, $\widehat{se} = 0.07$; and $\hat{P}_a = 0.63$, $\widehat{se} = 0.04$, respectively). Pooling all nests for all species results in an unconditional probability of detecting a nest within the transect of 0.78, $\widehat{se} = 0.02$. Mallards (and pintails) nest early in the season when most vegetation has little new growth and detection might be easier. Other species tend to nest later and may experience more concealment in the vegetation.

Estimates of P_a for nests of all species combined were higher in years when $w = 8.25$ feet ($\hat{P}_a = 0.84$, $\widehat{se} = 0.07$) compared to years when $w = 12$ feet ($\hat{P}_a = 0.78$, $\widehat{se} = 0.02$). In general, P_a is a function of w in that

$P_a = 1/\{w \cdot f(0)\}$; under large sample approximations, $cv(\hat{P}_a) = cv\{\hat{f}(0)\}$. However, fewer nests were found using the narrower transect. The narrow transects are inefficient as shown by the mean encounter rate (e.g. 0.867 and 1.137, respectively, for total nests). The values of w (either 8.25 or 12 feet)

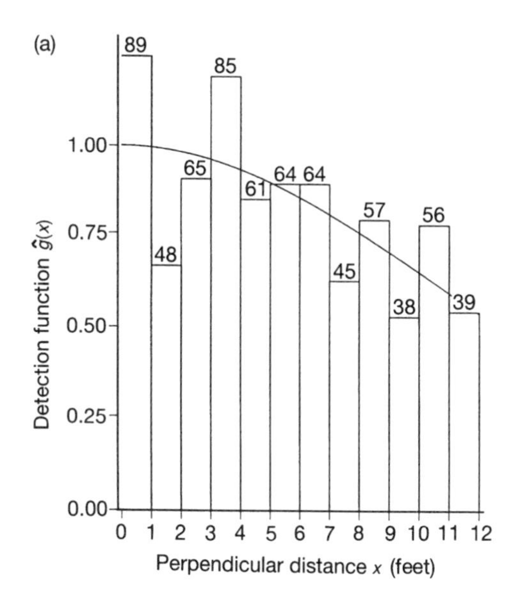

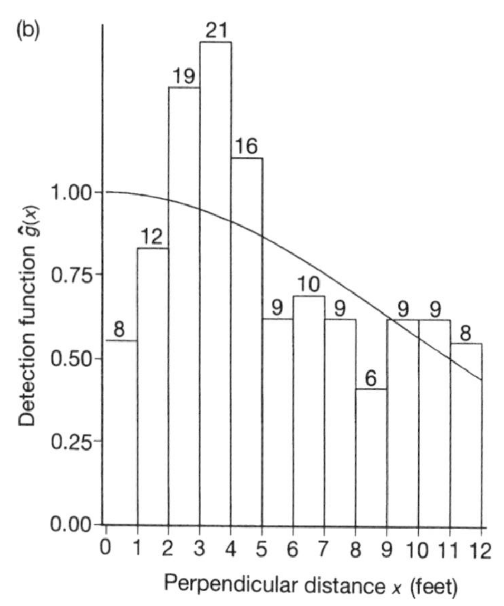

Fig. 8.5. *Continued*

were certainly too large to meet the assumptions of strip transect sampling and too small for good efficiency in line transect sampling. Using the data for mallard nests, pooled for the 1969–74 and 1986–87 period, $\hat{g}(8.25) \simeq 0.8$, thus the observer undoubtedly detected many nests beyond

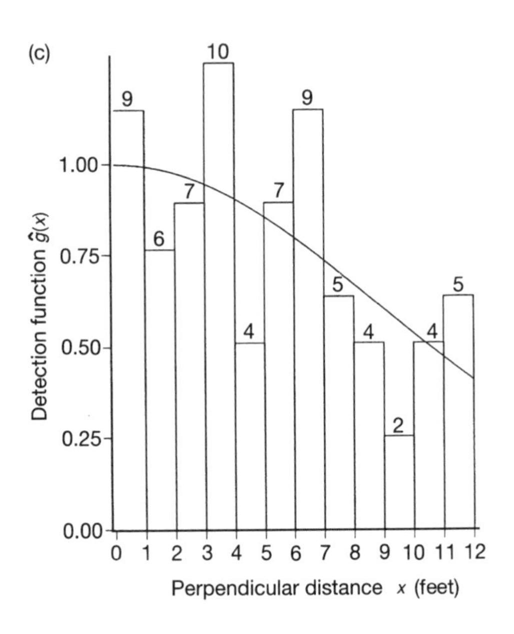

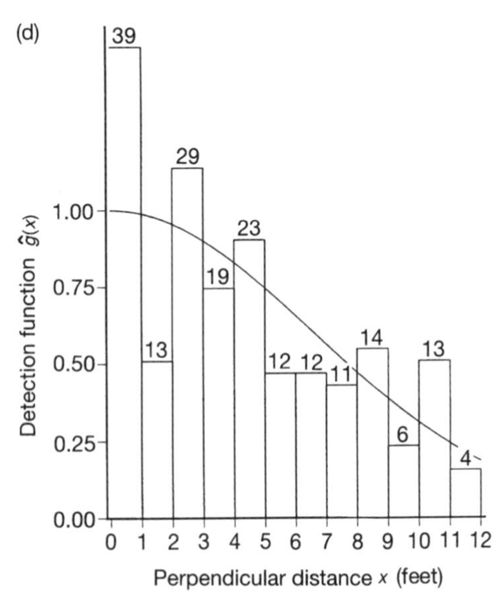

Fig. 8.5. *Continued*

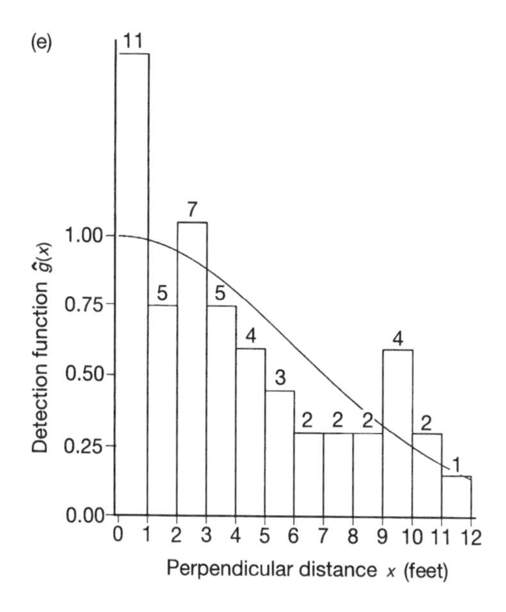

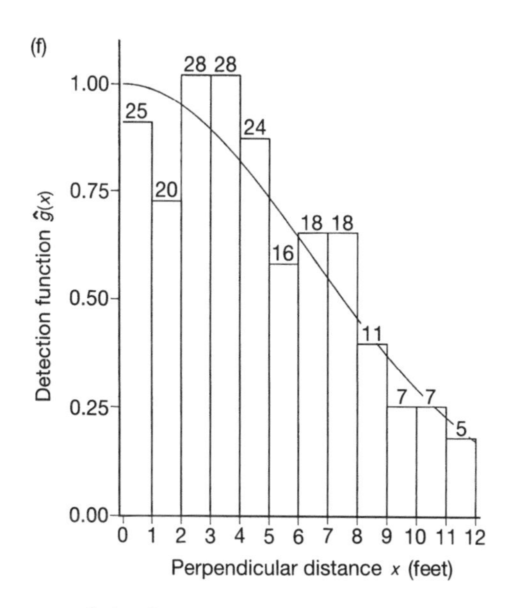

Fig. 8.5. Histograms of the distance data for the nests detected at the Monte Vista National Wildlife Refuge in Colorado, USA, during 1969–74 and 1986–87. In addition, estimated detection functions using the half-normal key function and Hermite polynomial adjustments are shown for nests of (a) mallard ($n = 711$), (b) pintail ($n = 136$), (c) gadwall ($n = 72$), (d) teal ($n = 195$), (e) shoveler ($n = 48$) and (f) unknown species ($n = 207$).

Table 8.5. Summary of statistics and estimates for survey of duck nests at the Monte Vista National Wildlife Refuge in Colorado during 1969–74 and 1986–87. See Fig. 8.5 for histograms of these data and estimated detection functions using the half-normal model and Hermite polynomial adjustment terms. Density is in nests/mile^2

Species	n	$\hat{P}_a$	n/L	$\hat{f}(0)$	$cv\{\hat{f}(0)\}$ (%)	$\hat{D}$	$cv(\hat{D})$ (%)*
All duck	1415	0.77	1.137	0.1079	2.6	323.8	3.7
Mallard	711	0.85	0.580	0.0901	3.8	149.6	5.4
Teal	195	0.64	0.157	0.1305	6.4	54.0	9.6
Pintail	136	0.78	0.109	0.1063	8.5	30.7	12.1
Gadwall	72	0.76	0.058	0.1094	11.5	16.7	16.5
Shoveler	48	0.60	0.039	0.1388	12.4	14.1	19.0
Unknown	207	0.63	0.166	0.1325	6.1	58.2	9.3

*Assuming $\widehat{var}(n) = 1.7n$.

w and could not record them. Detection near w is still good even for the wider transects ($\hat{g}(12) \simeq 0.6$), thus many nests were still readily detectable beyond the transect boundary. As only about 0.58 mallard nests were found per transect mile (Table 8.5), it is clear that increasing w would allow sample size to increase with little additional survey effort. Finding a nest is a relatively rare event and if more nests could be found, little additional time would be required to take and record the relevant measurements. It would be interesting to increase w to perhaps 15, 18 or even 20 feet in future surveys to improve efficiency for both the observers and the estimators. An additional adjustment term might be required in modelling $g(x)$ but an overall gain in the estimation process is likely. Observers would have to be cautioned to emphasize search on and near the centreline and not divert too much attention near w. Of course, it would be advantageous if heaping could be lessened and more accurate measurements were taken. Perpendicular distances might be remeasured as a test of accuracy when the fate of the nests is checked at a later time.

Mallard nests were found in adequate numbers to allow annual estimates of nest density to be made. Histograms of the distance data are shown in Fig. 8.6, with estimated detection functions $\hat{g}(x)$. Although the annual sample sizes were generally fairly adequate ($n_i > 60$ in every year), the histograms look 'rough' and certainly exhibit heaping and perhaps some careless measurements. Except in the early years of the survey, observers generally had little training and were possibly not that motivated.

Annual estimates of density from data such as those of Fig. 8.6 can be obtained from

$$\hat{D}_i = \frac{n_i \cdot \hat{f}_i(0)}{2L_i}, \quad \text{where } i = \text{year} \tag{8.2}$$

An attractive alternative exists if one is willing to make the assumption that the detection function $g_i(x)$, and hence $f_i(0)$, does not vary between years.

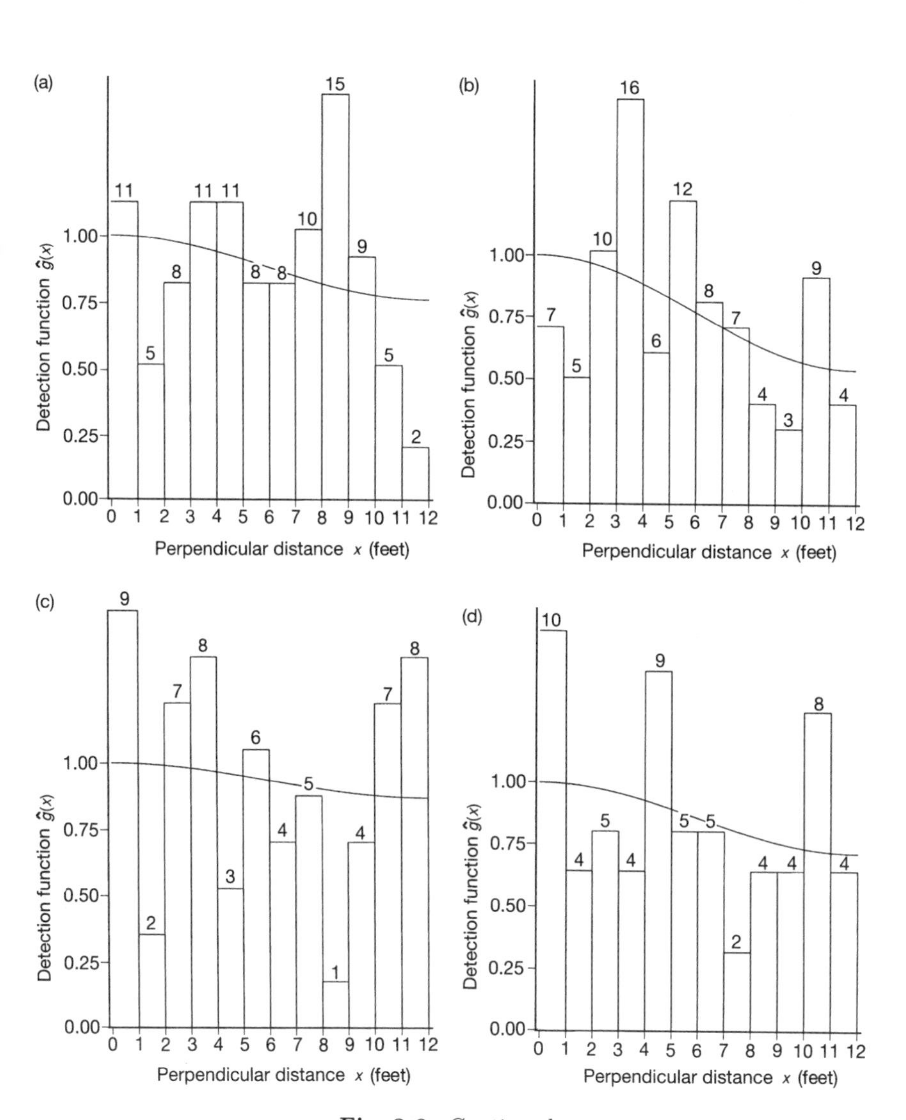

Fig. 8.6. *Continued*

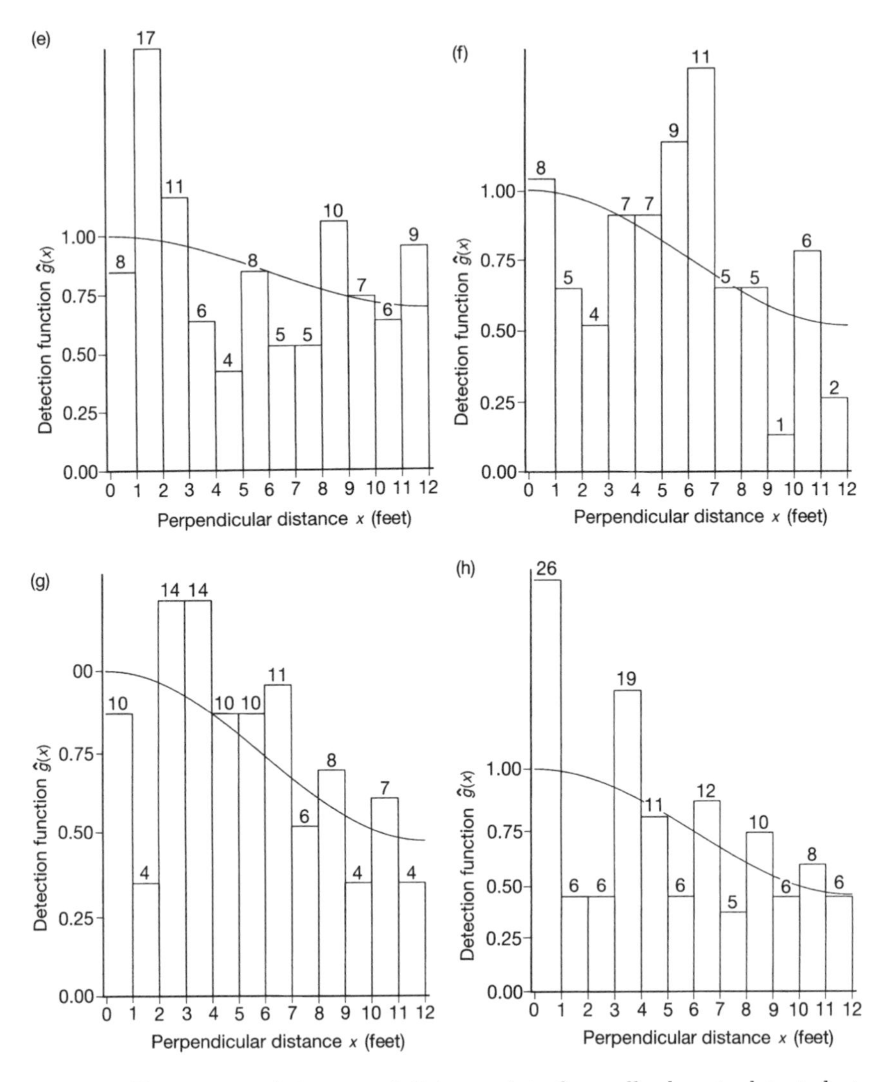

Fig. 8.6. Histograms of the annual distance data for mallard nests detected at the Monte Vista National Wildlife Refuge in Colorado, USA, during 1969–74 and 1986–87. The estimated detection function is also shown. The respective sample sizes (n) are (a) 103, (b) 91, (c) 64, (d) 64, (e) 96, (f) 70, (g) 102 and (h) 121.

This assumption seems biologically reasonable for the duck nest surveys as the vegetation varies little over time, and the search methods remained largely the same within the eight years analysed here. These reasons seem fairly compelling, but it is advisable to assess whether a common detection function can be assumed. A reasonable strategy is to fit the distance data, pooled over the eight years, to several good candidate models and select the model with the smallest AIC. These values for the mallard data under five models were:

Model	AIC	ΔAIC
Half-normal + cosine	3513.8	0.0
Uniform + cosine	3513.9	0.1
Uniform + polynomial	3514.3	0.5
Half-normal + Hermite	3513.8	0.0
Hazard-rate + cosine	3515.4	1.6

Any of the five models could be used in this case, with little difference among the models. AIC selects the single-parameter half-normal key function with no adjustment terms, although the uniform key function plus a cosine term (i.e. Fourier series) is second best by a trivial margin. The uniform + polynomial model might also be a satisfactory model, followed by the two-parameter hazard-rate model. For this example, the uniform + cosine (Fourier series) model will be used.

We now fit the eight individual distance data sets, using the same model (with the number of adjustment terms fixed) (Fig. 8.6), to give the AIC value for the model based on the pooled data together with the eight separate AIC values for the individual data sets. The sum of the eight AIC values is 3519.2, which exceeds 3513.9 ($\Delta\text{AIC} = 5.3$), the value obtained from pooled data for the Fourier series model above. This strongly favours pooling the distance data over the eight years to obtain a single $\hat{f}(0)$ and its standard error. Then, yearly estimates of density can be computed using the yearly sample size (n_i) and

$$\hat{D}_i = \frac{n_i \cdot \hat{f}(0)}{2L_i}, \quad \text{where } i = \text{year} \tag{8.3}$$

The estimate of $f(0)$ is specific to a given transect width, i.e. the value of $\hat{f}(0)$ for data from a transect with $w = 12$ feet cannot be used for years when the transect width was 8.25 feet. This approach seems appropriate for the mallard data, and it is nearly essential for species such as teal where yearly sample size would not support a reliable estimate of the year-specific detection function $g_i(x)$ (or, equivalently, $f_i(x)$). Data for the shoveler ($n = 48$) and gadwall ($n = 72$) support only an estimate of an average $f(0)$ over the eight year period, but this analysis approach allows annual estimates of density by using the year-specific sample sizes n_i. This general approach can be used for the analysis of other years of data where different transect widths were used.

Average estimates of density over the eight years under the five models were quite similar, ranging from 149.3 to 157.0 mallard nests per square mile. This might have been expected because the AIC values were all of similar magnitude. Models for $g(x)$ contained only one parameter (either a key or an adjustment parameter), except the hazard-rate model, with two parameters. Thus, two of the five models used only the half-normal key function.

8.4.3 *Nest detection in differing habitat types*

Despite large sample sizes, well distributed in time and space, the ability to examine differences in detectability by vegetation type was limited by the fact that baltic rush and greasewood made up approximately 68% and 15% of the vegetation on the refuge, respectively. Initially it was hypothesized that nest detectability would decline more rapidly with distance from the centreline in the tall, but often sparse, stands of greasewood when compared to the lower, more dense areas of rush. Instead, it became clear that the histogram of grouped distance data for nests found in greasewood indicated a mode well away from the transect centreline. It was hypothesized that observers would avoid the thorny greasewood (see Fig. 8.7) by walking off line and around these shrubs. Thus, nests at the base of these shrubs tended to go undetected near the transect centreline. Nests detected at the edge of greasewood clumps would be detected with near certainty while the observer was temporarily off the centreline (and thus avoiding the greasewood). Once such a nest was found its distance to the centreline was measured and recorded. Such temporary departures from the transect centreline could explain the odd distance data for the pintail nests (Fig. 8.5). Perhaps pintail were common nesters in greasewood types and, thus, many were missed near the centreline. Indeed, 24.2% of the pintail nests were found in greasewood; surely this percentage would be still higher if nests near the centreline in greasewood were all detected. Other species nested in greasewood types less frequently: mallard 15.6%, gadwall 19.5%, teal 6.9% and shoveler 2.6%. We tentatively conclude that observers were reluctant to enter the thorny greasewood type, and this resulted in nests being missed near the centreline.

An alternative explanation is that the observer measured the distance from his or her position to the nest and that pintail tended to nest at least 2 feet into the greasewood type. Then Fig. 8.5b would arise without missing any nests near the centreline; instead, the data would arise because the observer's path would go through habitat with a low pintail nest density. In any event, the presence of obstacles such as greasewood on the line must be dealt with effectively in the field survey or the analysis of the data can be problematic. We do not always advocate that the observer plunge through such cover types; instead, extra care in searching must be taken when an

Fig. 8.7. Stand of greasewood with extensive areas of bare ground typify many upland sites on the Monte Vista National Wildlife Refuge. Observers may tend to avoid walking through these thorny shrubs, thus biasing the data.

easier path is temporarily followed. For example, the observer could go around clumps of such vegetation both to the left and then to the right, searching the centreline more carefully. In any event, the measurements must be taken from the transect centreline, not to the observer who may be away from the centreline.

A definitive analysis of data such as those for the pintail nests is not possible. Approximate analyses that might be useful could be considered. First, one could fit a monotonically constrained function for $g(x)$, as is shown in Fig. 8.5b for the half-normal key function with Hermite polynomial adjustments. This is likely to result in an underestimate of density if a substantial number of nests near the centreline was undetected. However, in this particular case, one knows from several other, similar species in this survey that the shape of $g(x)$ has a broad shoulder, so that the procedure might be acceptable.

Second, one could use some arbitrary left-truncation and then estimate $f(0)$ and D using, for example, the uniform + cosine or half-normal + Hermite model. First, one could decide on a truncation point; 3 feet might be reasonable for the pintail nest data. Here the grouped distance data less than 3 feet could be discarded, the remaining data rescaled as if the fourth interval was actually the first interval, and proceed to estimate density in the usual way (Fig. 8.8a). This is likely to be similar to the first procedure because we have reason to suspect that the detection function for pintail nests is fairly flat. Still, in this case, some underestimation might be expected (unless $g(0) \simeq 1.0$ but nests close to zero tended to be recorded at around 3 feet; then overestimation might result).

Third, the left-truncation procedure of Alldredge and Gates (1985) could be employed, using the same truncation point. The result of this procedure is very dependent upon the model chosen and is often imprecise (Fig. 8.8b). In this example, where something is known about the distribution of distances of nests of other species of ducks, it seems likely that density of pintail nests is overestimated using this approach. Of course, any left truncation decreases sample size. The results of using the three approaches for the pintail nest data are summarized in Table 8.6 for the half-normal key function and Hermite polynomial adjustments.

The three estimates seem fairly reasonable for the pintail nest data, although one might prefer a density estimate near 30–32, rather than 35, unless the observer's path around greasewood types tended to sample areas of low pintail nest density. Considerable precision is lost in efforts to alleviate this problem; this is to be expected given the uncertainty introduced.

8.4.4 *Models for the detection function* $g(x)$

Various combinations of the key and adjustment functions provide flexibility in modelling the detection function $g(x)$. For data sets exhibiting a reasonable shoulder and meeting the other assumptions of distance sampling, the choice of model, among those recommended here, is relatively unimportant. Estimates of density and coefficients of variation are summarized in Table 8.7 for several reasonable models for nest data on mallard, gadwall, teal and shoveler for the Monte Vista data. The differences in estimated density are small relative to the coefficients of variation.

If the distance data are distributed in a more spiked form, the choice of model is more difficult and the estimate of density more tenuous. The models recommended here are likely to perform reasonably well, except in pathological cases. A model with an appreciably smaller χ^2 goodness of fit value, if constrained to be non-increasing, will tend to be better than other models with the same number of estimated parameters. However, if there is heaping or overdispersion in the data, goodness of fit tests are of relatively little help in model selection. Note that some lack of fit near w is of little consequence in comparing the fit of different models, whereas lack of fit close to the line can be problematic.

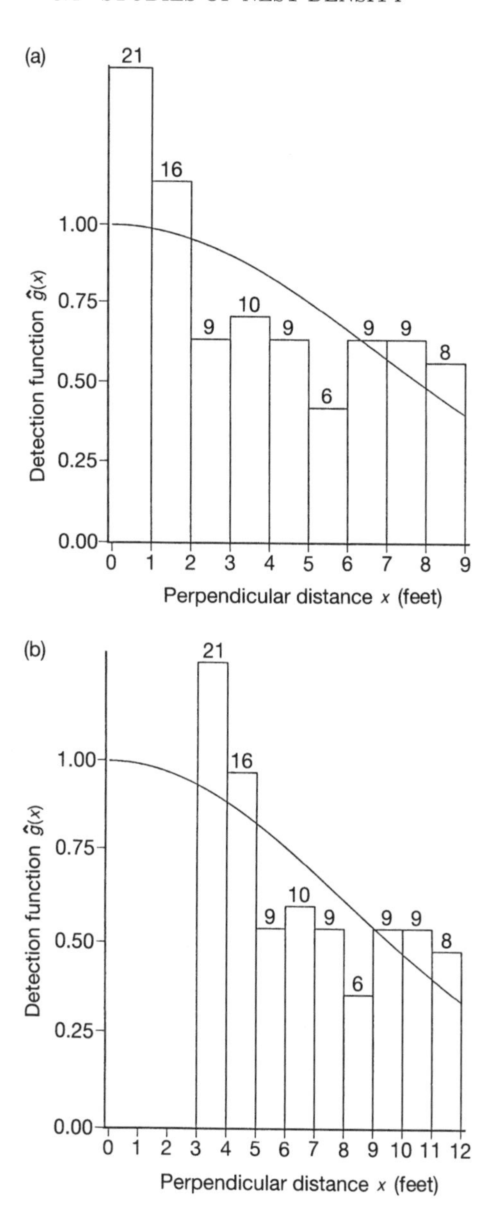

Fig. 8.8. Histograms of the distance data for pintail nests detected at the Monte Vista National Wildlife Refuge in Colorado, USA, during 1969–74 and 1986–87. Two estimates of $g(x)$ are shown using alternative ways to left-truncate the data and minimize the problems observed in the first 3–4 distance categories. Left-truncation and rescaling are shown in (a), and the Alldredge and Gates (1985) approach is shown in (b).

Table 8.6. Estimated density of pintail nests using three methods for allowing for the lack of detections near the trackline. 'Full data' means fitting a non-increasing detection function to the full data set; 'Left-truncate and rescale' means left-truncating all data within 3 feet of the trackline, and then analysing the data as if the line had been located 3 feet from the actual line; 'Full left-truncation' means truncating all data within 3 feet of the trackline, and extrapolating the detection function, fitted to these truncated data, back to the trackline

Method	n	$\hat{D}$	cv (%)
Full data	136	30.7	12.1
Left-truncate and rescale	97	29.8	14.2
Full left-truncation	97	35.0	17.9

Table 8.7. Summary of density estimates (above) and coefficients of variation (below) for five models of $g(x)$ and four duck species, 1969–74 and 1986–87

Key function	Adjustment terms	Gadwall ($n = 72$)	Teal ($n = 195$)	Shoveler ($n = 48$)	Mallard ($n = 711$)
Half-normal	Cosine	16.7	63.1	19.0	149.6
		16.5	12.4	22.4	5.4
Half-normal	Hermite	16.7	54.0	14.1	149.6
		16.5	9.6	19.0	5.4
Uniform	Cosine	17.4	54.8	14.0	155.2
		16.5	9.0	17.7	5.7
Uniform	Polynomial	15.9	61.5	12.2	147.3
		15.0	10.9	16.2	5.0
Hazard-rate	Cosine	16.2	67.0	17.9	147.6
		19.4	18.0	23.1	6.6

8.4.5 *Estimating trend in nest numbers*

Gilbert *et al.* (1996) provide estimates of nest density (nests per km^2) for all duck species combined for the years 1964–90. In years for which distance data were collected, densities were estimated using line transect sampling. For the remaining years, strip transects were conducted, and the strip counts were corrected based on line transect analysis of comparable data from other years. We use these to illustrate how temporal trends in density or abundance can be estimated. For short sequences of estimates, linear regression of the estimates, or their logarithms, on time is often used to assess whether there has been a 'significant' trend. Although this yields estimates of average change over a time period, they become increasingly less useful as the time span increases, as few populations show linear change (whether on a log scale or not) over an extended time period. We prefer therefore to estimate the pattern of change, using generalized additive modelling (Hastie and Tibshirani 1990).

We use the gam routine of Wood (2000). It has several advantages. One is that it has a stronger theoretical base than the methods of Hastie and Tibshirani (1990), yet is comparable in terms of computer resources. Another is that it incorporates an integrated approach, based on generalized cross-validation, for estimating the degrees of freedom of the smooth. The degrees of freedom control the extent to which the density estimates are smoothed. Suppose there are estimates of density for T years (not necessarily consecutive). No smoothing occurs when the degrees of freedom are equal to $T - 1$, which corresponds to fitting a generalized linear model with year as a factor at T levels; maximum smoothing occurs when degrees of freedom equal one, corresponding to fitting a generalized linear model with year as a continuous covariate, so that change is assumed linear on some scale. Wood's algorithm provides an objective means of selecting an intermediate option. The method is available as the function gam() from Wood's mgcv package, written for the software package R, both of which are free. The function uses one-dimensional penalized regression splines. Details are provided at http://www.ruwpa.st-and.ac.uk/simon/gam.html.

In our analysis, duck nest density is the response, and year is the covariate to be smoothed. We selected a gamma error distribution and a logarithmic link function. The estimated degrees of freedom were 5.87, and the resulting smooth is plotted in Fig. 8.9. There are clearly strong trends, with high nest densities in the early years of the surveys, reaching a maximum at around 1970. Densities fell quickly through the 1970s, reaching a minimum at around 1978, before increasing again to 1960s levels by the late 1980s. By 1990, numbers seemed to be falling again. Because of the large trends in these data, there is little need for further analysis; the confidence intervals in Fig. 8.9 indicate clearly that the decline and subsequent recovery are not just an artefact of sampling variation. For surveys with weaker trends, it may be useful to identify points in time at which the underlying trend changes significantly. The method of Fewster *et al.* (2000) allows significant change-points to be identified from a trend estimated using generalized additive modelling.

8.5 Fin whale abundance in the North Atlantic

Large-scale line transect surveys of the North Atlantic to assess whale abundance were carried out in 1987 and 1989 (North Atlantic Sightings Surveys, NASS-87 and NASS-89). We analyse here the fin whale (*Balaenoptera physalus*) data from the 1989 survey collected by Icelandic

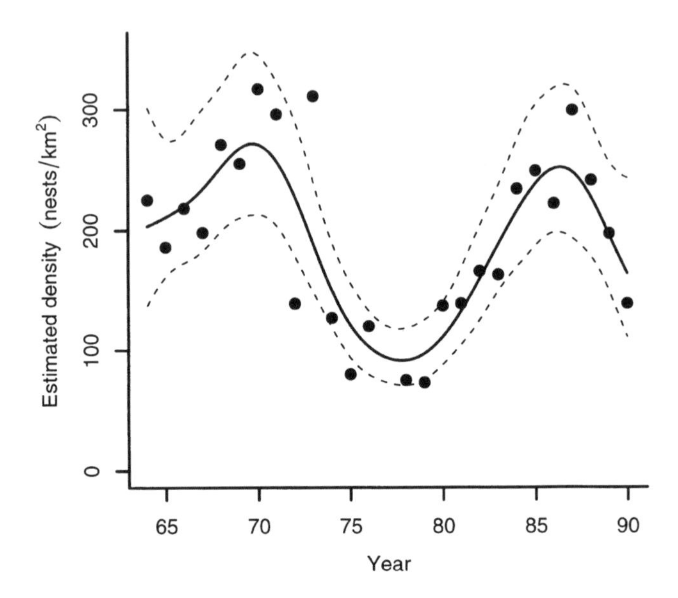

Fig. 8.9. Estimated trend in density of duck nests at the Monte Vista National Wildlife Refuge, 1964–90. Annual estimates of density are shown together with a smooth estimate of trend obtained by fitting a generalized additive model with a gamma error distribution and log link function. The dashed lines indicate approximate 95% confidence intervals for points on the curve.

vessels to illustrate the use of stratification. The analyses are extracted from Buckland *et al.* (1992b).

In 1989 four Icelandic vessels surveyed Icelandic and adjacent waters during July and August. The area covered was mostly within the East Greenland/Iceland stock boundaries for fin whales, and we consider here abundance estimation for that stock alone.

Sighting distances and angles were smeared (see Section 7.4.3.4) and assigned to perpendicular distance intervals, using smearing method 2 of Buckland and Anganuzzi (1988a), and the hazard-rate model was fitted to the group frequencies. Detections were often of more than one animal, so an analysis of clusters was carried out; average cluster (pod) size was roughly 1.5 whales. Several potential stratification factors were identified: geographic block, Beaufort (a scale for wind speed, generally determined from sea state), cloud cover, vessel and pod size. Ideally stratification should be by all of these factors, but sample size considerations preclude this. Variables Beaufort, cloud cover and pod size could be entered as covariates to avoid sample size difficulties, although it is then necessary to define a linear or generalized linear model between these effects and say effective strip half-width or encounter rate (see Buckland *et al.* in preparation), and confounding between say Beaufort and geographic location, and hence between Beaufort and whale density, is inevitable. For analysing minke whale data, Gunnlaugsson and Sigurjónsson (1990) used generalized linear modelling to estimate sighting efficiency in different Beaufort states during NASS-87. This approach also has shortcomings when Beaufort varies strongly with geographic location, if whale density also varies geographically. For example, encounter rate may be lower in high Beaufort simply because a disproportionate amount of rough weather encountered by survey vessels was in an area with low animal density. Geographic stratification reduces but does not eliminate this effect. The problem of estimating fin whale abundance is easier than that of estimating North Atlantic minke whale abundance since cues are more visible. We adopt a simpler approach here to determine stratification factors.

To assess say the effect of Beaufort, average pod size, encounter rate and effective strip half-width were estimated for each Beaufort category (0–6) in turn, pooling across all other possible stratification factors. Standard errors were calculated for each estimate, and z-tests carried out to assess whether there are significant differences in estimates at different Beauforts. Standard error for pod size was calculated as sample standard deviation divided by square root of sample size; for encounter rate, the rate per day was calculated, and the sample variance of these rates, weighted by daily effort, used as described for the empirical method of Section 3.6.2; and the standard error for effective strip half-width was obtained from likelihood methods, via the information matrix. The stratification factors are partially confounded with each other, and the above approach ignores interactions between them; analyses are supplemented here by knowledge of likely effects of the different factors on the three components of estimation to determine an appropriate analysis. Thus results from z-tests are not used blindly; if a pairwise test indicates that effective strip half-width is wider at Beaufort 4 than Beaufort 1, it would be considered spurious, because it is counter to the knowledge that detection is easier in low Beaufort, whereas if there was a trend towards narrower effective strip half-widths as Beaufort increases, stratification would be deemed necessary.

Suppose mean size of pods detected during Beaufort 0 is $\bar{s}_0$, and during Beaufort 1, $\bar{s}_1$. Denote their standard errors by $\widehat{se}(\bar{s}_0)$ and $\widehat{se}(\bar{s}_1)$, respectively. Then a z-test is carried out by calculating

$$z = \frac{\bar{s}_0 - \bar{s}_1}{\sqrt{\{\widehat{se}(\bar{s}_0)\}^2 + \{\widehat{se}(\bar{s}_1)\}^2}} \tag{8.4}$$

The distribution of z is approximately normal. Thus if $z > 1.96$ or $z < -1.96$, the mean pod sizes differ significantly at the 5% level ($p < 0.05$). Evidence against the null hypothesis that the mean pod size is the same in both sea states is strong if p is small, whereas if p is large, the data are consistent with the null hypothesis.

Table 8.8. Number of sightings (after truncation but before smearing), effective strip half-width, encounter rate and mean pod size by sea state, Icelandic fin whale data, NASS-89. Standard errors in parentheses. Values in the same column with different superscript letters differ significantly ($p < 0.05$)

Beaufort	Number of sightings n	Effective strip half-width (n.m.)		Encounter rate (pods/100 n.m.)		Mean pod size $\bar{s}$	
0	13	0.55	$(0.05)^a$	3.08	$(1.90)^a$	1.69	$(0.24)^{ab}$
1	42	2.37	$(0.22)^b$	3.47	$(0.54)^a$	1.48	$(0.15)^{ab}$
2	83	2.00	$(0.23)^{bc}$	4.19	$(1.12)^a$	1.54	$(0.08)^a$
3	78	1.60	$(0.20)^c$	4.13	$(0.47)^a$	1.63	$(0.11)^a$
4	44	1.17	$(0.29)^c$	2.55	$(0.38)^a$	1.25	$(0.10)^b$
5	33	0.49	$(0.19)^{ac}$	2.42	$(0.92)^a$	1.21	$(0.10)^b$
6	18	1.61	$(0.19)^c$	2.51	$(1.73)^a$	1.22	$(0.14)^b$

No significant differences in encounter rates by sea state (Beaufort 0–6) were found (Table 8.8). Mean pod size did not differ significantly for Beauforts 0–3 or for Beauforts 4–6, but there was strong evidence that the mean of recorded pod sizes in Beaufort ≥ 4 is smaller than for Beaufort 2 or 3. The effective strip half-width was significantly smaller for Beaufort 0 than for all other Beauforts except 5, and significantly larger for Beaufort 1 than for Beauforts 3, 4, 5 or 6 ($p < 0.05$). No other differences were significant at the 5% level, although the effective strip half-width was significantly smaller at Beaufort 5 than at Beauforts 2, 3 and 6 at the 10% level. The unexpected result for Beaufort 0 corresponds to a very small sample size (13); otherwise, there is an indication that effective strip half-width decreases with Beaufort, which is what we would expect. We estimate densities separately by low Beaufort (0–3) and high Beaufort (4–6), and average resulting estimates across Beaufort categories, weighting by effort. This analysis is valid provided the probability of detection on the centreline, $g(0)$, is unity for both Beaufort categories; the effective strip half-width need not be the same for both categories.

A similar analysis of cloud cover produced no significant differences, except that the encounter rate at cloud cover 3 was significantly higher than at cloud cover 2 ($p \simeq 0.01$), probably because relatively more cloud cover 3 occurred in areas of high fin whale density. If cloud cover 3 did increase detectability, effective strip half-width might be expected to increase, yet no pairwise comparisons provided any evidence of this ($p > 0.2$ for all six pairwise tests).

The geographic blocks defined for Icelandic surveys in 1989 are shown in Fig. 8.10. Highly significant differences between some blocks in encounter rate and mean pod size are unsurprising, and we stratify by block for each of these components of estimation. There are also several pairwise

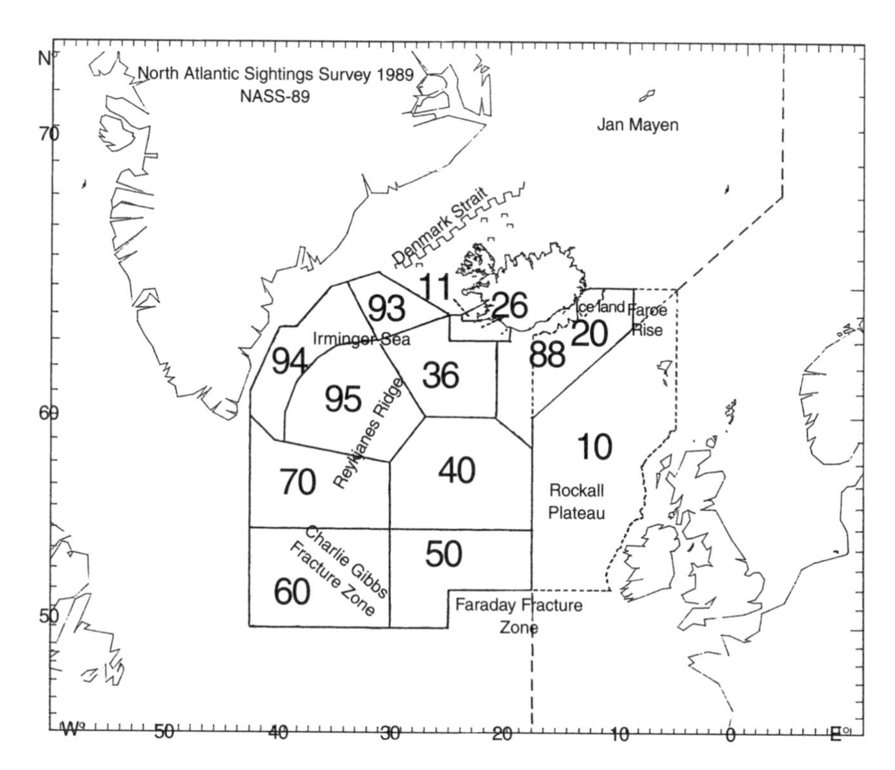

Fig. 8.10. Geographic blocks for which abundance of the East Greenland/ Iceland stock of fin whales is estimated from Icelandic 1989 data.

comparisons between blocks that indicate significant differences between effective strip half-widths. Blocks 40, 50 and 93 yield wide estimated effective strip half-widths, whereas the estimates for blocks 70 and 95 are small (Table 8.9). Given adequate sample size, stratification could be by block, as for encounter rate and mean pod size. However, effective strip half-width estimation is unreliable for small samples. There were only nine sightings in block 70 and five in block 95, rendering comparisons between them and other blocks of little value. Thus just the differences between block 94 and blocks 40 ($p \simeq 0.04$), 50 ($p \simeq 0.03$) and 93 ($p \simeq 0.05$) are genuine cause for concern. For estimating effective strip half-width, we choose here to stratify the area into two parts: south (blocks 40, 50, 60 and 70) and north, since this also effectively stratifies by vessel type (below).

The three components of estimation were also considered by vessel (Table 8.10). Most pairwise comparisons between vessels for encounter rate were significant, as were many of those for mean pod size. These differences arise largely because vessels operated in different blocks; there is strong confounding between vessel differences and block differences.

Table 8.9. Number of sightings (after truncation but before smearing), effective strip half-width, encounter rate and mean pod size by area, Icelandic fin whale data, NASS-89. Standard errors in parentheses. Blocks 11 (no sightings) and 26 (one sighting) are ignored. Values in the same column with different superscript letters differ significantly ($p < 0.05$)

Block	Number of sightings n	Effective strip half-width (n.m.)		Encounter rate (pods/100 n.m.)		Mean pod size $\bar{s}$	
36	54	0.94	$(0.51)^{abc}$	4.86	$(1.31)^{ab}$	1.35	$(0.10)^{ab}$
40	15	1.88	$(0.14)^{a}$	1.30	$(0.75)^{cd}$	1.13	$(0.14)^{a}$
50	23	2.07	$(0.35)^{a}$	2.03	$(1.06)^{bcd}$	1.35	$(0.13)^{ab}$
60	36	1.31	$(0.32)^{abc}$	3.26	$(1.40)^{abc}$	1.36	$(0.10)^{ab}$
70	9	0.68	$(0.18)^{c}$	1.18	$(0.51)^{cd}$	1.11	$(0.12)^{a}$
88	32	1.68	$(0.44)^{ab}$	2.57	$(0.45)^{bc}$	1.56	$(0.13)^{bc}$
93	70	1.87	$(0.18)^{a}$	16.39	$(1.96)^{e}$	1.69	$(0.12)^{c}$
94	66	1.14	$(0.21)^{bc}$	7.84	$(2.28)^{ae}$	1.53	$(0.12)^{bc}$
95	5	0.75	$(0.31)^{abc}$	0.24	$(0.20)^{d}$	1.20	$(0.22)^{abc}$

Table 8.10. Number of sightings (after truncation but before smearing), effective strip half-width, encounter rate and mean pod size by vessel, Icelandic fin whale data, NASS-89. Standard errors in parentheses. Values in the same column with different superscript letters differ significantly ($p < 0.05$)

Vessel	Number of sightings n	Effective strip half-width (n.m.)		Encounter rate (pods/100 n.m.)		Mean pod size $\bar{s}$	
Sk	43	1.09	$(0.43)^{a}$	1.79	$(0.47)^{a}$	1.26	$(0.08)^{a}$
AF	49	1.15	$(0.26)^{a}$	1.45	$(0.21)^{a}$	1.31	$(0.08)^{a}$
Hv8	83	1.43	$(0.33)^{a}$	4.58	$(0.47)^{b}$	1.43	$(0.08)^{ab}$
Hv9	136	1.37	$(0.17)^{a}$	8.02	$(2.46)^{b}$	1.61	$(0.08)^{b}$

If vessel differences in encounter rate in particular occurred because different vessels have different searching efficiencies, significant differences in effective strip half-width between vessels might be anticipated, yet none were close to significance ($p > 0.2$ in all pairwise tests). Given the similarity in effective strip half-widths across vessels, we pool distance data across vessels prior to analysis. Effective strip half-widths for the two research vessels (1.09 and 1.15 n.m.) were slightly smaller than for the whaling vessels (1.43 and 1.37 n.m.). Although these differences are not significant, the impact on the analyses of estimating the effective strip half-width for research vessels separately from that for whaling vessels was assessed, and found to be slight. Because all effort in southern blocks was carried out by research vessels and most effort in the northern blocks was by whaling vessels, the decision to estimate the effective strip half-width separately for

Table 8.11. Number of sightings (after truncation but before smearing), effective strip half-width and encounter rate by pod size, Icelandic fin whale data, NASS-89. Standard errors in parentheses. Values in the same column with different superscript letters differ significantly ($p < 0.05$)

Pod size	Number of sightings n	Effective strip half-width (n.m.)		Encounter rate (pods/100 n.m.)	
1	211	1.27	$(0.16)^a$	2.27	$(0.16)^a$
2	68	1.25	$(0.28)^a$	0.71	$(0.13)^b$
3	22	1.50	$(0.25)^a$	0.23	$(0.05)^c$
4	8	1.71	$(0.23)^a$	0.09	$(0.03)^d$
>4	2	—		—	

the northern and southern blocks, and to estimate encounter rate and mean pod size by individual block, in effect gives stratification by vessel type.

Effective strip half-width did not show significant differences by size of pod at the 5% level (Table 8.11), although there was a weak indication that the effective half-width was greater for pods of four or more animals than for single animals ($p \simeq 0.1$). Since 68% of sightings were of single animals, and a further 22% were of pairs, the effect of variation in detectability due to pod size on abundance estimates will be slight. However, stratification by pod size is likely to be more valid and was adopted. Estimated effective strip half-width is almost identical for single animals and for pairs, and very few pods of more than three animals were detected, so two strata were defined: small pods (one or two animals) and large pods (three or more). Small sample sizes forced one modification to the preferred method of analysis: the number of large pods was too small to allow estimation of effective strip half-width separately for high and low Beaufort, so that for large pods only, a pooled estimate of effective strip half-width across Beaufort categories was calculated.

To assess the impact of the decision to stratify by pod size on estimates, two further analyses were carried out. The first of these was exactly as above, except data were not stratified by pod size. In the second, the data were reanalysed with individual animals as the sighting unit. Thus a pod of size three between 0.75 and 1.0 n.m. perpendicular distance contributes a frequency count of three to that distance interval (before smearing). We do not recommend this approach in general, although it can be effective, if variances are estimated by robust methods. When estimates were summed across geographic blocks prior to combining across Beaufort categories, the preferred method of analysis (stratifying by pod size) gave a total estimate of 11 054 whales ($\widehat{se} = 1670$). Without stratification by pod size, the estimate was 11 702 whales ($\widehat{se} = 1896$). When individual whales were taken as the sampling unit, an estimate of 11 758 whales ($\widehat{se} = 1736$) was

Table 8.12. Abundance estimates by block, East Greenland/Iceland fin whale stock, 1989

Block	Number of sightings n	$\hat{D}$ (whales/ $10\,000\,\text{nm}^2$)	$\widehat{se}(\hat{D})$	Size of block (nm^2)	Abundance estimate $\hat{N}$	$\widehat{se}(\hat{N})$
36	54	270	72	44 172	1195	316
40	15	68	39	107 842	735	421
50	23	87	57	99 750	865	569
60	36	158	67	131 458	2071	879
70	9	74	58	88 571	658	517
88	32	129	38	59 848	770	230
93	70	873	220	21 761	1900	480
94	66	450	101	46 092	2073	467
95	5	16	14	69 396	111	95
All	323	155	25	668 891	10 378	1655

obtained. Note that this latter strategy gave a very similar standard error to the other methods, even though sightings of individual whales were not independent events, and sample size is thus artificially increased. This occurs because of the robust method of estimating the variance in encounter rate, found by calculating the sample variance of the rate per day, weighted by daily effort, used as described for the empirical method of Section 3.6.2. Nevertheless, this approach underestimates the variance in effective strip half-width, unless it is obtained by resampling methods.

Abundance estimates for the East Greenland/Iceland stock of fin whales are given by block in Table 8.12. The sum of these estimates does not equal the corresponding estimate of 11 054 whales given above, due to the effects of calculating a weighted average of high and low Beaufort estimates within each block instead of first combining across blocks. The two estimates would be equal if the proportion of effort at low Beaufort was the same in every block. Suppose that 50% of effort occurred at low Beaufort overall, but in a given block, just 5% of effort occurred at low Beaufort. The method of summing estimates across blocks before averaging across Beaufort categories would give equal weight to the low and high Beaufort estimates in this block, whereas the method of Table 8.12 would give the high Beaufort estimate 19 times the weight of the low Beaufort estimate. The latter method is more appropriate, so the final abundance estimate of 10 378 whales ($\widehat{se}$ = 1655) is obtained by weighting the low and high Beaufort estimates by respective effort in individual blocks.

This example shows that reliable abundance estimates may be obtained by geographic block even when sample size within a block is very small. In two of the blocks of Table 8.12, sample size was under 10, yet analysis was possible stratifying not only by block but also by pod size and

Beaufort category. Of the three components of estimation, only effective strip half-width (or equivalently, $f(0)$) cannot be reliably estimated when samples are small. If this parameter can be assumed to be constant across at least some of the stratification categories, small sample size problems are avoided. The method is far superior to prorating a total estimate, obtained by pooling data across blocks, between blocks according to their respective areas, which requires that density of animals is uniform across the entire surveyed area. Variance estimation requires some care, since the individual block estimates are not independent. Allowance must be made for the common component ($\hat{f}(0)$) of the respective estimates, using essentially the same method as was used to estimate the variance in the estimated change in density between surveys when a common $\hat{f}(0)$ is assumed (Section 3.6.5). For the relatively complex fin whale analyses, variances are found as follows.

Within a stratum, abundance N is estimated by

$$\hat{N} = \frac{n \cdot \hat{f}(0) \cdot \bar{s} \cdot A}{2L} \tag{8.5}$$

with

$$\widehat{var}(\hat{N}) = \hat{N}^2 \cdot \left[\frac{\widehat{var}(n)}{n^2} + \frac{\widehat{var}\{\hat{f}(0)\}}{\{\hat{f}(0)\}^2} + \frac{\widehat{var}(\bar{s})}{\bar{s}^2}\right] \tag{8.6}$$

where n is the number of sightings within 3 n.m. of the centreline in the stratum, $\hat{f}(0)$ the estimated probability density of perpendicular distances, evaluated at zero, $\bar{s}$ the mean pod size, L the distance covered while on effort and A the size of the area containing the population of N animals.

For a given block, the above yields independent estimates of $f(0)$, and hence of animal abundance, corresponding to small pods in low Beaufort $\hat{N}_{\text{sm,lo}}$, small pods in high Beaufort $\hat{N}_{\text{sm,hi}}$, and large pods $\hat{N}_{\text{la}}$ (unstratified by Beaufort). Then an estimate of abundance for animals in small pods is obtained by taking an average, weighted by effort carried out at low Beaufort (L_{lo}) and high (L_{hi}):

$$\hat{N}_{\text{sm}} = \frac{L_{\text{lo}} \cdot \hat{N}_{\text{sm,lo}} + L_{\text{hi}} \cdot \hat{N}_{\text{sm,hi}}}{L_{\text{lo}} + L_{\text{hi}}} \tag{8.7}$$

and

$$\widehat{var}(\hat{N}_{\text{sm}}) = \frac{L_{\text{lo}}^2 \cdot \widehat{var}(\hat{N}_{\text{sm,lo}}) + L_{\text{hi}}^2 \cdot \widehat{var}(\hat{N}_{\text{sm,hi}})}{(L_{\text{lo}} + L_{\text{hi}})^2} \tag{8.8}$$

An abundance estimate for the block is then

$$\hat{N}_{\text{bl}} = \hat{N}_{\text{sm}} + \hat{N}_{\text{la}} \tag{8.9}$$

with

$$\widehat{var}(\hat{N}_{\text{bl}}) = \widehat{var}(\hat{N}_{\text{sm}}) + \widehat{var}(\hat{N}_{\text{la}}) \tag{8.10}$$

Within say the northern blocks, for which $\hat{f}(0)$ estimates are in common, total abundance is estimated by

$$\hat{N}_N = \sum \hat{N}_{\text{bl}} \tag{8.11}$$

where summation is over all northern blocks. To estimate the variance of this estimate, note that $\hat{N}_{\text{bl}}$ may be expressed as

$$\begin{aligned}\hat{N}_{\text{bl}} &= \frac{L_{\text{lo}} \cdot \hat{N}_{\text{sm,lo}} + L_{\text{hi}} \cdot \hat{N}_{\text{sm,hi}}}{L_{\text{lo}} + L_{\text{hi}}} + \hat{N}_{\text{la}} \\ &= \frac{L_{\text{lo}} \cdot \hat{f}_{\text{sm,lo}}(0) \cdot \hat{M}_{\text{sm,lo}} + L_{\text{hi}} \cdot \hat{f}_{\text{sm,hi}}(0) \cdot \hat{M}_{\text{sm,hi}}}{L_{\text{lo}} + L_{\text{hi}}} + \hat{f}_{\text{la}}(0) \cdot \hat{M}_{\text{la}} \\ &= l_{\text{lo}} \cdot \hat{f}_{\text{sm,lo}}(0) \cdot \hat{M}_{\text{sm,lo}} + l_{\text{hi}} \cdot \hat{f}_{\text{sm,hi}}(0) \cdot \hat{M}_{\text{sm,hi}} + \hat{f}_{\text{la}}(0) \cdot \hat{M}_{\text{la}}\end{aligned} \tag{8.12}$$

where

$$l_{\text{lo}} = \frac{L_{\text{lo}}}{L_{\text{lo}} + L_{\text{hi}}} \qquad l_{\text{hi}} = \frac{L_{\text{hi}}}{L_{\text{lo}} + L_{\text{hi}}}$$

and

$$\hat{M}_{\text{sm,lo}} = \frac{n_{\text{sm,lo}} \cdot \bar{s}_{\text{sm,lo}} \cdot A}{2L_{\text{lo}}} \quad \text{evaluated for that block,}$$

and similarly for $\hat{M}_{\text{sm,hi}}$ and $\hat{M}_{\text{la}}$. The component $\hat{f}(0)$ is common across blocks, whereas the other components of the abundance estimate are not. Thus

$$\begin{aligned}\hat{N}_N = \hat{f}_{\text{sm,lo}}(0) \cdot \sum[l_{\text{lo}} \cdot \hat{M}_{\text{sm,lo}}] + \hat{f}_{\text{sm,hi}}(0) \cdot \sum[l_{\text{hi}} \cdot \hat{M}_{\text{sm,hi}}] \\ + \hat{f}_{\text{la}}(0) \cdot \sum \hat{M}_{\text{la}}\end{aligned} \tag{8.13}$$

where summation is over blocks. Denote the three terms in this expression by T_i, $i = 1, 2, 3$; these three terms are independent. Consider the final term,

$$T_3 = \hat{f}_{\text{la}}(0) \cdot \sum \hat{M}_{\text{la}} \tag{8.14}$$

This has variance

$$\widehat{var}(T_3) = T_3^2 \cdot \left[\frac{\widehat{var}\{\hat{f}_{\text{la}}(0)\}}{\{\hat{f}_{\text{la}}(0)\}^2} + \frac{\sum \widehat{var}(\hat{M}_{\text{la}})}{\{\sum \hat{M}_{\text{la}}\}^2} \right] \tag{8.15}$$

where

$$\widehat{var}(\hat{M}_{\text{la}}) = \hat{M}_{\text{la}}^2 \cdot \left[\frac{\widehat{var}(n_{\text{la}})}{n_{\text{la}}^2} + \frac{\widehat{var}(\bar{s}_{\text{la}})}{\bar{s}_{\text{la}}^2} \right] \quad \text{evaluated in each block} \tag{8.16}$$

Similarly,

$$\widehat{var}(T_1) = T_1^2 \cdot \left[\frac{\widehat{var}\{\hat{f}_{\mathrm{sm,lo}}(0)\}}{\{\hat{f}_{\mathrm{sm,lo}}(0)\}^2} + \frac{\sum l_{\mathrm{lo}}^2 \cdot \widehat{var}(\hat{M}_{\mathrm{sm,lo}})}{\{\sum l_{\mathrm{lo}} \cdot \hat{M}_{\mathrm{sm,lo}}\}^2} \right] \tag{8.17}$$

and likewise for $\widehat{var}(T_2)$. Finally,

$$\widehat{var}(\hat{N}_N) = \sum_{i=1}^{3} \widehat{var}(T_i) \tag{8.18}$$

If total abundance in the southern blocks is estimated by $\hat{N}_S$, it and its variance are estimated in the same way as for $\hat{N}_N$. Total abundance over the whole area is then the sum of these estimates, with variance equal to the sum of the respective variances, since $f(0)$ is estimated independently in the two areas. Applying the above methods, we obtain an abundance estimate of 10 378 fin whales, with standard error 1655. Assuming $\hat{N}$ is log-normal, the estimated 95% confidence interval for N is $(7607, 14\,158)$ animals (Section 3.6.1).

8.6 House wren densities in South Platte River bottomland

We use point transect data on house wrens (*Troglodytes aedon*) to illustrate model selection. The data were collected from 155 points, with between 14 and 16 points in each of ten 16-ha study blocks. The blocks were established in riparian vegetation along 30 km of South Platte River bottomland near Crook, Colorado. The study was described by Knopf (1986) and Sedgwick and Knopf (1987). The house wren was the most frequently recorded bird, and sample sizes were sufficient to allow estimation by block as well as across blocks. Thus, the option to stratify can also be examined.

The following models were tried: Fourier series (uniform key and up to four cosine adjustments); Hermite polynomial (half-normal key and up to four Hermite polynomial adjustments); half-normal and up to four cosine adjustments; and hazard-rate with at most two simple polynomial adjustments. Adjustment terms were tested for inclusion using the likelihood ratio test with a p-value of 0.05 and the Distance option 'look-ahead' set to two (see 'adjustment terms' in the 'detection function' tab under 'model definition' in Distance). Intervals for goodness of fit tests were set at 0.0, 7.5, 12.5, 17.5, 22.5, 27.5, 32.5, 42.5, 62.5 and 92.5 m. The largest detection distances were at 90 m. To assess the impact of truncation, the last two intervals were discarded. Thus, the truncation point was 42.5 m, corresponding to 10% truncation of observations. The intervals were chosen to avoid possible favoured rounding distances, such as 10 or 25 m. We recommend that goodness of fit is not used for model selection, but if it is, we recommend strongly that intervals are set using the diagnostics tab of

Table 8.13. Summary of results from fitting different models to house wren data. FS = Fourier series model (uniform key and cosine adjustments); HP = Hermite polynomial model (half-normal key and Hermite polynomial adjustments); HC = half-normal key and cosine adjustments; Hz = hazard-rate key and simple polynomial adjustments. The truncation distance of $w = 92.5$ m is larger than the largest recorded distance, so no data are truncated, and the value $w = 42.5$ m corresponds to truncation of 10% of detection distances. ΔAIC shows AIC$-$AIC$_{\text{min}}$, where AIC$_{\text{min}}$ (6624.8 for $w = 92.5$ m and 5523.8 for $w = 42.5$ m) is the AIC value corresponding to the best fitting model

Model	No. of adj. terms	Log-l'hood	χ^2	(df)	p-value	ΔAIC	$\hat{D}$	Log-based 95% confidence interval
Data untruncated ($w = 92.5$ m)								
FS	4	−3312.7	18.8	(3)	<0.001	8.6	6.72	(5.95, 7.58)
HP	3	−3308.4	10.8	(4)	0.03	0.0	8.28	(6.98, 9.82)
HC	3	−3308.4	10.7	(4)	0.03	5.1	8.47	(7.24, 9.91)
Hz	1	−3329.7	39.3	(4)	<0.001	40.7	6.05	(5.28, 6.93)
Data truncated at $w = 42.5$ m								
FS	3	−2760.0	7.0	(3)	0.07	2.2	9.05	(7.48, 10.95)
HP	1	−2762.0	12.1	(4)	0.02	4.2	7.84	(6.77, 9.07)
HC	1	−2760.4	7.6	(4)	0.11	1.0	9.01	(7.43, 10.92)
Hz	1	−2758.9	7.1	(3)	0.07	0.0	8.14	(6.44, 10.30)

the detection function section of the model definition in Distance 3.5 or 4. The default intervals used by Distance do not take account of rounding to favoured values, and may frequently give spurious significant test statistics.

A summary of results is given in Table 8.13. Note that the log-likelihood and the AIC are of no use for determining whether data should be truncated; values of these statistics for different models are only comparable if the same truncation point is selected. Density estimates from untruncated data in Table 8.13 are mostly smaller and more precise than those from truncated data. They are probably also more biased given that fits to the untruncated data are less good. The exception is the Hermite polynomial model, which provides the best of the four fits to the untruncated data, and the worst fit to the truncated data. The Fourier series and hazard-rate models perform particularly poorly on the untruncated data. The Fourier series model is not robust to poor choice of truncation point for both line and point transects, whereas the hazard-rate model appears to be robust when data are untruncated for line transects but not for point transects (Buckland 1985, 1987a). The fit of all but the Hermite polynomial model is improved by truncation, and density estimates are more similar under different models for truncated data. We therefore select the model that gives the largest log-likelihood and the smallest AIC value when applied to truncated data for further analyses. This model is the hazard-rate with simple polynomial adjustments.

Estimates stratified by block and by observer are shown in Table 8.14. Goodness of fit tests indicate that fits to eight of the ten blocks are very good, although the data from blocks 0 and 6 are less well-modelled. The effective detection radius is high for blocks 0 and 5, but is similar for all other blocks, at around 20 m. Densities vary appreciably between blocks. The final abundance estimate from the analysis stratified by block is very similar to that obtained from an unstratified analysis (Table 8.13, last row). The confidence interval is rather wider, reflecting the larger number of parameters that have been estimated. There seems little advantage here to stratification, unless estimates are required by block; this is likely to be true generally when effort per unit area is roughly the same in all strata. In this example, an option that would allow abundance to be estimated by block, without the loss in precision seen in Table 8.14, is to estimate the detection function (and hence the effective detection radius) from pooled data, while estimating encounter rate by block. However, the AIC values from the analyses of Table 8.14 sum to 5512.9, substantially less than 5523.8, the value obtained if data are pooled. Hence, if estimates by block are required, it is necessary to estimate the detection function separately by block.

Of more interest is the stratification by observer. Data from observers 1, 2 and 3 yield remarkably similar estimates. However, the first observer's data are modelled poorly. Inspection of the data and output from Distance shows that observer 1 preferentially rounded distances around, or rather over, 10 m to exactly 10 m, and distances between 25 and 40 m were predominantly recorded as, or close to, 30 m. Such rounding generally generates little bias, but intervals for goodness of fit testing need to be widened and reduced in number to obtain a reliable test when it is present. More serious is the apparent bias in the data of observer 4. The number of detections per point is rather greater than for the other observers, which is consistent with the higher effective detection radius, yet density is estimated to be well under half of that estimated from the data of each of the other observers. It is possible that observer 4 concentrated on detecting birds at greater distances, at the expense of missing many birds close to the point. More likely perhaps is that the effective detection radius was similar to that for the other observers, but that distances were overestimated by observer 4 by roughly 50%. This would be sufficient to explain the large difference in density estimates between observer 4 and the others. Whatever the cause, it seems clear that the data for observer 4 should be viewed with suspicion, whereas those for observers 2 and 3 appear to be most reliable. Our preferred analyses for these data use the hazard-rate model with up to two simple polynomial adjustments and truncation at 42.5 m, are unstratified by block (unless separate estimates by block are required), and discard

Table 8.14. Analyses of house wren data using the hazard-rate model with truncation at 42.5 m. Standard errors are given in parentheses; confidence intervals were calculated assuming a log-normal distribution for $\hat{D}$, and the Satterthwaite correction was not applied. *Estimated density found as the average of the block estimates; the corresponding standard error is found as (square root of the sum of squared standard errors for each block) divided by the number of blocks

Observer	Block	Effective detection radius $\hat{\rho}$	Encounter rate n/k	Estimated density	Log-based 95% confidence interval	Goodness of fit test p-value
All	0	30.4 (1.3)	0.64 (0.14)	2.21 (0.52)	(1.41, 3.47)	0.01
All	1	21.0 (3.2)	1.88 (0.24)	13.60 (4.45)	(7.28, 25.41)	0.42
All	2	22.7 (2.1)	2.17 (0.24)	13.35 (2.87)	(8.81, 20.24)	0.52
All	3	19.1 (10.8)	1.05 (0.14)	9.10 (10.40)	(1.52, 54.46)	0.46
All	4	20.4 (2.2)	1.71 (0.25)	13.05 (3.41)	(7.89, 21.61)	0.35
All	5	31.1 (1.9)	1.20 (0.20)	3.96 (0.83)	(2.64, 5.94)	0.39
All	6	23.5 (2.0)	1.22 (0.15)	7.00 (1.47)	(4.66, 10.51)	0.01
All	7	23.4 (5.1)	1.15 (0.6)	6.70 (3.06)	(2.86, 15.72)	0.35
All	8	23.5 (2.7)	1.16 (0.17)	6.69 (1.80)	(3.98, 11.25)	0.97
All	9	19.3 (4.9)	1.38 (0.22)	11.76 (6.22)	(4.44, 31.15)	0.31
All	All			8.38* (1.48)	(5.94, 11.83)	
1	All	21.8 (2.6)	1.26 (0.08)	8.44 (2.05)	(5.28, 13.49)	<0.001
2	All	18.8 (1.8)	1.06 (0.08)	9.56 (1.96)	(6.42, 14.24)	0.84
3	All	19.7 (1.6)	1.12 (0.09)	9.16 (1.67)	(6.43, 13.07)	0.15
4	All	33.0 (1.8)	1.40 (0.10)	4.11 (0.53)	(3.19, 5.30)	0.27
2 and 3	All	19.3 (1.2)	1.09 (0.07)	9.30 (1.28)	(7.10, 12.18)	0.31
1–3	All	20.1 (1.3)	1.15 (0.06)	9.06 (1.22)	(6.96, 11.79)	<0.001

data from observer 4. If there is concern about the poor fit of this model, the data of observer 1 should also be deleted. The resulting estimates with and without the data of observer 1 are shown in Table 8.14.

We have shown that poor model fits can be improved by truncating data. We now use the untruncated data to illustrate other strategies for improving the fit of a model. Other than truncation, the user of Distance has several options (see 'adjustment terms' in the 'detection function' tab of 'model definition'). First, the analyst has control over how many adjustment terms are tested before Distance concludes that no significant improvement in the fit has been obtained. It is not uncommon that a single adjustment term does not improve the fit of a model significantly, whereas the combined effect of it and a further term does yield a significant improvement. If 'look-ahead' is set equal to one (the default), the better model will not be found, whereas a value of two will allow Distance to select it, at the expense of slower run times. Second, the user can change the method by which Distance selects models using 'Selection method'. For method 'Sequential' (the default), it fits the lowest order term first, then adds successively higher order terms sequentially. If method 'Forward' is specified, Distance will test for inclusion of each term not yet in the model. If the term that gives the largest increase in the value of the likelihood yields a significant improvement in the fit, it is included, and Distance tests for inclusion of another term. This is analogous to a forward stepwise procedure in multiple regression. Method 'All' allows all possible combinations of adjustment terms to be tested, and that giving the minimum AIC value is selected. Third, the user may use AIC to select terms (the default), or specify the p-value for selecting between fits using likelihood ratio tests. For the runs considered in this example, only a p-value of 0.05 was used, but if this tended to fit too few terms, a larger value (e.g. 0.15, the default value) might be preferred. Under 'Models' in the 'Detection function' tab, the user has other options: the key function (half-normal, hazard-rate, uniform or negative exponential) may be changed, or a different type of adjustment term (simple or Hermite polynomial, or cosine) can be selected. The combinations of these options that were applied to the house wren data are listed in Table 8.15.

The results of Table 8.16 indicate a clear 'winner' among the models. The hazard-rate model with cosine adjustments and using selection methods 'forward' and 'all' both lead to a hazard-rate model with a single cosine adjustment of order 4. Only this model yields a goodness of fit statistic that is not significant at the 5% level, and its AIC value is 2.6 lower than the next best model. The density estimate is rather higher than that obtained from the favoured model on truncated data from above. Figure 8.11 shows that the fitted detection function has only a very narrow shoulder. For these data, use of cosine adjustment terms leads generally to a narrow shoulder, whereas polynomial adjustments to the hazard-rate model tend to preserve

Table 8.15. Models and model options used for fitting house wren data. Distance data were pooled across observers and blocks, and were untruncated ($w = 92.5$ m)

Model	Look-ahead	Selection method	Key function	Adjustment terms
1	1	Sequential	Hazard-rate	Cosine
2	2	Sequential	Hazard-rate	Cosine
3	1	Forward	Hazard-rate	Cosine
4	—	All	Hazard-rate	Cosine
5	1	Sequential	Hazard-rate	Simple polynomial
6	2	Sequential	Hazard-rate	Simple polynomial
7	1	Forward	Hazard-rate	Simple polynomial
8	—	All	Hazard-rate	Simple polynomial
9	1	Sequential	Uniform	Cosine
10	2	Sequential	Uniform	Cosine
11	1	Forward	Uniform	Cosine
12	—	All	Uniform	Cosine
13	1	Sequential	Uniform	Simple polynomial
14	2	Sequential	Uniform	Simple polynomial
15	1	Forward	Uniform	Simple polynomial
16	—	All	Uniform	Simple polynomial
17	1	Sequential	Half-normal	Cosine
18	2	Sequential	Half-normal	Cosine
19	1	Forward	Half-normal	Cosine
20	—	All	Half-normal	Cosine
21	1	Sequential	Half-normal	Hermite polynomial
22	2	Sequential	Half-normal	Hermite polynomial
23	1	Forward	Half-normal	Hermite polynomial
24	—	All	Half-normal	Hermite polynomial

a wider shoulder. In Fig. 8.12, the fitted detection function obtained by making simple polynomial adjustments to the hazard-rate key, together with selection option 'forward' or 'all', is shown. Although this model fits the data less well, its wider shoulder may be a better reflection of reality. It yields a density estimate rather lower than that from the favoured method for truncated data from above.

8.7 Songbird point transect surveys in Arapaho NWR

For this example, we use data supplied by F. L. Knopf from extensive songbird surveys of parts of the Arapaho National Wildlife Refuge, Colorado (Knopf *et al.* 1988). We consider counts carried out in June of 1980 and 1981, and analyse the six most numerous species, namely the yellow warbler (*Dendroica petechia*), brown-headed cowbird (*Molothrus ater*),

Table 8.16. Summary of results from fitting different models to house wren data. The models are defined in Table 8.15. ΔAIC shows AIC−AIC_{min}, where $AIC_{min} = 6619.3$ is the AIC value corresponding to the best fitting model

Model	No.of adj. terms	Log-l'hood	χ^2 (df)	p-value	ΔAIC	$\hat{D}$	Log-based 95% confidence interval
1	0	−3337.1	61.2 (6)	<0.001	59.0	5.56	(4.89, 6.32)
2	3	−3305.9	8.2 (3)	0.04	2.6	8.58	(7.33, 10.03)
3 & 4	1	−3306.7	9.4 (5)	0.10	0.0	8.74	(7.48, 10.22)
5 & 6	2	−3321.4	30.1 (4)	<0.001	31.5	6.72	(5.79, 7.79)
7 & 8	3	−3309.8	15.7 (3)	0.001	10.6	7.51	(6.31, 8.93)
9	3	−3332.0	59.9 (5)	<0.001	50.7	5.28	(4.75, 5.88)
10–12	5	−3308.1	13.2 (3)	0.004	7.0	7.46	(6.52, 8.54)
13	3	−3529.3	501.1 (5)	<0.001	445.3	2.65	(2.34, 3.00)
14	5	−3510.3	306.2 (3)	<0.001	411.4	3.24	(2.93, 3.58)
15	3	−3582.8	515.9 (5)	<0.001	552.2	2.56	(2.32, 2.82)
16	4	−3456.8	271.5 (4)	<0.001	302.4	3.39	(3.00, 3.82)
17 & 18	1	−3313.0	13.8 (6)	0.03	10.8	7.33	(6.48, 8.30)
19	2	−3310.4	11.2 (5)	0.05	7.4	8.38	(7.13, 9.84)
20	3	−3308.6	10.8 (4)	0.03	5.9	8.47	(7.24, 9.92)
21 & 23	0	−3327.3	39.6 (7)	<0.001	39.2	6.16	(5.51, 6.89)
22 & 24	2	−3312.8	13.9 (5)	0.02	12.4	7.33	(6.39, 8.40)

savannah sparrow (*Passerculus sandwichensis*), song sparrow (*Melospiza melodia*), red-winged blackbird (*Agelaius phoeniceus*) and American robin (*Turdus migratorius*). In 1980, three pastures, labelled Pastures 1, 2 and 3, were surveyed by one visit to each of 124, 126 and 123 points, respectively. In 1981, four pastures, 0, 1, 2 and 3, were surveyed during one visit to each of 100 points per pasture. All birds detected within 100 m of the point were noted and their locations were flagged, so that their distances could be measured to the nearest 10 cm. Although pastures varied in size, for the purposes of illustration, we assume that each was the same size.

Analyses were carried out adopting the half-normal key with cosine adjustments. This model combines the key of the Hermite polynomial model with the adjustments of the Fourier series model. It is computationally more efficient than the former model, and uses a more plausible key function than the latter. For yellow warbler, savannah sparrow and song sparrow, some fits were found to be poor, so the detection distances were truncated at 52.5 m. Other analyses are untruncated. The variance of the number of detections was estimated using the sample variance of number of detections per point (Section 3.6.2).

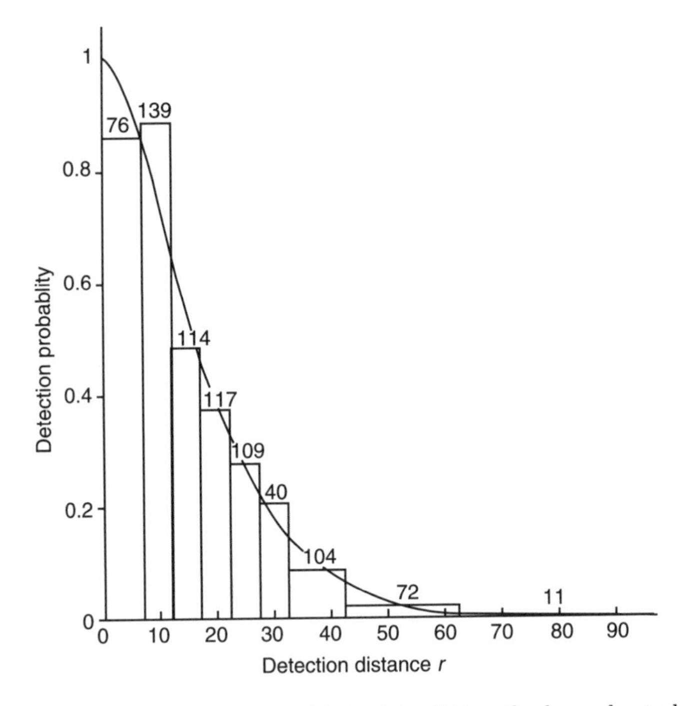

Fig. 8.11. The detection function obtained by fitting the hazard-rate key with cosine adjustments to untruncated house wren data.

Yellow warbler analyses are summarized in Table 8.17. Separate estimates were obtained by stratum (pasture). In all, 205 detections were made in 1980 and 342 in 1981. Although not apparent from the tabulated results, closer scrutiny of the data reveals that there is an excess of detections close to the point, with as many as 15 in 1981 being recorded within one metre of a point. Given the small area within a metre of a point, it is clear that an assumption has been violated. (At 6.25 birds/ha, and 400 points, the expected number of birds within one metre of a point is less than one.) In these surveys, field procedures for estimating detection distances were as follows. At the end of the count at a given point, the observer tied flags to the vegetation, to mark bird locations. The distance from the point to each flag was later measured to the nearest decimetre. It seems likely that for yellow warblers at least, birds typically moved around in the canopy during the count, and may not have been easily located. In the absence of a clear protocol for handling such species, and if the 'snapshot' method (Section 5.9) is not used, observers will have a tendency to record a detection distance too close to the point for birds moving around above them. The extent of the problem for these yellow warbler data is too great to allow reliable estimation of density. None of the other species analysed here seem to have been unduly affected by this problem.

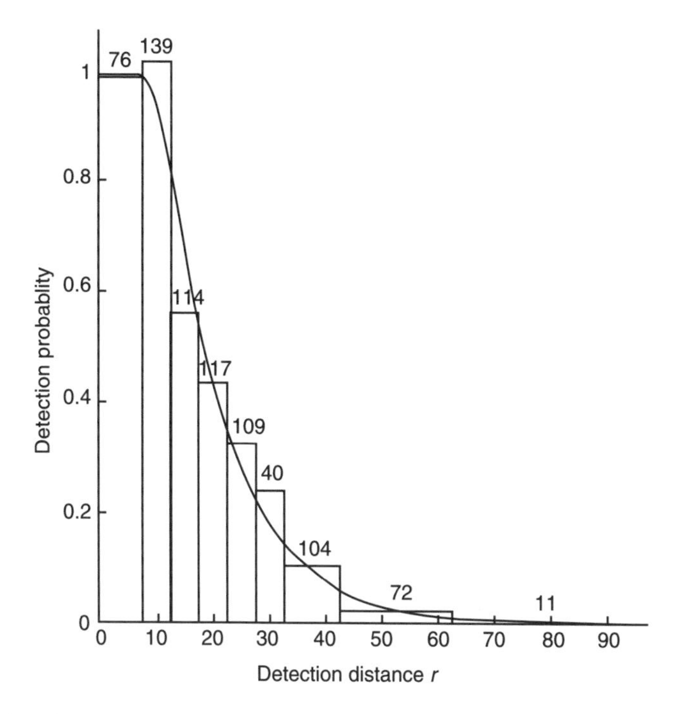

Fig. 8.12. The detection function obtained by fitting the hazard-rate key with simple polynomial adjustments to untruncated house wren data.

Analyses of the brown-headed cowbird counts (Table 8.18) are also not straightforward. First, count frequencies by distance from the point were highly variable in some pastures. Examination of the data showed that this was caused by detections of groups of birds. If more than one bird was recorded at exactly the same distance from the same point, we assume here that the birds comprised a single flock (cluster). For the first analyses, detection distance data were pooled across pastures, and a single set of estimates per year determined. The second set of analyses are by pasture. The estimates of $h(0)$ (or equivalently, of the effective radius of detection ρ) are imprecise, because the number of detections (as distinct from individual birds) per pasture was low, ranging from 21 (Pasture 1, 1980) to 50 (Pasture 3, 1980). Potentially more serious, bias may be high, as there is little information from which to select an appropriate model when sample size is small. Indeed, the effective radii for 1981 (Table 8.18) indicate appreciably more variability between pastures than can be explained by the standard errors. Either detectability of brown-headed cowbirds varied substantially between pastures or sample size was inadequate at least in some pastures for estimating the effective detection radius with low bias. If the latter explanation is more likely, then pasture estimates with

Table 8.17. Analyses of yellow warbler point transect data, Arapaho National Wildlife Refuge. Standard errors are given in parentheses. Estimated density for category 'all' is found as the average of the pasture estimates; the corresponding standard error is found as (square root of the sum of squared standard errors for each pasture) divided by the number of pastures

Year	Pasture	Effective detection radius $\hat{\rho}$	Encounter rate n/k	Estimated density	Log-based 95% confidence interval	Goodness of fit test p-value
1980	1	26.2	0.73	3.40	(1.97, 5.87)	0.67
		(3.5)	(0.06)	(0.97)		
	2	27.1	0.45	1.96	(1.38, 2.79)	0.42
		(1.7)	(0.06)	(0.36)		
	3	17.0	0.37	4.02	(2.42, 6.67)	0.80
		(1.6)	(0.07)	(1.06)		
	All			3.12	(2.30, 4.23)	
				(0.49)		
1981	0	17.7	0.71	7.18	(4.87, 10.58)	0.60
		(1.4)	(0.09)	(1.44)		
	1	20.8	0.98	7.22	(4.52, 11.54)	0.07
		(2.4)	(0.08)	(1.75)		
	2	26.0	0.69	3.25	(2.36, 4.47)	0.13
		(1.5)	(0.08)	(0.53)		
	3	16.4	0.62	7.34	(5.03, 10.71)	0.19
		(1.2)	(0.08)	(1.43)		
	All			6.25	(5.05, 7.74)	
				(0.68)		

higher precision and lower bias may be obtained by estimating the effective radius from data pooled across pastures (first section of Table 8.18), and estimating other parameters individually by pasture (second section of Table 8.18). This assumes that detectability does not vary with pasture, and utilizes the fact that average cluster size and expected number of detections can be estimated with low bias from small samples, whereas the effective radius often cannot. The approach is described in Section 3.7.

The third section of Table 8.18 shows the estimates obtained from the above approach, assuming the pastures were equal in area. Note how much some of the pasture estimates differ from those found by estimating the effective radius of detection within each pasture. Note also that the estimates for all pastures combined are the same as those for which all data were pooled (first section of Table 8.18). If exactly the same effort (points per unit area) is expended in each stratum, the two methods are equivalent. However, the current method (1) allows separate estimates by stratum, (2) is still valid if effort differs by stratum and (3) is preferable to the fully stratified analysis if sample sizes are too small to estimate effective

Table 8.18. Analyses of brown-headed cowbird point transect data, Arapaho National Wildlife Refuge. The first set of results was obtained by carrying out an unstratified analysis, the second set by stratifying by pasture, the third set by estimating $h(0)$ from unstratified distance data and other components separately by pasture, the fourth set by stratifying by cluster size, and the fifth set by correcting for size bias in mean cluster size. Standard errors are given in parentheses

Year	Pasture	Effective detection radius $\hat{\rho}$	Encounter rate n/k	Mean cluster size $\bar{s}$	Estimated density	Log-based 95% conf. interval	Goodness of fit test p-value
Unstratified							
1980	All	38.2	0.26	1.72	0.99	(0.73, 1.35)	0.44
		(2.0)	(0.03)	(0.11)	(0.16)		
1981	All	34.8	0.35	1.73	1.59	(1.18, 2.13)	0.44
		(1.9)	(0.03)	(0.11)	(0.24)		
Stratified by pasture							
1980	1	36.6	0.17	1.71	0.69	(0.34, 1.41)	0.46
		(5.0)	(0.04)	(0.21)	(0.26)		
	2	38.7	0.21	2.04	0.93	(0.50, 1.70)	0.60
		(4.2)	(0.04)	(0.28)	(0.29)		
	3	38.5	0.41	1.56	1.36	(0.92, 2.02)	0.55
		(2.5)	(0.06)	(0.12)	(0.28)		
	All				0.99	(0.72, 1.36)	
					(0.16)		
1981	0	25.5	0.28	1.39	1.91	(1.03, 3.54)	0.23
		(3.2)	(0.05)	(0.14)	(0.62)		
	1	56.9	0.33	2.03	0.66	(0.38, 1.14)	0.59
		(5.1)	(0.06)	(0.26)	(0.19)		
	2	42.1	0.42	1.76	1.33	(0.86, 2.06)	0.44
		(2.8)	(0.06)	(0.18)	(0.30)		
	3	24.9	0.36	1.69	3.12	(1.64, 5.94)	0.66
		(3.4)	(0.05)	(0.25)	(1.05)		
	All				1.75	(1.23, 2.49)	
					(0.32)		
Stratified by pasture, except for $h(0)$							
1980	1				0.63	(0.39, 1.13)	
					(0.18)		
	2				0.95	(0.61, 1.59)	
					(0.24)		
	3				1.39	(0.97, 2.00)	
					(0.26)		
	All				0.99	(0.73, 1.35)	
					(0.16)		
1981	0				1.03	(0.67, 1.68)	
					(0.25)		
	1				1.76	(1.11, 2.90)	
					(0.44)		

Table 8.18. *Continued*

Year	Pasture	Effective detection radius $\hat{\rho}$	Encounter rate n/k	Mean cluster size $\bar{s}$	Estimated density	Log-based 95% conf. interval	Goodness of fit test p-value
	2				1.95 (0.43)	(1.28, 2.98)	
	3				1.61 (0.38)	(1.07, 2.63)	
	All				1.59 (0.24)	(1.18, 2.13)	
Stratified by cluster size							
Single birds							
1980	All	36.5 (2.9)	0.16 (0.02)	1.00 (0.08)	0.37	(0.25, 0.56)	0.07
1981	All	32.5 (2.1)	0.22 (0.02)	1.00 (0.11)	0.65	(0.47, 0.90)	0.19
Clusters (two or more birds)							
1980	All	40.5 (3.0)	0.11 (0.02)	2.78 (0.15)	0.58 (0.13)	(0.37, 0.89)	0.49
1981	All	39.4 (4.3)	0.13 (0.02)	2.92 (0.20)	0.80 (0.21)	(0.48, 1.32)	0.30
All birds							
1980	All				0.95 (0.15)	(0.70, 1.29)	
1981	All				1.44 (0.23)	(1.05, 1.97)	
Mean cluster size corrected for size bias							
1980	All	38.2 (2.0)	0.26 (0.03)	1.68 (0.10)	0.97 (0.16)	(0.71, 1.33)	0.44
1981	All	34.8 (1.9)	0.35 (0.03)	1.49 (0.07)	1.36 (0.20)	(1.02, 1.81)	0.44

detection radii reliably by stratum, although it assumes that detectability does not vary across strata.

The fourth section of Table 8.18 shows another method of analysing these data. In this case, data were stratified by cluster size (Quinn 1979, 1985). Detections were divided into two categories: single birds and at least two birds. The results suggest that clusters are more detectable than single birds, although the overall estimates of density differ very little from those obtained above.

The final section of Table 8.18 shows adjusted mean cluster size, estimated by regressing logarithm of cluster size on probability of detection, and from this regression, estimating mean cluster size when probability of detection is one. For each cluster, its probability of detection was

Table 8.19. Analyses of savannah sparrow point transect data, Arapaho National Wildlife Refuge. Standard errors are given in parentheses

Year	Pasture	Effective detection radius $\hat{\rho}$	Encounter rate n/k	Estimated density	Log-based 95% confidence interval	Goodness of fit test p-value
1980	1	33.4	0.48	1.36	(0.89, 2.06)	0.02
		(2.6)	(0.07)	(0.29)		
	2	27.1	0.95	4.12	(3.14, 5.40)	0.17
		(1.3)	(0.10)	(0.57)		
	3	31.2	0.72	2.37	(1.73, 3.26)	0.86
		(2.0)	(0.07)	(0.39)		
	All			2.62	(2.18, 3.17)	
				(0.25)		
1981	0	27.0	0.31	1.36	(0.80, 2.29)	0.87
		(2.8)	(0.05)	(0.37)		
	1	47.0	0.32	0.46	(0.24, 0.89)	0.12
		(6.8)	(0.06)	(0.16)		
	2	31.6	0.51	1.63	(1.02, 2.58)	0.96
		(2.8)	(0.08)	(0.39)		
	3	34.6	0.48	1.27	(0.77, 2.11)	0.66
		(3.7)	(0.07)	(0.33)		
	All			1.18	(0.90, 1.54)	
				(0.16)		

estimated by substituting its detection distance into the fitted detection function from the analyses of the first section of Table 8.18. The correlation between log cluster size and detection probability was not significant for 1980 ($r = -0.022$, df $= 96$, $p > 0.1$), and estimation was barely affected. For 1981, the correlation was significant ($r = -0.193$, df $= 137$, $p < 0.05$), and estimated cluster size was reduced (1.49 birds per cluster, compared with 1.73 birds per cluster for the detected clusters).

The analyses for savannah and song sparrows presented no special difficulties, and the estimates are given in Tables 8.19 and 8.20, respectively. For red-winged blackbirds (Table 8.21), it was again necessary to analyse the detections as clusters, although average cluster size was small, so bias arising from possible greater detectability of groups of two or more birds would be small. Data were insufficient to stratify, either by pasture or by cluster size.

The final analyses from this example are of American robin (Table 8.22). Again, sample sizes were too small to stratify, but the data presented no additional problems. A single cosine adjustment to the half-normal fit was selected for both the 1980 and the 1981 data. To show the effects of allowing for estimation of the number of adjustments required on the confidence interval for density, both years' data were reanalysed selecting

Table 8.20. Analyses of song sparrow point transect data, Arapaho National Wildlife Refuge. Standard errors are given in parentheses

Year	Pasture	Effective detection radius $\hat{\rho}$	Encounter rate n/k	Estimated density	Log-based 95% confidence interval	Goodness of fit test p-value
1980	1	27.0	0.38	1.66	(1.12, 2.45)	0.79
		(1.9)	(0.05)	(0.33)		
	2	23.7	0.40	2.29	(1.51, 3.47)	0.56
		(1.8)	(0.06)	(0.49)		
	3	23.8	0.41	2.34	(1.59, 3.42)	0.04
		(1.6)	(0.06)	(0.46)		
	All			2.10	(1.66, 2.65)	
				(0.25)		
1981	0	31.5	0.43	1.38	(0.83, 2.30)	0.23
		(3.4)	(0.06)	(0.36)		
	1	32.1	0.47	1.45	(0.94, 2.25)	0.99
		(2.8)	(0.07)	(0.33)		
	2	24.6	0.39	2.05	(1.26, 3.32)	0.12
		(2.5)	(0.06)	(0.51)		
	3	23.9	0.29	1.62	(0.98, 2.67)	0.16
		(2.2)	(0.05)	(0.42)		
	All			1.63	(1.27, 2.08)	
				(0.21)		

Table 8.21. Analyses of red-winged blackbird point transect data, Arapaho National Wildlife Refuge. Standard errors are given in parentheses

Year	Pasture	Effective detection radius $\hat{\rho}$	Encounter rate n/k	Mean cluster size $\bar{s}$	Estimated density	Log-based 95% conf. interval	Goodness of fit test p-value
1980	All	27.9	0.18	1.29	0.93	(0.63, 1.37)	0.38
		(1.6)	(0.03)	(0.09)	(0.19)		
1981	All	32.9	0.17	1.20	0.61	(0.39, 0.97)	0.39
		(3.2)	(0.02)	(0.07)	(0.14)		

the bootstrap option for estimating the variance of $\hat{h}(0)$. The resulting confidence intervals were (0.15, 0.58) and (0.18, 0.64), respectively, wider than the intervals of Table 8.22, as might be expected.

The density estimates of Tables 8.17–8.22 are consistently higher than those of Knopf *et al.* (1988). They used the Fourier series model on squared distances, as recommended by Burnham *et al.* (1980). We no longer recommend this approach, as it can lead to underestimation of density (Buckland 1987a), so the differences between their estimates and those given here might be anticipated.

Table 8.22. Analyses of American robin point transect data, Arapaho National Wildlife Refuge. Standard errors are given in parentheses

Year	Pasture	Effective detection radius $\hat{\rho}$	Encounter rate n/k	Estimated density	Log-based 95% confidence interval	Goodness of fit test p-value
1980	All	30.2	0.09	0.30	(0.18, 0.50)	0.22
		(2.8)	(0.02)	(0.08)		
1981	All	36.1	0.14	0.34	(0.22, 0.52)	0.87
		(3.2)	(0.02)	(0.08)		

8.8 Assessing the effects of habitat on density

The design of line and point transect surveys was discussed in Chapter 7. Suppose estimates of object density are required by habitat. The study area should first be stratified by habitat type. Surveys may then be designed within each stratum as described in Chapter 7. A belt of width w, where w corresponds to the distance within which say 85–90% of detections are expected to lie, might be defined just within the border of each stratum. If line or point transects are constrained so that they do not lie within the belt, then differences in density between habitats will be easier to detect. If such provision is not made, density will be underestimated in habitat types holding high densities, and overestimated in habitats with low densities.

Sometimes, comparisons of density between uniform blocks of habitat and habitat edge are of interest. Point transects are more suited to such comparisons than line transects. As before, habitat edges should be determined by stratifying the area by habitat categories. Points should then be positioned randomly, or systematically (say every 200 m) with a random starting point, along each edge. Detections at these edge points should be recorded according to which side of the edge (i.e. which habitat type) they are in, to allow edge versus centre comparisons within the same habitat type. Thus there would be two analyses of edge points, each taking the fraction of the circle surveyed to be one half. Alternatively, data from each side can be pooled to obtain an estimate of 'edge' density to compare with densities in either or both habitat types, found as described above.

Note that density at a distance w from the edge may be appreciably different from that at the edge itself. This will not invalidate the above analysis, unless the trend in density is very great, although the estimated detection function will be biased. Provided the assumption $g(0) = 1$ holds, the method will give a valid estimate of density at the edge. For similar reasons, points from which density away from the edge is estimated could be taken to be at least a distance w from the edge, rather than say at least $2w$, although a larger value might be preferred for other reasons; for example, a value equal to the maximum likely territory diameter could be chosen.

In reality, habitat may be too patchy and heterogeneous to divide a study area into a small number of strata. In this case, density might be better considered as a function of habitat characteristics. One approach to this would be to include habitat information as covariates (Buckland *et al.* in preparation), so that the surface representing object density is modelled. We use as an example a different approach. The following is summarized from Bibby and Buckland (1987).

We consider here binomial count data collected during 1983 for a study into bird populations of recently restocked conifer plantations throughout north Wales. In total, 326 points were covered, divided among 62 forestry plots that had been restocked between 1972 and 1981. Further details are given by Bibby *et al.* (1985).

Each detected bird was recorded as to whether it was within or beyond 30 m of the point. The half-normal binomial model of Section 6.6.1 was applied to these data, together with a linear model, for which analytic results are also available, and a single parameter hazard-rate model, with power parameter b set equal to 3.3, fitted by numerical methods (Buckland 1987a). Table 8.23 (reproduced from Buckland 1987a) shows that the linear model consistently yields higher estimates of densities than does the half-normal model, which in turn yields higher estimates than the hazard-rate model in most cases. Standard errors of these estimates are similar for all three models. Note that all three models give very similar relative densities between species. For example, the ratio of willow warbler density to wren density is estimated as 2.15, 2.13 and 2.19 under the half-normal, linear and hazard-rate models, respectively. Indeed, the three models give virtually the same ranking of species by density. This suggests that binomial counts may be effective for estimating relative density, but yield potentially biased estimates of absolute density. Total counts, which fail to take account of variability in detectability between species or between habitats, give a markedly different ordering of species. Counts of birds within 30 m of the point give a better indication of relative densities, although the ordering of goldcrest (*Regulus regulus*) and tree pipit (*Anthus trivialis*), and of blackcap (*Sylvia atricapilla*) and redpoll (*Carduelis flammea*), is reversed relative to the density estimates.

Table 8.23 shows that estimates of $r_{1/2}$ agree remarkably well under the three models, considering their widely differing shapes, indicating that the estimates provide a useful guide to the relative detectability of species. Variation within a species in values of $\hat{\rho}$ suggests that estimation of ρ may be less robust than that of $r_{1/2}$.

Various aspects of the habitat within a 30-m radius of each point were recorded, and a principal components analysis carried out. The first component was identified as succession. The plantations had been restocked

Table 8.23. Analyses of binomial count data on songbirds from Welsh restocked conifer plantations under three models. Standard errors are given below estimates. n_1 is the number of birds detected within $c_1 = 30$ m of the point, and n_2 is the number beyond 30 m. Scientific names are given in Appendix A

			Linear model			Half-normal model			Hazard-rate model		
Species	n_1	n_2	$\hat{D}$	$\hat{r}_{1/2}$	$\hat{\rho}$	$\hat{D}$	$\hat{r}_{1/2}$	$\hat{\rho}$	$\hat{D}$	$\hat{r}_{1/2}$	$\hat{\rho}$
Willow	421	504	6.09	32.1	38.5	6.65	31.9	36.8	5.23	32.2	43.3
warbler			0.37	0.8	1.0	0.38	0.7	0.9	0.34	1.0	1.3
Wren	208	347	2.83	36.4	43.8	3.12	36.1	41.7	2.39	37.3	50.1
			0.22	1.3	1.5	0.23	1.2	1.4	0.18	1.4	1.9
Goldcrest	108	57	1.90	24.2	29.1	1.96	24.8	28.6	1.99	21.7	29.1
			0.26	1.2	1.5	0.25	1.1	1.2	0.38	1.9	2.5
Tree	127	235	1.70	38.0	45.6	1.88	37.6	43.3	1.44	39.0	52.3
pipit			0.18	1.7	2.0	0.19	1.6	1.9	0.15	1.8	2.5
Robin	78	89	1.14	31.5	37.8	1.24	31.4	36.2	0.98	31.5	42.4
			0.16	1.8	2.2	0.16	1.7	2.0	0.15	2.2	3.0
Chaffinch	73	141	0.97	38.7	46.4	1.07	38.2	44.1	0.82	39.7	53.3
			0.13	2.3	2.7	0.15	2.2	2.5	0.11	2.4	3.3
Garden	58	87	0.80	34.9	42.0	0.88	34.6	40.0	0.68	35.6	47.8
warbler			0.13	2.3	2.8	0.13	2.2	2.6	0.11	2.6	3.5
Siskin	36	74	0.47	39.7	47.6	0.52	39.2	45.3	0.40	40.7	54.7
			0.10	3.3	4.0	0.11	3.2	3.7	0.08	3.5	4.7
White-	33	48	0.46	34.5	41.5	0.51	34.2	39.5	0.39	35.1	47.2
throat			0.10	3.0	3.7	0.10	2.9	3.3	0.08	3.5	4.6
Coal tit	29	38	0.41	33.2	39.8	0.45	33.0	38.1	0.35	33.6	45.1
			0.10	3.1	3.7	0.10	2.9	3.4	0.09	3.7	4.9
Dunnock	28	32	0.41	31.5	37.8	0.45	31.4	36.3	0.35	31.6	42.4
			0.10	3.0	3.6	0.10	2.8	3.3	0.09	3.7	4.9
Song	27	79	0.34	46.1	55.3	0.38	45.5	52.5	0.30	47.4	63.6
thrush			0.08	4.4	5.3	0.08	4.4	5.1	0.06	4.6	6.1
Long-	18	12	0.30	26.1	31.3	0.31	26.5	30.5	0.29	24.4	32.8
tailed tit			0.10	3.2	3.8	0.10	2.8	3.3	0.12	4.6	6.2
Blackbird	15	40	0.19	44.3	53.2	0.21	43.7	50.4	0.16	45.5	61.1
			0.06	5.7	6.9	0.06	5.6	6.5	0.04	5.9	7.9
Blackcap	10	6	0.17	25.2	30.3	0.18	25.7	29.7	0.17	23.2	31.1
			0.07	4.1	5.0	0.07	3.7	4.2	0.10	6.2	8.3
Redpoll	12	20	0.16	36.4	43.8	0.18	36.1	41.6	0.14	37.3	50.0
			0.07	5.3	6.4	0.08	5.1	5.9	0.06	5.8	7.8
Chiffchaff	9	27	0.11	46.6	55.9	0.12	46.0	53.1	0.10	47.9	64.3
			0.04	7.8	9.4	0.05	7.7	8.9	0.03	8.0	10.8
Mistle	6	41	0.07	67.6	81.2	0.08	67.1	77.5	0.06	70.9	95.2
thrush			0.03	13.8	16.6	0.04	14.0	16.2	0.03	16.5	22.1

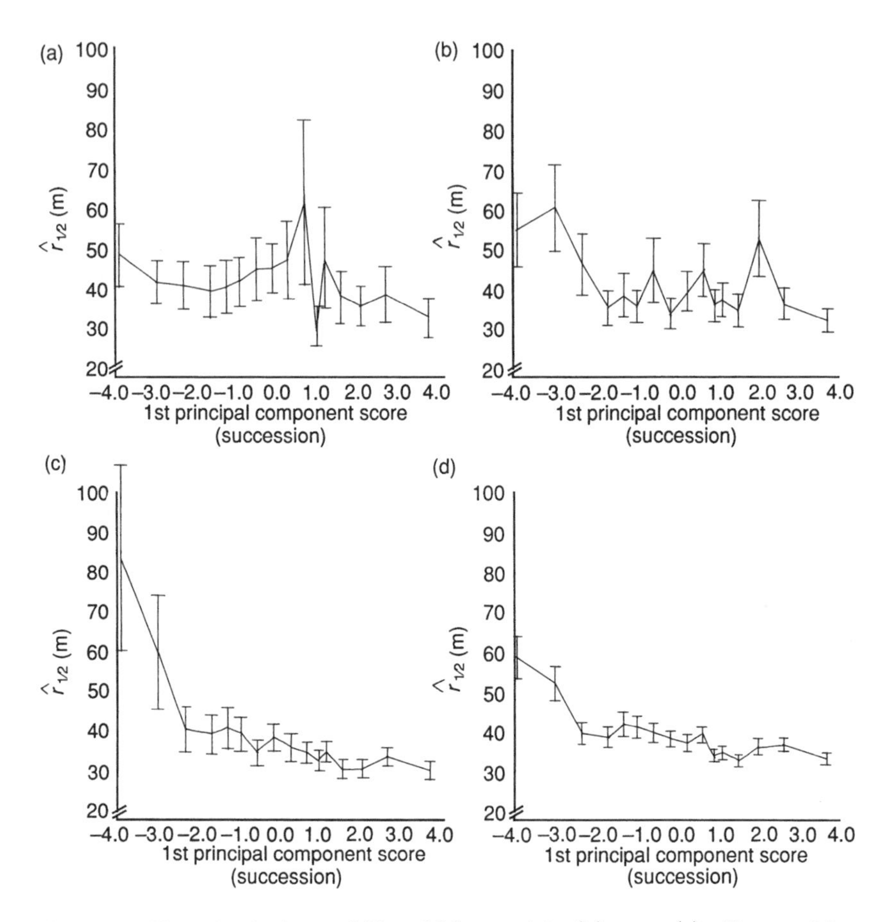

Fig. 8.13. Variation in detectability of (a) tree pipit, (b) wren, (c) willow warbler and (d) all species (pooled data) with habitat succession in conifer plantations aged between 2 and 11 years.

between two and eleven years previously, so the environment ranged from open to very dense. Birds of each species were recorded according to whether they were within or beyond 30 m of the point. The binomial half-normal model for the detection function was assumed, and $r_{1/2}$, the distance at which probability of detection is one half, was used as a measure of detectability. Three species, the tree pipit, wren (*Troglodytes troglodytes*) and willow warbler (*Phylloscopus trochilus*) were present in sufficient numbers at each stage of development to examine their change in detectability and density with succession in habitat. Figure 8.13 shows the estimated change in detectability with succession for these three species, and for all species combined. Both the wren and the willow warbler appear to be more detectable in the very early stages of succession. The pattern

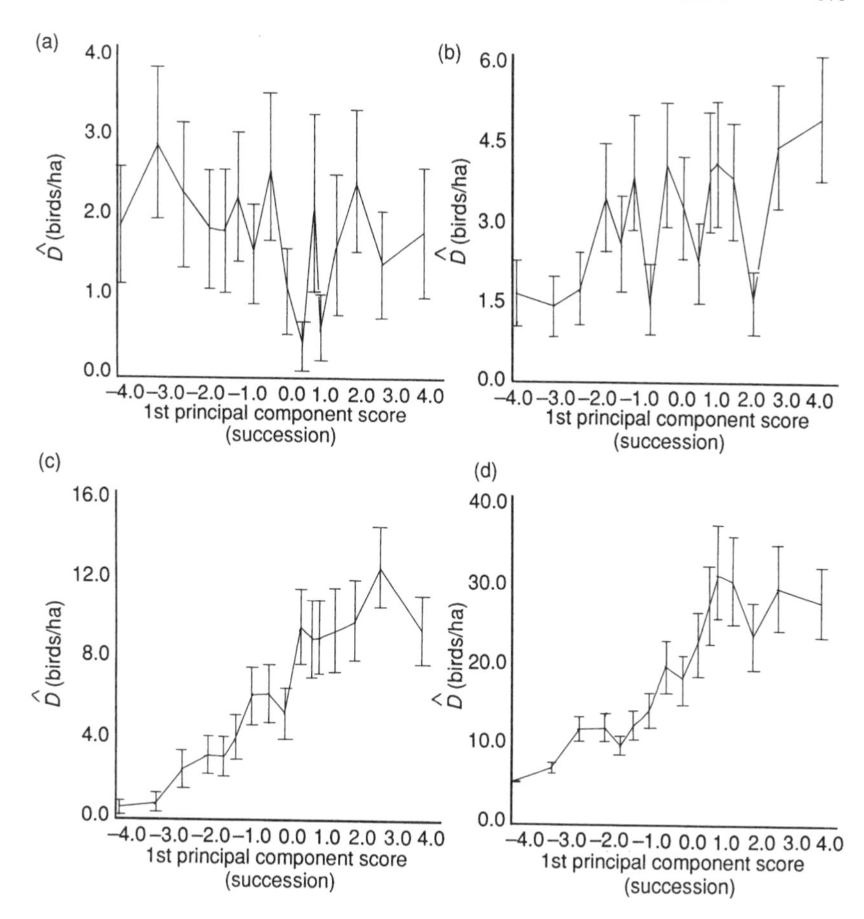

Fig. 8.14. Variation in density of (a) tree pipit, (b) wren, (c) willow warbler and (d) all species (pooled data) with habitat succession in conifer plantations aged between 2 and 11 years.

for the tree pipit is less clear. Analyses of the combined data set show a similar pattern to those for wren and willow warbler.

To measure trends in bird density with succession, it is therefore necessary to adjust for greater detectability in more open habitats. In Fig. 8.14, estimated change in density with succession is shown for the same three species and for all species combined. Both the wren and the willow warbler show a trend to higher densities in the older plantations. The plot for all species combined shows roughly a fivefold increase in density for eleven-year-old restocks relative to two-year-old. If unadjusted counts of birds are used as measures of relative abundance, this increase is estimated to be just 1.4-fold, indicating the importance of adjusting counts for detectability.

Principal components analysis was used in the above because it proved effective at reducing the dimensionality of the habitat variables. The second component represented a trend from a more diverse habitat, with herbaceous plants and regenerating broadleaf trees, through to pure coniferous stand with little undergrowth. If the only aspects of interest were variation in detectability and density with succession, the analysis could have been simplified by replacing the first principal component by stand age.

Bibliography

*Publications referenced in the text are indicated by an asterisk

*Akaike, H. (1973) Information theory and an extension of the maximum likelihood principle. In *International Symposium on Information Theory*, 2nd edn (eds. B.N. Petran and F. Csàaki), Akadèemiai Kiadi, Budapest, Hungary, pp. 267–81.

*Akaike, H. (1985) Prediction and entropy. In *A Celebration of Statistics* (eds. A.C. Atkinson and S.E. Fienberg), Springer-Verlag, Berlin, pp. 1–24.

Akin, J.A. (1998) Fourier series estimation of ground skink population density. *Copeia*, 519–22.

*Alldredge, J.R. and Gates, C.E. (1985) Line transect estimators for left-truncated distributions. *Biometrics*, **41**, 273–80.

*Alpizar-Jara, R. and Pollock, K.H. (1996) A combination line transect and capture recapture sampling model for multiple observers in aerial surveys. *Environmental and Ecological Statistics*, **3**, 311–27.

*Alpizar-Jara, R. and Pollock, K.H. (1999) Combining line transect and capture–recapture for mark-resighting studies. In *Marine Mammal Survey and Assessment Methods* (eds. G.W. Garner, S.C. Amstrup, J.L. Laake, B.F.J. Manly, L.L. McDonald and D.G. Robertson), Balkema, Rotterdam, pp. 99–114.

Andersen, D.E., Rongstad, O.J. and Mytton, W.R. (1985) Line transect analysis of raptor abundance along roads. *Wildlife Society Bulletin*, **13**, 533–9.

*Anderson, B.W. and Ohmart, R.D. (1981) Comparisons of avian census results using variable distance transect and variable circular plot techniques. In *Estimating Numbers of Terrestrial Birds. Studies in Avian Biology No. 6* (eds. C.J. Ralph and J.M. Scott), Cooper Ornithological Society, pp. 186–92.

*Anderson, D.R., Burnham, K.P. and Crain, B.R. (1978) A log-linear model approach to estimation of population size using the line-transect sampling method. *Ecology*, **59**, 190–3.

*Anderson, D.R., Burnham, K.P. and Crain, B.R. (1979a) Line transect estimation of population size: the exponential case with grouped data. *Communications in Statistics – Theory and Methods*, **A8**, 487–507.

*Anderson, D.R., Burnham, K.P. and Crain, B.R. (1980) Some comments on Anderson and Pospahala's correction of bias in line transect sampling. *Biometrical Journal*, **22**, 513–24.

*Anderson, D.R., Burnham, K.P. and Crain, B.R. (1985a) Estimating population size and density using line transect sampling. *Biometrical Journal*, **7**, 723–31.

*Anderson, D.R., Burnham, K.P. and Crain, B.R. (1985b) Some mathematical models for line transect sampling. *Biometrical Journal*, **7**, 741–52.

*Anderson, D.R., Burnham, K.P., Lubow, B.C., Thomas, L., Corn, P.S., Medica, P.A. and Marlow, R.W. (in press) Field trials of line transect methods applied to estimation of desert tortoise abundance. *Journal of Wildlife Management*, **65**.

*Anderson, D.R., Burnham, K.P., White, G.C. and Otis, D.L. (1983) Density estimation of small-mammal populations using a trapping web and distance sampling methods. *Ecology*, **64**, 674–80.

*Anderson, D.R., Laake, J.L., Crain, B.R. and Burnham, K.P. (1979b) Guidelines for line transect sampling of biological populations. *Journal of Wildlife Management*, **43**, 70–8.

*Anderson, D.R. and Pospahala, R.S. (1970) Correction of bias in belt transects of immotile objects. *Journal of Wildlife Management*, **34**, 141–6.

*Anderson, D.R. and Southwell, C. (1995) Estimates of macropod density from line transect surveys relative to analyst expertise. *Journal of Wildlife Management*, **59**, 852–7.

Anganuzzi, A.A. and Buckland, S.T. (1989) Reducing bias in estimated trends from dolphin abundance indices derived from tuna vessel data. *Report of the International Whaling Commission*, **39**, 323–34.

Anganuzzi, A.A., Buckland, S.T. and Cattanach, K.L. (1991) Relative abundance of dolphins associated with tuna in the eastern tropical Pacific, estimated from tuna vessel sightings data for 1988 and 1989. *Report of the International Whaling Commission*, **41**, 497–506.

Anthony, R.M. and Stehn, R.A. (1994) Navigating aerial transects with a laptop computer map. *Wildlife Society Bulletin*, **22**, 674–6.

Arendt, W.J., Gibbons, D.W. and Gray, G. (1999) Status of the volcanically threatened Montserrat Oriole *Icterus oberi* and other forest birds in Montserrat, West Indies. *Bird Conservation International*, **9**, 351–72.

*Avery, T.E. (1967) *Forest Measurements*. McGraw-Hill, New York.

Baldi, A. and Kisbenedek, T. (1999) Species-specific distribution of reed-nesting passerine birds across reed-bed edges: effects of spatial scale and edge type. *Acta Zoologica Academiae Scientiarum Hungaricae*, **45**, 97–114.

Ballance, L.T. and Pitman, R.L. (1998) Cetaceans of the western tropical Indian Ocean: Distribution, relative abundance, and comparisons with cetacean communities of two other tropical ecosystems. *Marine Mammal Science*, **14**, 429–59.

Balph, M.H., Stoddart, S.L. and Balph D.H. (1977) A simple technique for analyzing bird transect counts. *Auk*, **94**, 606–7.

*Barabesi, L. (2000) Local likelihood density estimation in line transect sampling. *Environmetrics*, **11**, 413–22.

*Barabesi, L. and Fattorini, L. (1993) Asymptotic properties of orthogonal series density estimators in line transect sampling with grouped data. *Metron*, **51**, 59–81.

*Barabesi, L. and Fattorini, L. (1994) A note on bandwidth selections for kernel density estimators at endpoints. *Metron*, **52**, 43–56.

Barabesi, L. and Fattorini, L. (1998) The use of replicated plot, line and point sampling for estimating species abundance and ecological diversity. *Environmental and Ecological Statistics*, **5**, 353–70.

Barlow, J. (1988) Harbor porpoise (*Phocoena phocoena*) abundance estimation for California, Oregon and Washington: I. Ship surveys. *Fishery Bulletin*, **86**, 417–31.

Barlow, J. (1994) Abundance of large whales in California coastal waters: a comparison of ship surveys in 1979/80 and in 1991. *Report of the International Whaling Commission*, **44**, 399–406.

*Barlow, J. (1997) *Preliminary estimates of cetacean abundance off California, Oregon, and Washington based on a 1996 ship survey and comparisons of passing and closing modes.* Southwest Fisheries Science Center Administrative Report LJ-97-11, 25pp.

Barlow, J. (1999) Trackline detection probability for long-diving whales. In *Marine Mammal Survey and Assessment Methods* (eds. G.W. Garner, S.C. Amstrup, J.L. Laake, B.F.J. Manly, L.L. McDonald and D.G. Robertson), Balkema, Rotterdam, pp. 209–21.

Barlow, J., Gerrodette, T. and Silber, G. (1997) First estimates of vaquita abundance. *Marine Mammal Science*, **13**, 44–58.

Barlow, J., Oliver, C.W., Jackson, T.D. and Taylor, B.L. (1988) Harbor porpoise (*Phocoena phocoena*) abundance estimation for California, Oregon, and Washington: II. Aerial surveys. *Fishery Bulletin*, **86**, 433–44.

*Barlow, R.E., Bartholomew, D.J., Bremner, J.M. and Brunk, H.D. (1972) *Statistical Inference under Order Restrictions.* Wiley, New York.

Barnes, A., Hill, G.J.E. and Wilson, G.R. (1986) Correcting for incomplete sighting in aerial surveys of kangaroos. *Australian Wildlife Research*, **13**, 339–48.

*Barnes, R.F.W., Blom, A., Alers, M.P.T. and Barnes, K.L. (1995) An estimate of the numbers of forest elephants in Gabon. *Journal of Tropical Ecology*, **11**, 27–37.

Bart, J. and Herrick, J. (1984) Diurnal timing of bird surveys. *Auk*, **101**, 384–7.

*Bart, J. and Schoultz, J.D. (1984) Reliability of singing bird surveys: changes in observer efficiency with avian density. *Auk*, **101**, 307–18.

*Batcheler, C.L. (1975) Development of a distance method for deer census from pellet groups. *Journal of Wildlife Management*, **39**, 641–52.

*Beasom, S.L., Hood, J.C. and Cain, J.R. (1981) The effect of strip width on helicopter censusing of deer. *Journal of Range Management*, **34**, 36–7.

*Beavers, S.C. and Ramsey, F.L. (1998) Detectability analysis in transect surveys. *Journal of Wildlife Management*, **62**, 948–57.

Becker, B.H., Beissinger, S.R. and Carter, H.R. (1997) At-sea density monitoring of marbled Murrelets in central California: methodological considerations. *Condor*, **99**, 743–55.

Bell, H.L. and Ferrier, S. (1985) The reliability of estimates of density from transect counts. *Corella*, **9**, 3–13.

*Bengtson, J.L., Blix, A.S., Boyd, I.L., Cameron, M.F., Hanson, M.B. and Laake, J.L. (1995) Antarctic pack-ice seal research, February and March 1995. *Antarctic Journal*, **30**, 191–3.

*Bergstedt, R.A. and Anderson, D.R. (1990) Evaluation of line transect sampling based on remotely sensed data from underwater video. *Transactions of the American Fisheries Society*, **119**, 86–91.

*Best, L.B. (1981) Seasonal changes in detection of individual bird species. In *Estimating Numbers of Terrestrial Birds. Studies in Avian Biology No. 6* (eds. C.J. Ralph and J.M. Scott), Cooper Ornithological Society, pp. 252–61.

*Best, P.B. and Butterworth, D.S. (1980) Report of the Southern Hemisphere minke whale assessment cruise, 1978/79. *Report of the International Whaling Commission*, **30**, 257–83.

*Bibby, C.J. and Buckland, S.T. (1987) Bias of bird census results due to detectability varying with habitat. *Acta Ecologica*, **8**, 103–12.

Bibby, C.J., Burgess, N.D. and Hill, D.A. (1992) *Bird Census Techniques.* Academic Press, London.

*Bibby, C.J., Burgess, N.D., Hill, D.A. and Mustoe, S. (2000) *Bird Census Techniques*, 2nd edn, Academic Press, London and San Diego.

*Bibby, C.J., Phillips, B.N. and Seddon, A.J.E. (1985) Birds of restocked conifer plantations in Wales. *Journal of Applied Ecology*, **22**, 619–33.

Blake, J.G., Hanowski, J.M., Niemi, G.J. and Collins, P.T. (1991) Hourly variation in transect counts of birds. *Ornis Fennica*, **68**, 139–47.

Blake, J.G., Hanowski, J.M., Niemi, G.J. and Collins, P.T. (1994) Annual variation in bird populations of mixed-conifer northern hardwood forests. *Condor*, **96**, 381–99.

Bodkin, J.L. and Udevitz, M.S. (1999) An aerial survey method to estimate sea otter abundance. In *Marine Mammal Survey and Assessment Methods* (eds. G.W. Garner, S.C. Amstrup, J.L. Laake, B.F.J. Manly, L.L. McDonald and D.G. Robertson), Balkema, Rotterdam, pp. 13–26.

*Bollinger, E.K., Gavin, T.A. and McIntyre, D.C. (1988) Comparison of transects and circular-plots for estimating bobolink densities. *Journal of Wildlife Management*, **52**, 777–86.

Bonnell, M.L. and Ford, R.G. (1987) California sea lion distribution: a statistical analysis of aerial transect data. *Journal of Wildlife Management*, **51**, 13–20.

Borchers, D.L. (1994) Methods of estimating mean school size from IWC sightings survey data. *Report of the International Whaling Commission*, **44**, 429–37.

*Borchers, D.L. (1999) Composite mark-recapture line transect surveys. In *Marine Mammal Survey and Assessment Methods* (eds. G.W. Garner, S.C. Amstrup, J.L. Laake, B.F.J. Manly, L.L. McDonald and D.G. Robertson), Balkema, Rotterdam, pp. 115–26.

*Borchers, D.L., Buckland, S.T., Goedhart, P.W., Clarke, E.D. and Hedley, S.L. (1998a) Horvitz-Thompson estimators for double-platform line transect surveys. *Biometrics*, **54**, 1221–37.

*Borchers, D.L., Zucchini, W. and Fewster, R.M. (1998b) Mark-recapture models for line transect surveys. *Biometrics*, **54**, 1207–20.

Bottenberg, H., Litsinger, J.A. and Kenmore, P.E. (1992) A line transect survey method for rice tungro virus. In *Proceedings, Third International Conference on Plant Protection in the Tropics*, Volume V, Kuala Lumpur, Malaysia, March 1990.

Brandt, C.A. and Rickard, W.H. (1992) Effects of survey frequency on bird density estimates in the shrub-steppe environment. *Northwest Science*, **66**, 172–82.

Brennan, L.A. and Block, W.M. (1986) Line transect estimates of mountain quail density. *Journal of Wildlife Management*, **50**, 373–7.

Briggs, K.T., Tyler, W.B. and Lewis, D.B. (1985) Aerial surveys for seabirds: methodological experiments. *Journal of Wildlife Management*, **49**, 412–7.

Brockelman, W.Y. (1980) The use of the line transect sampling method for forest primates. In *Tropical Ecology and Development* (ed. J.I. Furtado), The International Society of Tropical Ecology, Kuala Lumpur, Malaysia, pp. 367–71.

Broome, L.S., Bishop, K.D. and Anderson, D.R. (1984) Population density and habitat use by *Megapodius freyinet eremita* in West New Britain. *Australian Wildlife Research*, **11**, 161–71.

Brown, B.M. and Cowling, A. (1998) Clustering and abundance estimation for Neyman-Scott models and line transect surveys. *Biometrika*, **85**, 427–38.

Brown, J.A. and Boyce, M.S. (1998) Line transect sampling of Karner blue butterflies (*Lycaeides melissa samuelis*). *Environmental and Ecological Statistics*, **5**, 81–91.

*Brunk, H.D. (1978) Univariate density estimation by orthogonal series. *Biometrika*, **65**, 521–8.

*Buckland, S.T. (1980) A modified analysis of the Jolly–Seber capture–recapture model. *Biometrics*, **36**, 419–35.

*Buckland, S.T. (1982) A note on the Fourier series model for analysing line transect data. *Biometrics*, **38**, 469–77.

*Buckland, S.T. (1984) Monte Carlo confidence intervals. *Biometrics*, **40**, 811–7.

*Buckland, S.T. (1985) Perpendicular distance models for line transect sampling. *Biometrics*, **41**, 177–95.

*Buckland, S.T. (1987a) On the variable circular plot method of estimating animal density. *Biometrics*, **43**, 363–84.

*Buckland, S.T. (1987b) An assessment of the performance of line transect models for fitting IWC/IDCR cruise data, 1978/79 to 1984/85. *Report of the International Whaling Commission*, **37**, 277–9.

Buckland, S.T. (1987c) Estimation of minke whale numbers from the 1984/85 IWC/IDCR Antarctic sightings data. *Report of the International Whaling Commission*, **37**, 263–8.

*Buckland, S.T. (1992a) Fitting density functions using polynomials. *Applied Statistics*, **41**, 63–76.

*Buckland, S.T. (1992b) Maximum likelihood fitting of Hermite and simple polynomial densities. *Applied Statistics*, **41**, 241–66.

Buckland, S.T. (1992c) Effects of heterogeneity on estimation of probability of detection on the trackline. *Report of the International Whaling Commission*, **42**, 569–73.

*Buckland, S.T., Anderson, D.R., Burnham, K.P. and Laake, J.L. (1993a) *Distance Sampling: Estimating Abundance of Biological Populations.* Chapman and Hall, London.

*Buckland, S.T., Anderson, D.R., Burnham, K.P., Laake, J.L., Borchers, D. L. and Thomas, L. (eds) (in preparation) *Advanced Distance Sampling*, Oxford University Press, Oxford.

*Buckland, S.T. and Anganuzzi, A.A. (1988a) Comparison of smearing methods in the analysis of minke sightings data from IWC/IDCR Antarctic cruises. *Report of the International Whaling Commission*, **38**, 257–63.

Buckland, S.T. and Anganuzzi, A.A. (1988b) Estimated trends in abundance of dolphins associated with tuna in the eastern tropical Pacific. *Report of the International Whaling Commission*, **38**, 411–37.

Buckland, S.T., Bloch, D., Cattanach, K.L., Gunnlaugsson, Th., Hoydal, K., Lens, S. and Sigurjónsson, J. (1993b) Distribution and abundance of long-finned pilot whales in the North Atlantic, estimated from NASS-87 and NASS-89 data. In *Biology of Northern Hemisphere Pilot Whales*, International Whaling Commission, Cambridge, pp. 33–49.

*Buckland, S.T., Burnham, K.P. and Augustin, N.H. (1997) Model selection: an integral part of inference. *Biometrics*, **53**, 603–18.

Buckland, S.T., Cattanach, K.L. and Anganuzzi, A.A. (1992a) Estimating trends in abundance of dolphins associated with tuna in the eastern tropical Pacific Ocean, using sightings data collected on commercial tuna vessels. *Fishery Bulletin*, **90**, 1–12.

*Buckland, S.T., Cattanach, K.L. and Gunnlaugsson, Th. (1992b) Fin whale abundance in the North Atlantic, estimated from Icelandic and Faroese NASS-87 and NASS-89 data. *Report of the International Whaling Commission*, **42**, 645–651.

*Buckland, S.T. and Garthwaite, P.H. (1990) Estimating confidence intervals by the Robbins–Monro search process. *Applied Statistics*, **39**, 413–24.

*Buckland, S.T., Goudie, I.B.J. and Borchers, D.L. (2000) Wildlife population assessment: past developments and future directions. *Biometrics*, **56**, 1–12.

*Buckland, S.T., Macmillan, D.C., Duff, E.I. and Hanley, N. (1999) Estimating mean willingness to pay from dichotomous choice contingent valuation studies. *The Statistician*, **48**, 109–24.

*Buckland, S.T. and Turnock, B.J. (1992) A robust line transect method. *Biometrics*, **48**, 901–9.

Buford, E.W. and Capen, D.E. (1999) Abundance and productivity of forest songbirds in a managed, unfragmented landscape in Vermont. *Journal of Wildlife Management*, **63**, 180–8.

Buford, E.W., Capen, D.E. and Williams, B.K. (1996) Distance sampling to estimate fledgling brood density of forest birds. *Canadian Field Naturalist*, **110**, 642–8.

*Burdick, D.L. (1979) On estimation of the number of porpoise schools. Publication 79–2, San Diego State University, San Diego, CA, USA.

*Burnham, K.P. (1979) A parametric generalization of the Hayne estimator for line transect sampling. *Biometrics*, **35**, 587–95.

Burnham, K.P. (1981) Summarizing remarks: environmental influences. In *Estimating Numbers of Terrestrial Birds. Studies in Avian Biology No. 6* (eds. C.J. Ralph and J.M. Scott), Cooper Ornithological Society, pp. 324–5.

*Burnham, K.P. and Anderson, D.R. (1976) Mathematical models for non-parametric inferences from line transect data. *Biometrics*, **32**, 325–36.

*Burnham, K.P. and Anderson, D.R. (1984) The need for distance data in transect counts. *Journal of Wildlife Management*, **48**, 1248–54.

Burnham, K.P. and Anderson, D.R. (1992) Data-based selection of an appropriate biological model: the key to modern data analysis. In *Wildlife 2001: Populations* (eds. D. R. McCullough and R. H. Barrett), Elsevier Science Publishers, London, pp. 16–30.

*Burnham, K.P. and Anderson, D.R. (1998) *Model Selection and Inference: A Practical Information–Theoretic Approach.* Springer, New York.

*Burnham, K.P., Anderson, D.R. and Laake, J.L. (1979) Robust estimation from line transect data. *Journal of Wildlife Management*, **43**, 992–6.

*Burnham, K.P., Anderson, D.R. and Laake, J.L. (1980) Estimation of density from line transect sampling of biological populations. *Wildlife Monographs*, **72**, 1–202.

*Burnham, K.P., Anderson, D.R. and Laake, J.L. (1981) Line transect estimation of bird population density using a Fourier series. In *Estimating Numbers of Terrestrial Birds. Studies in Avian Biology No. 6* (eds. C.J. Ralph and J.M. Scott), Cooper Ornithological Society, pp. 466–82.

*Burnham, K.P., Anderson, D.R. and Laake, J.L. (1985) Efficiency and bias in strip and line transect sampling. *Journal of Wildlife Management*, **49**, 1012–8.

*Burnham, K.P., Anderson, D.R., White, G.C., Brownie, C. and Pollock, K.H. (1987) Design and analysis methods for fish survival experiments based on release–recapture. American Fisheries Society, Monograph No. 5.

*Butterworth, D.S. (1982a) A possible basis for choosing a functional form for the distribution of sightings with right angle distance: some preliminary ideas. *Report of the International Whaling Commission*, **32**, 555–8.

*Butterworth, D.S. (1982b) On the functional form used for $g(y)$ for minke whale sightings, and bias in its estimation due to measurement inaccuracies. *Report of the International Whaling Commission*, **32**, 883–8.

Butterworth, D.S. (1986) A note on the analysis of the 1980/81 variable speed experiment. *Report of the International Whaling Commission*, **36**, 485–9.

*Butterworth, D.S. and Best, P.B. (1982) Report of the Southern Hemisphere minke whale assessment cruise, 1980/81. *Report of the International Whaling Commission*, **32**, 835–74.

Butterworth, D.S., Best, P.B. and Basson, M. (1982) Results of analysis of sighting experiments carried out during the 1980/81 Southern Hemisphere minke whale assessment cruise. *Report of the International Whaling Commission*, **32**, 819–34.

*Butterworth, D.S., Best, P.B. and Hembree, D. (1984) Analysis of experiments carried out during the 1981/82 IWC/IDCR Antarctic minke whale assessment cruise in Area II. *Report of the International Whaling Commission*, **34**, 365–92.

Butterworth, D.S. and Borchers, D.L. (1988) Estimates of $g(0)$ for minke schools from the results of the independent observer experiment on the 1985/86 and 1986/87 IWC/IDCR Antarctic assessment cruises. *Report of the International Whaling Commission*, **38**, 301–13.

*Byth, K. (1982) On robust distance-based intensity estimators. *Biometrics*, **38**, 127–35.

Byth, K. and Ripley, B.D. (1980) On sampling spatial patterns by distance methods. *Biometrics*, **36**, 279–84.

*Carretta, J.V., Forney, K.A. and Laake, J.L. (1998) Abundance of southern California coastal bottlenose dolphins estimated from tandem aerial surveys. *Marine Mammal Science*, **14**, 655–75.

*Carroll, J.R. and Ruppert, D. (1988) *Transformation and Weighting in Regression*. Chapman and Hall, London.

Casagrande, D.G. and Beissinger, S.R. (1997) Evaluation of four methods for estimating parrot population size. *Condor*, **99**, 445–57.

Cassey, P. and McArdle, B.H. (1999) An assessment of distance sampling techniques for estimating animal abundance. *Environmetrics*, **10**, 261–78.

Cassey, P. and Ussher, G.T. (1999) Estimating abundance of tuatara. *Biological Conservation*, **88**, 361–6.

Catt, D.C., Baines, D., Picozzi, N., Moss, R. and Summers, R.W. (1998) Abundance and distribution of capercaillie *Tetrao urogallus* in Scotland 1992–1994. *Biological Conservation*, **85**, 257–67.

Caughley, G. (1972) Improving the estimates from inaccurate censuses. *Journal of Wildlife Management*, **36**, 135–40.

Caughley, G. (1974) Bias in aerial survey. *Journal of Wildlife Management*, **38**, 921–33.

Caughley, G., Sinclair, R. and Scott-Kemmis, D. (1976) Experiments in aerial survey. *Journal of Wildlife Management*, **40**, 290–300.

*Chafota, J. (1988) *Effect of measurement errors in estimating density from line transect sampling.* MS Paper, Colorado State University, Ft Collins, 47pp.

Chapman, C.A. and Lambert, J.E. (2000) Habitat alteration and the conservation of African primates: case study of Kibale National Park, Uganda. *American Journal of Primatology*, **50**, 169–85.

*Chatfield, C. (1988) *Problem Solving: a Statistician's Guide.* Chapman and Hall, London.

*Chatfield, C. (1991) Avoiding statistical pitfalls. *Statistical Science*, **6**, 240–68.

*Chen, S.X. (1996a) A kernel estimate for the density of a biological population by using line transect sampling. *Applied Statistics*, **45**, 135–50.

*Chen, S.X. (1996b) Studying school size effects in line transect sampling using the kernel method. *Biometrics*, **52**, 1283–94.

*Chen, S.X. (1998) Measurement errors in line transect surveys. *Biometrics*, **54**, 899–908.

*Chen, S.X. (1999) Estimation in independent observer line transect surveys for clustered populations. *Biometrics*, **55**, 754–9.

Chen, S.X. (2000) Animal abundance estimation in independent observer line transect surveys. *Environmental and Ecological Statistics*, **7**, 285–99.

Chen, S.X. and Lloyd, C.J. (2000) A nonparametric approach to the analysis of two-stage mark-recapture experiments. *Biometrika*, **87**, 633–49.

Cherenkov, S.E. (1998) Accuracy of one-visit censuses of forest passerine birds during a breeding season. *Zoologichesky Zhurnal*, **77**, 474–85.

Chiarello, A.G. (1999) Effects of fragmentation of the Atlantic forest on mammal communities in south-eastern Brazil. *Biological Conservation*, **89**, 71–82.

Childs, J.E., Robinson, L.E., Sadek, R., Madden, A., Miranda, M.E. and Miranda, N.L. (1998) Density estimates of rural dog populations and an assessment of marking methods during a rabies vaccination campaign in the Philippines. *Preventive Veterinary Medicine*, **33**, 207–18.

Clancy, T.F., Pople, A.R. and Gibson, L.A. (1997) Comparison of helicopter line transects with walked line transects for estimating densities of kangaroos. *Wildlife Research*, **24**, 397–409.

*Clark, P.J. and Evans, F.C. (1954) Distance to nearest neighbour as a measure of spatial relationships in populations. *Ecology*, **35**, 23–30.

*Cochran, W.G. (1977) *Sampling Techniques*, 3rd edn. Wiley, New York.

Conroy, M.J., Nichols, J.D. and Asanza, E.R. (1997) Métodos cuantitativos contemporáneos para entender y manejar poblaciones y comunidades animales [Contemporary quantitative methods to understand and manage animal populations and communities]. *Interciencia*, **22**, 247–58.

*Cook, R.D. and Jacobsen, J.O. (1979) A design for estimating visibility bias in aerial surveys. *Biometrics*, **35**, 735–42.

*Cooke, J.G. (1985) Notes on the estimation of whale density from line transects. *Report of the International Whaling Commission*, **35**, 319–23.

Cooke, J.G. (1987) Estimation of the population of minke whales in Antarctic Area IVW in 1984/85. *Report of the International Whaling Commission*, **37**, 273–6.

Cooke, J.G. (1997) An implementation of a surfacing-based approach to abundance estimation of minke whales from shipborne surveys. *Report of the International Whaling Commission*, **47**, 513–28.

Corn, J.L. and Conroy, M.J. (1998) Estimation of density of mongooses with capture-recapture and distance sampling. *Journal of Mammalogy*, **79**, 1009–15.

Cottam, G. and Curtis, J.T. (1956) The use of distance measures in phytosociological sampling. *Ecology*, **37**, 451–60.

Coulson, G. (1993) Use of heterogeneous habitat by the western gray kangaroo, *Macropus fuliginosus*. *Wildlife Research*, **20**, 137–49.

*Coulson, G.M. and Raines, J.A. (1985) Methods for small-scale surveys of grey kangaroo populations. *Australian Wildlife Research*, **12**, 119–25.

Cowling, A. (1998) Spatial methods for line transect surveys. *Biometrics*, **54**, 828–39.

*Cox, D.R. (1969) Some sampling problems in technology. In *New Developments in Survey Sampling* (eds. N.L. Johnson and H. Smith, Jr.), Wiley–Interscience, New York, USA, pp. 506–27.

*Cox, D.R. and Snell, E.J. (1989) *Analysis of Binary Data*, 2nd edn. Chapman and Hall, London.

*Cox, T.F. (1976) The robust estimation of the density of a forest stand using a new conditioned distance method. *Biometrika*, **63**, 493–500.

Crain, B.R. (1974) Estimation of distributions using orthogonal expansions. *The Annals of Statistics*, **2**, 454–63.

Crain, B.R. (1998) Some comments on line transect grouped data analysis. *Ecological Modelling*, **109**, 243–9.

Crain, B.R. (1998) Window sensitivity functions for line transect sampling. *Environmental Management*, **22**, 471–81.

*Crain, B.R., Burnham, K.P., Anderson, D.R. and Laake, J.L. (1978) *A Fourier Series Estimator of Population Density for Line Transect Sampling.* Utah State University Press, Logan, UT, USA.

*Crain, B.R., Burnham, K.P., Anderson, D.R. and Laake, J.L. (1979) Nonparametric estimation of population density for line transect sampling using Fourier series. *Biometrical Journal*, **21**, 731–48.

*Dahlheim, M., York, A., Towell, R., Waite, J. and Breiwick, J. (2000) Harbor porpoise (*Phocoena phocoena*) abundance in Alaska: Bristol Bay to Southeast Alaska, 1991-1993. *Marine Mammal Science*, **16**, 28–45.

Daniel, A., Holechek, J., Valdez, R., Tembo, A., Saiwana, L., Fusco, M. and Cardenas, M. (1993) Jackrabbit densities on fair and good condition Chihuahuan desert range. *Journal of Range Management*, **46**, 524–8.

*Davison, A.C. and Hinkley, D.V. (1997) *Bootstrap Methods and their Application.* Cambridge University Press, Cambridge.

*Davison, A.C., Hinkley, D.V. and Schechtman, E. (1986) Efficient bootstrap simulation. *Biometrika*, **73**, 555–66.

*Dawson, D.G. (1981) Counting birds for a relative measure (index) of density. In *Estimating Numbers of Terrestrial Birds. Studies in Avian Biology No. 6* (eds. C.J. Ralph and J.M. Scott), Cooper Ornithological Society, pp. 12–6.

DeJong, M.J. and Emlen, J.T. (1985) The shape of the auditory detection function and its implications for songbird censusing. *Journal of Field Ornithology*, **56**, 213–23.

*Dempster, A.P., Laird, N.M. and Rubin, D.B. (1977) Maximum likelihood from incomplete data via the EM algorithm (with Discussion). *Journal of the Royal Statistical Society, Series B*, **39**, 1–39.

*DeSante, D.F. (1981) A field test of the variable circular-plot censusing technique in a California coastal scrub breeding bird community. In *Estimating Numbers of Terrestrial Birds. Studies in Avian Biology No. 6* (eds. C.J. Ralph and J.M. Scott), Cooper Ornithological Society, pp. 177–85.

*DeSante, D.F. (1986) A field test of the variable circular-plot censusing method in a Sierran subalpine forest habitat. *Condor*, **88**, 129–42.

DeVries, P.G. (1979a) Line intersect sampling – statistical theory, applications, and suggestions for extended use in ecological inventory. In *Sampling Biological Populations* (eds. R.M. Cormack, G.P. Patil and D.S. Robson), International Co-operative Publishing House, Fairland, MD, USA, pp. 1–70.

DeVries, P.G. (1979b) A generalization of the Hayne-type estimator as an application of line intercept sampling. *Biometrics*, **35**, 743–8.

Devy, M.S., Ganesh, T. and Davidar, P. (1998) Patterns of butterfly distribution in the Andaman islands: implications for conservation. *Acta Oecologica*, **19**, 527–34.

DeYoung, C.A., Guthery, F.S., Beasom, S.L., Coughlin, S.P. and Heffelfinger, J.R. (1989) Improving estimates of white-tailed deer abundance from helicopter surveys. *Wildlife Society Bulletin*, **17**, 275–9.

Diaz, J.A. and Carrascal, L.M. (1991) Regional distribution of a Mediterranean lizard – influence of habitat cues and prey abundance. *Journal of Biogeography*, **18**, 291–7.

*Dice, L.R. (1938) Some census methods for mammals. *Journal of Wildlife Management*, **2**, 119–30.

Diggle, P.J. (1975) Robust density estimation using distance methods. *Biometrika*, **62**, 39–48.

Diggle, P.J. (1977) A note on robust density estimation for spatial point patterns. *Biometrika*, **64**, 91–5.

*Diggle, P.J. (1983) *Statistical Analysis of Spatial Point Patterns.* Academic Press, London.

Dobkin, D.S. and Rich, A.C. (1998) Comparison of line transect, spot map, and point count surveys for birds in riparian habitats of the Great Basin. *Journal of Field Ornithology*, **69**, 430–43.

Dodd, C.K. (1990) Line transect estimation of Red Hills salamander burrow density using a Fourier series. *Copeia*, 555–7.

Dohl, T.P., Bonnell, M.L. and Ford, R.G. (1986) Distribution and abundance of common dolphin, *Delphinus delphis*, in the Southern California Bight: a quantitative assessment based upon aerial transect data. *Fishery Bulletin*, **84**, 333–43.

Doi, T. (1971) Further development of sighting theory on whales. *Bulletin of Tokai Regional Fisheries Research Laboratory*, **68**, 1–22.

Doi, T. (1974) Further development of whale sighting theory. In *The Whale Problem: a Status Report* (ed. W.E. Schevill), Harvard University Press, Cambridge, MA, USA, pp. 359–68.

Doi, T., Kasamatsu, F. and Nakano, T. (1982) A simulation study on sighting survey of minke whales in the Antarctic. *Report of the International Whaling Commission*, **32**, 919–28.

Doi, T., Kasamatsu, F. and Nakano, T. (1983) Further simulation studies on sighting by introducing both concentration of sighting effort by angle and aggregations of minke whales in the Antarctic. *Report of the International Whaling Commission*, **33**, 403–12.

*Drummer, T.D. (1985) *Size-bias in line transect sampling.* PhD Thesis, University of Wyoming, Laramie, WY, USA. Available from University Microfilms International, 300 N. Zeeb Road, Ann Arbor, MI 48106.

Drummer, T.D. (1990) Estimation of proportions and ratios from line transect data. *Communications in Statistics – Theory and Methods*, **19**, 3069–91.

*Drummer, T.D. (1991) SIZETRAN: analysis of size-biased line transect data. *Wildlife Society Bulletin*, **19**, 117–8.

Drummer, T.D. (1999) Planning abundance estimation surveys when detectability is < 1.0. In *Marine Mammal Survey and Assessment Methods* (eds. G.W. Garner, S.C. Amstrup, J.L. Laake, B.F.J. Manly, L.L. McDonald and D.G. Robertson), Balkema, Rotterdam, pp. 67–73.

*Drummer, T.D., Degange, A.R., Pank, L.L. and McDonald, L.L. (1990) Adjusting for group size influence in line transect sampling. *Journal of Wildlife Management*, **54**, 511–4.

*Drummer, T.D. and McDonald, L.L. (1987) Size bias in line transect sampling. *Biometrics*, **43**, 13–21.

Duffy, D.C. and Schneider, D.C. (1984) A comparison of two transect methods of counting birds at sea. *Cormorant*, **12**, 95–8.

*Eberhardt, L.L. (1967) Some developments in 'distance sampling'. *Biometrics*, **23**, 207–16.

*Eberhardt, L.L. (1968) A preliminary appraisal of line transect. *Journal of Wildlife Management*, **32**, 82–8.

*Eberhardt, L.L. (1978a) Transect methods for population studies. *Journal of Wildlife Management*, **42**, 1–31.

*Eberhardt, L.L. (1978b) Appraising variability in population studies. *Journal of Wildlife Management*, **42**, 207–38.

*Eberhardt, L.L. (1979) Line-transects based on right-angle distances. *Journal of Wildlife Management*, **43**, 768–74.

Eberhardt, L.L., Chapman, D.G. and Gilbert, J.R. (1979) A review of marine mammal census methods. *Wildlife Monographs*, **63**, 1–46.

*Edwards, D.K., Dorsey, G.L. and Crawford, J.A. (1981) A comparison of three avian census methods. In *Estimating Numbers of Terrestrial Birds. Studies in Avian Biology No. 6* (eds. C.J. Ralph and J.M. Scott), Cooper Ornithological Society, pp. 170–6.

*Efron, B. (1979) Bootstrap methods: another look at the jackknife. *Annals of Statistics*, **7**, 1–16.

*Efron, B. (1981) Nonparametric standard errors and confidence intervals (with discussion). *Canadian Journal of Statistics*, **9**, 139–72.

*Efron, B. and Tibshirani, R.J. (1993) *An Introduction to the Bootstrap*. Chapman and Hall, London.

*Emlen, J.T. (1971) Population densities of birds derived from transect counts. *Auk*, **88**, 323–42.

*Emlen, J.T. (1977) Estimating breeding season bird densities from transect counts. *Auk*, **94**, 455–68.

Emlen, J.T. and DeJong, M.J. (1981) The application of song detection threshold distance to census operations. In *Estimating Numbers of Terrestrial Birds. Studies in Avian Biology No. 6* (eds. C.J. Ralph and J.M. Scott), Cooper Ornithological Society, pp. 346–52.

Ensign, W.E., Angermeier, P.L. and Dolloff, C.A. (1995) Use of line transect methods to estimate abundance of benthic stream fishes. *Canadian Journal of Fisheries and Aquatic Sciences*, **52**, 213–22.

Epperly, S.P., Braun, J. and Chester, A.J. (1995) Aerial surveys for sea turtles in North Carolina inshore waters. *Fishery Bulletin*, **93**, 254–61.

*Erickson, A.W., Siniff, D.B. and Harwood, J. (1993) Estimation of population sizes. In *Antarctic Seals: Research Methods and Techniques* (ed. R.M. Laws), Cambridge University Press, Cambridge, pp. 29–45.

Erwin, R.M. (1982) Observer variability in estimating numbers: an experiment. *Journal of Field Ornithology*, **53**, 159–67.

Estades, C.F. and Temple, S.A. (1999) Deciduous-forest bird communities in a fragmented landscape dominated by exotic pine plantations. *Ecological Applications*, **9**, 573–85.

Estes, J.A. and Gilbert, J.R. (1978) Evaluation of an aerial survey of Pacific walruses. *Journal of the Fisheries Research Board of Canada*, **35**, 1130–40.

*Fancy, S.G. (1997) A new approach for analyzing bird densities from variable circular-plot counts. *Pacific Science*, **51**, 107–14.

Farina, A. (1995) Distribution and dynamics of birds in a rural sub-Mediterranean landscape. *Landscape and Urban Planning*, **31**, 269–80.

Fashing, P.J. and Cords, M. (2000) Diurnal primate densities and biomass in the Kakamega Forest: an evaluation of census methods and a comparison with other forests. *American Journal of Primatology*, **50**, 139–52.

*Fay, J.M. (1991) An elephant (*Loxodonta africana*) survey using dung counts in the forests of the Central African Republic. *Journal of Tropical Ecology*, **7**, 25–36.

*Fay, J.M. and Agnagna, M. (1991) A population survey of forest elephants (*Loxodonta africana cyclotis*) in northern Congo. *African Journal of Ecology*, **29**, 177–87.

*Fewster, R.M., Buckland, S.T., Siriwardena, G.M., Baillie, S.R. and Wilson, J.D. (2000) Analysis of population trends for farmland birds using generalized additive models. *Ecology*, **81**, 1970–84.

Fitzgerald, S.M. and Tanner, G.W. (1992) Avian community response to fire and mechanical shrub control in south Florida. *Journal of Range Management*, **45**, 396–400.

Fleming, K.K., and Giuliano, W.M. (1998) Effect of border-edge cuts on birds at woodlot edges in southwestern Pennsylvania. *Journal of Wildlife Management*, **62**, 1430–7.

Folkard, N.F.G. and Smith, J.N.M. (1995) Evidence for bottom up effects in the boreal forest: do passerine birds respond to large scale experimental fertilization? *Canadian Journal of Zoology*, **73**, 2231–7.

Foote, K.G. and Stefansson, G. (1993) Definition of the problem of estimating fish abundance over an area from acoustic line transect measurements of density. *ICES Journal of Marine Science*, **50**, 369–81.

Forbes, A.R., Mueller, J.M., Mitchell, R.B., Dabbert, C.B. and Wester, D.B. (2000) Accuracy of red imported fire ant mound density estimates. *Southwestern Entomologist*, **25**, 109–12.

*Forbes, S.A. (1907) An ornithological cross-section of Illinois in autumn. *Illinois Natural History Survey Bulletin*, **7**, 305–35.

*Forbes, S.A. and Gross, A.O. (1921) The orchard birds of an Illinois summer. *Illinois Natural History Survey Bulletin*, **14**, 1–8.

Forcada, J., Disciara, G.N. and Fabbri, F. (1995) Abundance of fin whales and striped dolphins summering in the Corso-Ligurian Basin. *Mammalia*, **59**, 127–40.

Forcada, J. and Hammond, P. (1998) Geographical variation in abundance of striped and common dolphins of the western Mediterranean. *Journal of Sea Research*, **39**, 313–25.

Forney, K.A. and Barlow, J. (1998) Seasonal patterns in the abundance and distribution of California cetaceans, 1991–1992. *Marine Mammal Science*, **14**, 460–89.

*Franzreb, K.E. (1976) Comparison of variable strip transect and spot-map methods for censusing avian populations in a mixed-coniferous forest. *Condor*, **78**, 260–2.

*Franzreb, K.E. (1981) Determination of avian densities using the variable-strip and fixed-width transect surveying methods. In *Estimating the Numbers of Terrestrial Birds. Studies in Avian Biology No. 6* (eds. C.J. Ralph and J.M. Scott), Cooper Ornithological Society, pp. 139–45.

Fuller, R.J. and Langslow, D.R. (1984) Estimating numbers of birds by point counts: how long should counts last? *Bird Study*, **31**, 195–202.

Gaillard, J.M., Boutin, J.-M. and Van Laere, G. (1993) The use of line transects for estimating the population density of roe deer – a feasibility study. *Revue d'Ecologie – la Terre et la Vie*, **48**, 73–85.

Gannier, A. (1997) Estimation of summer abundance of the fin whale *Balaenoptera physalus* (Linne, 1758) in the Liguro-Provencal Basin (West Mediterranean). *Revue d'Ecologie – la Terre et la Vie*, **52**, 69–86.

Gannier, A. (1998) Estimation of summer abundance of the striped dolphin *Stenella coeruleoalba* (Meyen, 1833) in the future northwestern Mediterranean international marine sanctuary. *Revue d'Ecologie – la Terre et la Vie*, **53**, 255–72.

*Garrett-Logan, N. and Smith, T. (1997) A hand-held pen-based computer system for marine mammal sighting surveys. *Marine Mammal Science*, **13**, 694–700.

*Garthwaite, P.H. and Buckland, S.T. (1992) Generating Monte Carlo confidence intervals by the Robbins–Monro search process. *Applied Statistics*, **41**, 159–71.

Gaston, A.J., Collins, B.L. and Diamond, A.W. (1987) The 'snapshot' count for estimating densities of flying seabirds during boat transects: a cautionary comment. *Auk*, **104**, 336–8.

*Gates, C.E. (1969) Simulation study of estimators for the line transect sampling method. *Biometrics*, **25**, 317–28.

*Gates, C.E. (1979) Line transect and related issues. In *Sampling Biological Populations* (eds. R.M. Cormack, G.P. Patil and D.S. Robson), International Co-operative Publishing House, Fairland, MD, USA, pp. 71–154.

*Gates, C.E. (1980) LINETRAN, a general computer program for analyzing line transect data. *Journal of Wildlife Management*, **44**, 658–61.

*Gates, C.E., Evans, W., Gober, D.R., Guthery, F.S. and Grant, W.E. (1985) Line transect estimation of animal densities from large data sets. In *Game Harvest Management* (eds. S.L. Beasom and S.F. Roberson), Caesar Kleberg Wildlife Research Institute, Texas A&I University, Kingsville, TX, USA, pp. 37–50.

*Gates, C.E., Marshall, W.H. and Olson, D.P. (1968) Line transect method of estimating grouse population densities. *Biometrics*, **24**, 135–45.

*Gates, C.E. and Smith, P.W. (1980) An implementation of the Burnham–Anderson distribution free method of estimating wildlife densities from line transect data. *Biometrics*, **36**, 155–60.

Geimsdell, J.J.R. and Westley, S.B. (eds.) (1979) *Low-Level Aerial Survey Techniques*. International Livestock Centre for Africa, Addis Ababa, Ethiopia.

Gelatt, T.S. and Siniff, D.B. (1999) Line transect survey of crabeater seals in the Amundsen-Bellingshausen Seas, 1994. *Wildlife Society Bulletin*, **27**, 330–6.

*Gerard, P.D. and Schucany, W.R. (1999) Local bandwidth selection for kernel estimation of population densities with line transect sampling. *Biometrics*, **55**, 769–73.

*Gilbert, D.W., Anderson, D.R., Ringelman, J.K. and Szymczak, M.R. (1996) Response of nesting ducks to habitat management on the Monte Vista National Wildlife Refuge, Colorado. *Wildlife Monographs*, **131**, 1–44.

*Gill, R.M.A., Thomas, M.L. and Stocker, D. (1997) The use of portable thermal imaging for estimating deer population density in forest habitats. *Journal of Applied Ecology*, **34**, 1273–86.

Gillings, S., Fuller, R.J. and Henderson, A.C.B. (1998) Avian community composition and patterns of bird distribution within birch-heath mosaics in north-east Scotland. *Ornis Fennica*, **75**, 27–37.

Gogan, P.J., Thompson, S.C., Pierce, W. and Barrett, R.H. (1986) Line-transect censuses of fallow and black-tailed deer on the Point Reyes Peninsula. *California Game and Fish*, **72**, 47–61.

Gotmark, F. and Post, P. (1996) Prey selection by sparrowhawks, *Accipiter nisus*: relative predation risk for breeding passerine birds in relation to their size, ecology and behaviour. *Philosophical Transactions of the Royal Society of London B*, **351**, 1559–77.

Granholm, S.L. (1983) Bias in density estimates due to movement of birds. *Condor*, **85**, 243–8.

Granjon, L., Cosson, J.F., Judas, J. and Ringuet, S. (1996) Influence of tropical rainforest fragmentation on mammal communities in French Guiana: short-term effects. *Acta Oecologica*, **17**, 673–84.

*Gray, H.L. and Schucany, W.R. (1972) *The Generalized Jackknife Statistic.* Marcel Dekker, New York, 308pp.

Green, G.A., Brueggeman, J.J., Bowlby, C.E., Grotefendt, R.A., Bonnell, M.L. and Balcomb, K.T. (1992) Cetacean distribution and abundance off Oregon and Washington, 1989–1990. Chapter I. In *Oregon and Washington Marine Mammal and Seabird Surveys* (ed. J.J. Brueggeman), final report prepared by Ebasco Environmental, Bellevue, WA, and Ecological Consulting, Inc., Portland, OR, for the Minerals Management Service, Pacific OCS Region. OCS Study MMS 91–0072.

Gregory, R.D. and Baillie, S.R. (1998) Large-scale habitat use of some declining British birds. *Journal of Applied Ecology*, **35**, 785–99.

*Grigg, G.C., Pople, A.R. and Beard, L.A. (1997) Application of an ultralight aircraft to aerial surveys of kangaroos on grazing properties. *Wildlife Research*, **24**, 359–72.

Gross, J.E., Stoddart, L.C. and Wagner, F.H. (1974) Demographic analysis of a northern Utah jackrabbit population. *Wildlife Monographs*, **40**, 1–68.

*Guenzel, R.J. (1997) *Estimating pronghorn abundance using aerial line transect sampling.* Wyoming Game and Fish Dept., 5400 Bishop Blvd., Cheyenne, WY, 174pp.

Guix, J.C., Martin, M. and Manosa, S. (1999) Conservation status of parrot populations in an Atlantic rainforest area of southeastern Brazil. *Biodiversity and Conservation*, **8**, 1079–88.

*Gunnlaugsson, Th. and Sigurjónsson, J. (1990) NASS-87: estimation of whale abundance based on observations made on board Icelandic and Faroese survey vessels. *Report of the International Whaling Commission*, **40**, 571–80.

*Guthery, F.S. (1988) Line transect sampling of bobwhite density on rangeland: evaluation and recommendations. *Wildlife Society Bulletin*, **16**, 193–203.

Gutzwiller, K.J. and Marcum, H.A. (1997) Bird reactions to observer clothing color: implications for distance-sampling techniques. *Journal of Wildlife Management*, **61**, 935–47.

*Hahn, H.C. (1949) A method of censusing deer and its application in the Edwards Plateau of Texas. *Final Report for Texas Federal Aid Project 25-R, July 1, 1946 to March 30, 1948.*

*Hain, J.H.W., Ellis, S.L., Kenney, R.D. and Slay, C.K. (1999) Sightability of right whales in coastal waters of the southeastern United States with implications for the aerial monitoring program. In *Marine Mammal Survey and Assessment Methods* (eds. G.W. Garner, S.C. Amstrup, J.L. Laake, B.F.J. Manly, L.L. McDonald and D.G. Robertson), Balkema, Rotterdam, pp. 191–207.

Hamel, P.B. (1984) Comparison of variable circular-plot and spot-map censusing methods in temperate forest. *Ornis Scandinavica*, **15**, 266–74.

Hamel, P.B. (1990) Response to Tomiatojc and Verner. *Auk*, **107**, 451–3.

Hamel, P.B., Smith, W.P. and Wahl, J.W. (1993) Wintering bird populations of fragmented forest habitat in the Central Basin, Tennessee. *Biological Conservation*, **66**, 107–15.

*Hammond, P.S. (1984) An investigation into the effects of different techniques of smearing the IWC/IDCR minke whale sighting data and the use of different models to estimate density of schools. *Report of the International Whaling Commission*, **34**, 301–7.

*Hammond, P.S. and Laake, J.L. (1983) Trends in estimates of abundance of dolphins (*Stenella* spp. and *Delphinus delphis*) involved in the purse-seine fishery for tunas in the eastern Pacific Ocean. *Report of the International Whaling Commission*, **33**, 565–88.

Hammond, P.S. and Laake, J.L. (1984) Estimates of sperm whale density in the eastern tropical Pacific, 1974–1982. *Report of the International Whaling Commission*, **34**, 255–8.

Hanowski, J.M., Niemi, G.J. and Blake, J.G. (1990) Statistical perspectives and experimental design when counting birds on line transects. *Condor*, **92**, 326–35.

Harden, R.H., Muir, R.J. and Milledge, D.R. (1986) An evaluation of the strip transect method for censusing bird communities in forests. *Australian Wildlife Research*, **13**, 203–11.

Harmata, A.R., Podruzny, K.M., Zelenak, J.R. and Morrison, M.L. (2000) Passage rates and timing of bird migration in Montana. *American Midland Naturalist*, **143**, 30–40.

Harris, R.B. (1986) Reliability of trend lines obtained from variable counts. *Journal of Wildlife Management*, **50**, 165–71.

Hashimoto, C. (1995) Population census of the chimpanzees in the Kalinzu Forest, Uganda – comparison between methods with nest counts. *Primates*, **36**, 477–88.

*Hashmi, D. (in preparation) A method to estimate flux across line transects.

*Hastie, T.J. and Tibshirani, R.J. (1990) *Generalized Additive Models.* Chapman and Hall, London.

Haukioja, E. (1968) Reliability of the line survey method in bird census with reference to reed bunting and sedge warbler. *Ornis Fennica*, **45**, 105–13.

*Hayes, R.J. (1977) *A critical review of line transect methods.* MSc Thesis, University of Edinburgh, Scotland.

*Hayes, R.J. and Buckland, S.T. (1983) Radial-distance models for the line-transect method. *Biometrics*, **39**, 29–42.

*Hayne, D.W. (1949) An examination of the strip census method for estimating animal populations. *Journal of Wildlife Management*, **13**, 145–57.

Healy, W.M. and Welsh, C.J.E. (1992) Evaluating line transects to monitor gray squirrel populations. *Wildlife Society Bulletin*, **20**, 83–90.

Heckman, N. and Rice, J. (1997) Line transects of two-dimensional random fields: estimation and design. *Canadian Journal of Statistics*, **25**, 481–501.

Hedley, S.L., Buckland, S.T. and Borchers, D.L. (1999) Spatial modelling from line transect data. *Journal of Cetacean Research and Management*, **1**, 255–64.

Heide-Jørgensen, M.P., Mosbech, A., Teilmann, J., Benke, H. and Schultz, W. (1992) Harbor porpoise (*Phocoena phocoena*) densities obtained from aerial surveys north of Fyn and in the Bay of Kiel. *Ophelia*, **35**, 133–46.

Hein, E.W. (1997) Demonstration of line transect methodologies to estimate urban gray squirrel density. *Environmental Management*, **21**, 943–7.

Heitjan, D.F. (1989) Inference from grouped continuous data: a review. *Statistical Science*, **4**, 164–83.

Hemingway, P. (1971) Field trials of the line transect method of sampling large populations of herbivores. In *The Scientific Management of Animal and Plant Communities for Conservation* (eds. E. Duffey and A.S. Watts), Blackwell Scientific Publications, Oxford, England, pp. 405–11.

Heydon, M.J. and Bulloh, P. (1997) Mousedeer densities in a tropical rainforest: the impact of selective logging. *Journal of Applied Ecology*, **34**, 484–96.

*Heydon, M.J., Reynolds, J.C. and Short, M.J. (2000) Variation in abundance of foxes (*Vulpes vulpes*) between three regions of rural Britain, in relation to landscape and other variables. *Journal of Zoology*, **251**, 253–64.

*Hiby, A.R. (1982) Using average number of whales in view to estimate population density. *Report of the International Whaling Commission*, **32**, 562–5.

*Hiby, A.R. (1985) An approach to estimating population densities of great whales from sighting surveys. *IMA Journal of Mathematics Applied in Medicine and Biology*, **2**, 201–20.

*Hiby, A.R. (1986) Results of a hazard rate model relevant to experiments on the 1984/85 IDCR minke whale assessment cruise. *Report of the International Whaling Commission*, **36**, 497–8.

*Hiby, A.R. and Hammond, P.S. (1989) Survey techniques for estimating current abundance and monitoring trends in abundance of cetaceans. In *The Comprehensive Assessment of Whale Stocks: the early years* (ed. G.P. Donovan), International Whaling Commission, Cambridge, pp. 47–80.

*Hiby, A.R., Martin, A.R. and Fairfield, F. (1984) IDCR cruise/aerial survey in the north Atlantic 1982: aerial survey. *Report of the International Whaling Commission*, **34**, 633–44.

*Hiby, A.R. and Ward, A.J. (1986a) Simulation trials of a cue-counting technique for censusing whale populations. *Report of the International Whaling Commission*, **36**, 471–2.

*Hiby, A.R. and Ward, A.J. (1986b) Analysis of cue-counting and blow rate estimation experiments carried out during the 1984/85 IDCR minke whale assessment cruise. *Report of the International Whaling Commission*, **36**, 473–6.

*Hiby, A.R., Ward, A.J. and Lovell, P. (1989) Analysis of the 1987 north Atlantic sightings survey: aerial survey results. *Report of the International Whaling Commission*, **39**, 447–55.

Hiby, L. (1999) The objective identification of duplicate sightings in aerial survey for porpoise. In *Marine Mammal Survey and Assessment Methods* (eds. G.W. Garner, S.C. Amstrup, J.L. Laake, B.F.J. Manly, L.L. McDonald and D.G. Robertson), Balkema, Rotterdam, pp. 179–89.

*Hiby, L. and Krishna, M.B. (in press) Line transect sampling from a curving path. *Biometrics*, **57**.

*Hiby, L. and Lovell, P. (1998) Using aircraft in tandem formation to estimate abundance of harbour porpoise. *Biometrics*, **54**, 1280–9.

Hilden, O. and Järvinen, A. (1989) Efficiency of the line-transect method in mountain birch forest. *Annales Zoologici Fennici*, **26**, 185–90.

Hill, K., Padwe, J., Bejyvagi, C., Bepurangi, A., Jakugi, F., Tykuarangi, R. and Tykuarangi, T. (1997) Impact of hunting on large vertebrates in the Mbaracayu reserve, Paraguay. *Conservation Biology*, **11**, 1339–53.

Högmander, H. (1991) A random field approach to transect counts of wildlife populations. *Biometrical Journal*, **33**, 1013–23.

*Holgate, P. (1964) The efficiency of nearest neighbour estimators. *Biometrics*, **20**, 647–9.

Holt, R.S. and Cologne, J. (1987) Factors affecting line transect estimates of dolphin school density. *Journal of Wildlife Management*, **51**, 836–43.

*Holt, R.S. and Powers, J.E. (1982) *Abundance Estimation of Dolphin Stocks Involved in the Eastern Tropical Pacific Yellowfin Tuna Fishery Determined From Aerial and Ship Surveys to 1979*, United States Department of Commerce, NOAA Technical Memorandum NOAA-TM-NMFS-SWFC-23, 95pp.

*Hone, J. (1986) Accuracy of the multiple regression method for estimating density in transect counts. *Australian Wildlife Research*, **13**, 121–6.

*Hone, J. (1988) A test of the accuracy of line and strip transect estimators in aerial survey. *Australian Wildlife Research*, **15**, 493–7.

*Hurvich, C.M. and Tsai, C.L. (1989) Regression and time series model selection in small samples. *Biometrika*, **76**, 297–307.

*Hurvich, C.M. and Tsai, C.L. (1995) Model selection for extended quasi-likelihood models in small samples. *Biometrics*, **51**, 1077–84.

Hutto, R.L., Pletschet, S.M. and Hendricks, P. (1986) A fixed radius point count method for nonbreeding and breeding season use. *Auk*, **103**, 593–602.

Jaramillo-Legorreta, A.M., Rojas-Bracho, L. and Gerrodette, T. (1999) A new abundance estimate for vaquitas: first step for recovery. *Marine Mammal Science*, **15**, 957–73.

Järvinen, O. (1976) Estimating relative densities of breeding birds by the line transect method. II. Comparison between two methods. *Ornis Scandinavica*, **7**, 43–8.

Järvinen, O. (1978) Estimating relative densities of land birds by point counts. *Annales Zoologici Fennici*, **15**, 290–3.

*Järvinen, O. (1978) Species-specific efficiency in line transects. *Ornis Scandinavica*, **9**, 164–7.

*Järvinen, O. and Väisänen, R.A. (1975) Estimating relative densities of breeding birds by the line transect method. *Oikos*, **26**, 316–22.

Järvinen, O. and Väisänen, R.A. (1976) Between-year component of diversity in communities of breeding land birds. *Oikos*, **27**, 34–9.

Järvinen, O. and Väisänen, R.A. (1976) Estimating relative densities of breeding birds by the line transect method. IV. Geographical constancy of the proportion of main belt observations. *Ornis Fennica*, **53**, 87–90.

Järvinen, O. and Väisänen, R.A. (1976) Finnish line transect censuses. *Ornis Fennica*, **53**, 115–8.

Järvinen, O. and Väisänen, R.A. (1983) Confidence limits for estimates of population density in line transects. *Ornis Scandinavica*, **14**, 129–34.

Järvinen, O. and Väisänen, R.A. (1983) Correction coefficients for line transect censuses of breeding birds. *Ornis Fennica*, **60**, 97–104.

Järvinen, O., Väisänen, R.A. and Haila, Y. (1976) Estimating relative densities of breeding birds by the line transect method. III. Temporal constancy of the proportion of main belt observations. *Ornis Fennica*, **53**, 40–5.

Jarvis, A.M. and Robertson, A. (1999) Predicting population sizes and priority conservation areas for 10 endemic Namibian bird species. *Biological Conservation*, **88**, 121–31.

Jefferson, T.A. (1996) Estimates of abundance of cetaceans in offshore waters of the northwestern Gulf of Mexico, 1992–1993. *Southwest Naturalist*, **41**, 279–87.

*Jenkins, R.K.B., Brady, L.D., Huston, K., Kauffmann, J.L.D., Rabearivony, J., Raveloson, G. and Rowcliffe, J.M. (1999) The population status of chameleons within Ranomafana National Park, Madagascar, and recommendations for future monitoring. *Oryx*, **33**, 38–46.

Jensen, A.L. (1996) Subsampling with line transects for estimation of animal abundance. *Environmetrics*, **7**, 283–9.

Jett, D.A. and Nichols, J.D. (1987) A field comparison of nested grid and trapping web density estimators. *Journal of Mammalogy*, **68**, 888–92.

Johnsingh, A.J.T. and Joshua, J. (1994) Avifauna in 3 vegetation types on Mundanthurai Plateau, south India. *Journal of Tropical Ecology*, **10**, 323–35.

*Johnson, B.K. and Lindzey, F.G. (unpublished) *Guidelines for Estimating Pronghorn Numbers Using Line Transects*, Wyoming Cooperative Fish and Wildlife Research Unit, University of Wyoming, Laramie, 15pp. plus appendices.

Johnson, B.K., Lindzey, F.G. and Guenzel, R.J. (1991) Use of aerial line transect surveys to estimate pronghorn populations in Wyoming. *Wildlife Society Bulletin*, **19**, 315–21.

*Johnson, E.G. and Routledge, R.D. (1985) The line transect method: a nonparametric estimator based on shape restrictions. *Biometrics*, **41**, 669–79.

Johnson, F.A., Pollock, K.H. and Montalbano, F., III. (1989) Visibility bias in aerial surveys of mottled ducks. *Wildlife Society Bulletin*, **17**, 222–7.

Jones, J., McLeish, W.J. and Robertson, R.J. (2000) Density influences census technique accuracy for Cerulean Warblers in eastern Ontario. *Journal of Field Ornithology*, **71**, 46-56.

Kaiser, L. (1983) Unbiased estimation in line-intercept sampling. *Biometrics*, **39**, 965–76.

Karanth, K.U. and Sunquist, M.E. (1992) Population structure, density and biomass of large herbivores in the tropical forests of Nagarahole, India. *Journal of Tropical Ecology*, **8**, 21–35.

Karanth, K.U. and Sunquist, M.E. (1995) Prey selection by tiger, leopard and dhole in tropical forests. *Journal of Animal Ecology*, **64**, 439–50.

Karunamuni, R.J. and Quinn, T.J. (1995) Bayesian estimation of animal abundance for line transect sampling. *Biometrics*, **51**, 1325–37.

Kasamatsu, F. and Joyce, G.G. (1995) Current status of odontocetes in the Antarctic. *Antarctic Science*, **7**, 365–79.

*Kelker, G.H. (1945) *Measurement and interpretation of forces that determine populations of managed deer.* PhD Dissertation. University of Michigan, Ann Arbor, MI, USA.

Kelley, J.R. (1996) Line transect sampling for estimating breeding wood duck density in forested wetlands. *Wildlife Society Bulletin*, **24**, 32–6.

Khan, J.A. (1995) Conservation and management of Gir Lion Sanctuary and National Park, Gujarat, India. *Biological Conservation*, **73**, 183–8.

*Kie, J.G. and Boroski, B.B. (1995) Using spotlight counts to estimate mule deer population size and trends. *California Fish and Game*, **81**, 55–70.

Kingsley, M.C.S. and Reeves, R.R. (1998) Aerial surveys of cetaceans in the Gulf of St. Lawrence in 1995 and 1996. *Canadian Journal of Zoology*, **76**, 1529–50.

Kishino, H., Kasamatsu, F. and Toda, T. (1988) On the double line transect method. *Report of the International Whaling Commission*, **38**, 273–9.

Kishino, H., Kato, H., Kasamatsu, F. and Fujise, Y. (1991) Detection of heterogeneity and estimation of population characteristics from the field survey data: 1987/88 Japanese feasibility study of the Southern Hemisphere minke whales. *Annals of the Institute of Statistical Mathematics*, **43**, 435–53.

Kitahara, M. and Fujii, K. (1994) Biodiversity and community structure of temperate butterfly species within a gradient of human disturbance – an analysis based on the concept of generalist vs specialist strategies. *Researches on Population Ecology*, **36**, 187–99.

*Klavitter, J.L. (2000) *Survey methodology, abundance and demography of the endangered Hawaiian hawk: is delisting warranted?* MS Thesis. University of Washington, Seattle WA, 102pp.

*Knopf, F.L. (1986) Changing landscapes and the cosmopolitism of the eastern Colorado avifauna. *Wildlife Society Bulletin*, **14**, 132–42.

*Knopf, F.L., Sedgwick, J.A. and Cannon, R.W. (1988) Guild structure of a riparian avifauna relative to seasonal cattle grazing. *Journal of Wildlife Management*, **52**, 280–90.

Koopman, B.O. (1956) The theory of search II. Target detection. *Operations Research*, **4**, 503–31.

*Koopman, B.O. (1980) *Search and Screening: General Principles with Historical Applications.* Pergamon Press, New York, USA, 369pp.

Koster, S.H. (1985) *An evaluation of line transect census methods in West African wooded savanna.* PhD Dissertation. Michigan State University, Ann Arbor, MI, USA, 207pp.

Kovner, J.L. and Patil, S.A. (1974) Properties of estimators of wildlife population density for the line transect method. *Biometrics*, **30**, 225–30.

Kuitunen, M., Rossi, E. and Stenroos, A. (1998) Do highways influence density of land birds? *Environmental Management*, **22**, 297–302.

Kulbicki, M. and Sarramegna, S. (1999) Comparison of density estimates derived from strip transect and distance sampling for underwater visual censuses: a case study of Chaetodontidae and Pomacanthidae. *Aquatic Living Resources*, **12**, 315–25.

*Kullback, S. and Leibler, R.A. (1951) On information and sufficiency. *Annals of Mathematical Statistics*, **22**, 79–86.

*Laake, J.L. (1978) *Line transect estimators robust to animal movement.* MS Thesis. Utah State University, Logan, UT, USA, 55pp.

Laake, J.L. (1981) Abundance estimation of dolphins in the eastern Pacific with line transect sampling – a comparison of the techniques and suggestions for future research. In *Report of the Workshop on Tuna–Dolphin Interactions* (ed. P.S. Hammond), Inter-American Tropical Tuna Commission Special Report Number 4, pp. 56–95.

*Laake, J.L. (1999) Distance sampling with independent observers: reducing bias from heterogeneity by weakening the conditional independence assumption. In *Marine Mammal Survey and Assessment Methods* (eds. G.W. Garner, S.C. Amstrup, J.L. Laake, B.F.J. Manly, L.L. McDonald and D.G. Robertson), Balkema, Rotterdam, pp. 137–48.

*Laake, J.L., Buckland, S.T., Anderson, D.R. and Burnham, K.P. (1993) *DISTANCE User's Guide.* Colorado Cooperative Fish and Wildlife Research Unit, Colorado State University, Fort Collins, CO 80523, USA.

*Laake, J.L., Burnham, K.P. and Anderson, D.R. (1979) *User's Manual for Program TRANSECT.* Utah State University Press, Logan, UT, USA.

*Laake, J.L., Calambokidis, J., Osmek, S.D. and Rugh, D.J. (1997) Probability of detecting harbor porpoise from aerial surveys: estimating $g(0)$. *Journal of Wildlife Management*, **61**, 63–75.

Lacki, M.J., Hummer, J.W. and Fitzgerald, J.L. (1994) Application of line transects for estimating population density of the endangered copperbelly water snake in southern Indiana. *Journal of Herpetology*, **28**, 241–5.

Langbein, J., Hutchings, M.R., Harris, S., Stoate, C., Tapper, S.C. and Wray, S. (1999) Techniques for assessing the abundance of Brown Hares *Lepus europaeus*. *Mammal Review*, **29**, 93–116.

Lauerman, L.M.L., Kaufmann, R.S. and Smith, K.L. (1996) Distribution and abundance of epibenthic megafauna at a long time-series station in the abyssal northeast Pacific. *Deep Sea Research I*, **43**, 1075–103.

Leatherwood, S., Gilbert, J.R. and Chapman, D.G. (1978) An evaluation of some techniques for aerial censuses of bottlenosed dolphins. *Journal of Wildife Management*, **42**, 239–50.

Leatherwood, S. and Show, I.T. Jr. (1982) Effects of varying altitude on aerial surveys of bottlenose dolphins. *Report of the International Whaling Commission*, **32**, 569–75.

Leatherwood, S., Show, I.T. Jr., Reeves, R.R. and Wright, M.B. (1982) Proposed modification of transect models to estimate population size from aircraft with obstructed downward visibility. *Report of the International Whaling Commission*, **32**, 577–9.

*Leopold, A. (1933) *Game Management.* Charles Schribner's Sons, New York.

Leopold, M.F., vanderWerf, B., Ries, E.H. and Reijnders, P.J.H. (1997) The importance of the North Sea for winter dispersal of harbour seals *Phoca vitulina* from the Wadden Sea. *Biological Conservation*, **81**, 97–102.

*Lerczak, J.A. and Hobbs, R.C. (1998) Calculating sighting distances from angle readings during ship-board, aerial and shore-based marine mammal surveys. *Marine Mammal Science*, **14**, 590–8.

LeResche, R.E. and Rausch, R.A. (1974) Accuracy and precision of aerial moose censusing. *Journal of Wildlife Management*, **38**, 175–82.

Link, W.A. and Barker, R.J. (1994) Density estimation using the trapping web design – a geometric analysis. *Biometrics*, **50**, 733–45.

*Link, W.A. and Sauer, J.R. (1997) Estimation of population trajectories from count data. *Biometrics*, **53**, 488–497.

Lochmiller, R.L., Boggs, J.F., McMurry, S.T., Leslie, D.M. and Engle, D.M. (1991) Response of cottontail rabbit populations to herbicide and fire applications on Cross Timbers rangeland. *Journal of Range Management*, **44**, 150–5.

Lucas, H.A. and Seber, G.A.F. (1977) Estimating coverage and particle density using line intercept method. *Biometrika*, **64**, 618–22.

Lynch, T.B. and Rusydi, R. (1999) Distance sampling for forest inventory in Indonesian teak plantations. *Forest Ecology and Management*, **113**, 215–21.

Mack, Y.P. (1998) Testing for the shoulder condition in transect sampling. *Communications in Statistics – Theory and Methods*, **27**, 423–32.

*Mack, Y.P. and Quang, P.X. (1998) Kernel methods in line and point transect sampling. *Biometrics*, **54**, 606–19.

*Mack, Y.P., Quang, P.X. and Zhang, S. (1999) Kernel estimation in transect sampling without the shoulder condition. *Communications in Statistics – Theory and Methods*, **28**, 2277–96.

Mackintosh, N.A. and Brown, S.G. (1956) Estimates of the southern population of the larger baleen whales. *The Norwegian Gazette*, **45**, 469.

Mandujano, S. and Gallina, S. (1995) Comparison of deer censusing methods in tropical dry forest. *Wildlife Society Bulletin*, **23**, 180–6.

*Manly, B.F.J. (1997) *Randomization, Bootstrap and Monte Carlo Methods in Biology*, 2nd edn. Chapman and Hall, London.

*Manly, B.F.J., McDonald, L.L. and Garner, G.W. (1996) Maximum likelihood estimation for the double-count method with independent observers. *Journal of Agricultural, Biological, and Environmental Statistics*, **1**, 170–89.

*Marques, F.F.C. and Buckland, S.T. (in preparation) Incorporating covariates into standard line transect analyses.

*Marques, F.F.C., Buckland, S.T., Goffin, D., Dixon, C.E., Borchers, D.L., Mayle, B.A. and Peace, A.J. (2001) Estimating deer abundance from line transect surveys of dung: sika deer in southern Scotland. *Journal of Applied Ecology*, **38**, 349–63.

*Marsden, S.J. (1999) Estimation of parrot and hornbill densities using a point count distance sampling method. *Ibis*, **141**, 377–90.

Mattson, D.J. and Reinhart, D.P. (1996) Indicators of red squirrel (*Tamiasciurus hudsonicus*) abundance in the whitebark pine zone. *Great Basin Naturalist*, **56**, 272–5.

*Mayfield, H.F. (1981) Problems in estimating population size through counts of singing males. In *Estimating Numbers of Terrestrial Birds. Studies in Avian Biology No. 6* (eds. C.J. Ralph and J.M. Scott), Cooper Ornithological Society, pp. 220–4.

*McCullagh, P. and Nelder, J.A. (1989) *Generalized Linear Models*, 2nd edn, Chapman and Hall, London.

McDonald, L.L. (1980) Line-intercept sampling for attributes other than coverage and density. *Journal of Wildlife Management*, **44**, 530–3.

McDonald, L.L., Garner, G.W. and Robertson, D.G. (1999) Comparison of aerial survey procedures for estimating polar bear density: results of pilot studies in Northern Alaska. In *Marine Mammal Survey and Assessment Methods* (eds. G.W. Garner, S.C. Amstrup, J.L. Laake, B.F.J. Manly, L.L. McDonald and D.G. Robertson), Balkema, Rotterdam, pp. 37–51.

McIntyre, G.A. (1953) Estimation of plant density using line transects. *Journal of Ecology*, **41**, 319–30.

McIntyre, N.E. (1995) Methamidophos application effects on *Pasimachus elongatus* (coleoptera, carabidae) – an update. *Environmental Entomology*, **24**, 559–63.

Mercey, K.A. and Jayaraman, K. (1999) Predicting the variation in detection function in line transect sampling through random parameter model. *Environmental and Ecological Statistics*, **6**, 341–50.

Meriggi, A. and Verri, A. (1990) Population dynamics and habitat selection of the European hare on poplar monocultures in northern Italy. *Acta Theriologica*, **35**, 69–76.

*Miller, R.G. (1974) The jackknife – a review. *Biometrika*, **61**, 1–15.

*Milliken, G.A. and Johnson, D.E. (1984) *Analysis of Messy Data*. Lifetime Learning Publications, Belmont, California.

Mills, J.N., Yates, T.L., Ksiazek, T.G., Peters, C.J. and Childs, J.E. (1999) Long-term studies of hantavirus reservoir populations in the southwestern United States: Rationale, potential, and methods. *Emerging Infectious Diseases*, **5**, 95–101.

*Morgan, D.G. (1986) *Estimating Vertebrate Population Densities by Line Transect Methods.* Occasional Papers Number 11, Melbourne College, Melbourne, Australia.

Morrison, M.L., Mannan, R.W. and Dorsey, G.L. (1981) Effects of number of circular plots on estimates of avian density and species richness. In *Estimating Numbers of Terrestrial Birds. Studies in Avian Biology No. 6* (eds. C.J. Ralph and J.M. Scott), Cooper Ornithological Society, pp. 405–8.

Morrison, M.L. and Marcot, B.G. (1984) Expanded use of the variable circular-plot census method. *Wilson Bulletin*, **96**, 313–5.

Moskat, C. and Baldi, A. (1999) The importance of edge effect in line transect censuses applied in marshland habitats. *Ornis Fennica*, **76**, 33–40.

Muchaal, P.K. and Ngandjui, G. (1999) Impact of village hunting on wildlife populations in the western Dia Reserve, Cameroon. *Conservation Biology*, **13**, 385–96.

Mugangu, T.E., Hunter, M.L. and Gilbert, J.R. (1995) Food, water, and predation: a study of habitat selection by buffalo in Virunga National Park, Zaire. *Mammalia*, **59**, 349–62.

Myrberget, S. (1976) Field tests of line transect census methods for grouse. *Norwegian Journal of Zoology*, **24**, 307–17.

*Nice, M.M. and Nice, L.B. (1921) The roadside census. *Wilson Bulletin*, **33**, 113–23.

Nichols, J.D., Hines, J.E., Sauer, J.R., Fallon, F.W., Fallon, J.E. and Heglund, P.J. (2000) A double-observer approach for estimating detection probability and abundance from point counts. *Auk*, **117**, 393–408.

Nichols, J.D., Tomlinson, R.E. and Waggerman, G. (1986) Estimating nest detection probabilities for white-winged dove nest transects in Tamaulipas, Mexico. *Auk*, **103**, 825–8.

O'Connell, V.M. and Carlile, D.W. (1993) Habitat-specific density of adult yelloweye rockfish *Sebastes ruberrimus* in the eastern Gulf of Alaska. *Fishery Bulletin*, **91**, 304–9.

Olawsky, C.D. and Smith, L.M. (1991) Lesser prairie-chicken densities on tebuthiuron-treated and untreated sand shinnery oak rangelands. *Journal of Range Management*, **44**, 364–8.

Oliveira, P., Jones, M., Caires, D. and Menezes, D. (1999) Population trends and status of the Madeira Laurel Pigeon *Columba trocaz*. *Bird Conservation International*, **9**, 387–95.

*O'Meara, T.E. (1981) A field test of two density estimators for transect data. In *Estimating the Numbers of Terrestrial Birds. Studies in Avian Biology No. 6* (eds. C.J. Ralph and J.M. Scott), Cooper Ornithological Society, pp. 193–6.

*Otis, D.L., Burnham, K.P., White, G.C. and Anderson, D.R. (1978) Statistical inference from capture data on closed animal populations. *Wildlife Monographs*, **62**, 1–135.

Otten, A. and deVries, P.G. (1984) On line-transect estimates for population density, based on elliptic flushing curves. *Biometrics*, **40**, 1145–50.

*Otto, M.C. and Pollock, K.H. (1990) Size bias in line transect sampling: a field test. *Biometrics*, **46**, 239–45.

*Overton, W.S. and Davis, D.E. (1969) Estimating the number of animals in wildlife populations. In *Wildlife Management Techniques* (ed. R.H. Giles, Jr.), The Wildlife Society, Washington, DC, pp. 405–55.

Owiunji, I. (2000) Changes in avian communities of Budongo Forest Reserve after 70 years of selective logging. *Ostrich*, **71**, 216–9.

Owiunji, I. and Plumptre, A.J. (1998) Bird communities in logged and unlogged compartments in Budongo Forest, Uganda. *Forest Ecology and Management*, **108**, 115–26.

Palka, D. and Pollard, J. (1999) Adaptive line transect survey for harbor porpoises. In *Marine Mammal Survey and Assessment Methods* (eds. G.W. Garner, S.C. Amstrup, J.L. Laake, B.F.J. Manly, L.L. McDonald and D.G. Robertson), Balkema, Rotterdam, pp. 3–11.

Parker, K.R. (1979) Density estimation by variable area transect. *Journal of Wildlife Management*, **43**, 484–92.

Parmenter, C.A., Yates, T.L., Parmenter, R.R. and Dunnum, J.L. (1999) Statistical sensitivity for detection of spatial and temporal patterns in rodent population densities. *Emerging Infectious Diseases*, **5**, 118–25.

*Parmenter, R.R., MacMahon, J.A. and Anderson, D.R. (1989) Animal density estimation using a trapping web design: field validation exteriments. *Ecology*, **70**, 169–79.

*Parmenter, R.R., Yates, T.L., Anderson, D.R., Burnham, K.P., Dunnum, J.L., Franklin, A., Friggens, M.T., Lubow, B., Miller, M., Olson, G.S., Parmenter, C.A., Pollard, J., Rexstad, E., Shenk, T., Stanley, T.R. and White, G.C. (in preparation) Small mammal density estimation: a field comparison of grid-based versus web-based density estimators using enclosed rodent populations.

*Patil, G.P. and Ord, J.K. (1976) On size-biased sampling and related form-invariant weighted distributions. *Sankhyā B*, **38**, 48–61.

*Patil, G.P. and Rao, C.R. (1978) Weighted distribution and size-biased sampling with applications to wildlife populations and human families. *Biometrics*, **34**, 179–89.

*Patil, G.P., Taillie, C. and Wigley, R.L. (1979a) Transect sampling methods and their application to deep-sea red crab. In *Environmental Biomonitoring, Assessment, Prediction, and Management – Certain Case Studies and Related Quantitative Issues* (eds. J. Cairns, Jr, G.P. Patil and W.E. Waters), International Co-operative Publishing House, Fairland, MD, USA, pp. 51–75.

*Patil, S.A., Burnham, K.P. and Kovner, J.L. (1979b) Nonparametric estimation of plant density by the distance method. *Biometrics*, **35**, 597–604.

*Patil, S.A., Kovner, J.L. and Burnham, K.P. (1982) Optimum nonparametric estimation of population density based on ordered distances. *Biometrics*, **38**, 243–8.

*Peel, M.J.S. and Bothma, J.D. (1995) Comparison of the accuracy of four methods commonly used to count impala. *South African Journal of Wildlife Research*, **25**, 41–3.

Pelletier, L. and Krebs, C.J. (1997) Line transect sampling for estimating ptarmigan (*Lagopus* spp.) density. *Canadian Journal of Zoology*, **75**, 1185–92.

Peres, C.A. (1996) Population status of white-lipped *Tayassu pecari* and collared peccaries *T. tajacu* in hunted and unhunted Amazonian forests. *Biological Conservation*, **77**, 115–23.

Peres, C.A. (1997) Effects of habitat quality and hunting pressure on arboreal folivore densities in neotropical forests: a case study of howler monkeys (*Alouatta* spp.) *Folia Primatologica*, **68**, 199–222.

Peres, C.A. (1997) Primate community structure at twenty western Amazonian flooded and unflooded forests. *Journal of Tropical Ecology*, **13**, 381–405.

Peres, C.A. (2000) Effects of subsistence hunting on vertebrate community structure in Amazonian forests. *Conservation Biology*, **14**, 240–53.

Perez, J.M., Granados, J.E. and Soriguer, R.C. (1994) Population dynamic of the Spanish ibex *Capra pyrenaica* in Sierra Nevada Natural Park (southern Spain). *Acta Theriologica*, **39**, 289–94.

Persson, O. (1971) The robustness of estimating density by distance measurements. In *Statistical Ecology*, Vol. 2 (eds. G.P. Patil, E.C. Pielou and W.E. Waters), Pennsylvania State University Press, University Park, pp. 175–90.

Peterhofs, E. and Priednieks, J. (1989) Problems in applying the line-transect method without repeated counts when the breeding season is long. *Annales Zoologici Fennici*, **26**, 181–4.

Philibert, H., Wobeser, G. and Clark, R.G. (1993) Counting dead birds – examination of methods. *Journal of Wildlife Diseases*, **29**, 284–9.

*Pierce, B.L. (2000) *A non-linear spotlight line transect method for estimating white-tailed deer population densities.* MS Thesis, Southwest Texas State University, TX, USA, 86pp.

Pinder, L. (1996) Marsh deer *Blastocerus dichotomus* population estimate in the Parana River, Brazil. *Biological Conservation*, **75**, 87–91.

Pinkowski, B. (1990) Performance of Fourier series in line transect simulations. *Simulation*, **54**, 211–2.

*Plumptre, A.J. (2000) Monitoring mammal populations with line transect techniques in African forests. *Journal of Applied Ecology*, **37**, 356–68.

*Plumptre, A.J. and Harris, S. (1995) Estimating the biomass of large mammalian herbivores in a tropical montane forest – a method of fecal counting that avoids assuming a steady-state system. *Journal of Applied Ecology*, **32**, 111–20.

*Plumptre, A.J. and Reynolds, V. (1994) The effect of selective logging on the primate populations in the Budongo Forest Reserve, Uganda. *Journal of Applied Ecology*, **31**, 631–41.

*Plumptre, A.J. and Reynolds, V. (1996) Censusing chimpanzees in the Budongo Forest, Uganda. *International Journal of Primatology*, **17**, 85–99.

*Plumptre, A.J. and Reynolds, V. (1997) Nesting behavior of chimpanzees: implications for censuses. *International Journal of Primatology*, **18**, 475–85.

Pojar, T.M., Bowden, D.C. and Gill, R.B. (1995) Aerial counting experiments to estimate pronghorn density and herd structure. *Journal of Wildlife Management*, **59**, 117–28.

Pollard, J.H. (1971) On distance estimators of density in randomly distributed forests. *Biometrics*, **27**, 991–1002.

*Pollard, J.H. and Buckland, S.T. (1997) A strategy for adaptive sampling in shipboard line transect surveys. *Report of the International Whaling Commission*, **47**, 921–31.

*Pollock, K.P. (1978) A family of density estimators for line-transect sampling. *Biometrics*, **34**, 475–8.

Pollock, K.P. and Kendall, W.L. (1987) Visibility bias in aerial surveys: a review of estimation procedures. *Journal of Wildlife Management*, **51**, 502–9.

Pontes, A.R.M. (1999) Environmental determinants of primate abundance in Maraca Island, Roraima, Brazilian Amazonia. *Journal of Zoology*, **247**, 189–99.

Pople, A.R., Cairns, S.C., Clancy, T.F., Grigg, G.C., Beard, L.A. and Southwell, C.J. (1998) An assessment of the accuracy of kangaroo surveys using fixed-wing aircraft. *Wildlife Research*, **25**, 315–26.

Puertas, P. and Bodmer, R.E. (1993) Conservation of a high diversity primate assemblage. *Biodiversity and Conservation*, **2**, 586–93.

Quang, P.X. (1990) Confidence intervals for densities in line transect sampling. *Biometrics*, **46**, 459–72.

*Quang, P.X. (1991) A nonparametric approach to size-biased line transect sampling. *Biometrics*, **47**, 269–79.

*Quang, P.X. (1993) Nonparametric estimators for variable circular plot surveys. *Biometrics*, **49**, 837–52.

Quang, P.X. and Becker, E.F. (1996) Line transect sampling under varying conditions with application to aerial surveys. *Ecology*, **77**, 1297–302.

*Quang, P.X. and Becker, E.F. (1997) Combining line transect and double count sampling techniques for aerial surveys. *Journal of Agricultural, Biological, and Environmental Statistics*, **2**, 230–42.

*Quang, P.X. and Becker, E.F. (1999) Aerial survey sampling of contour transects using double-count and covariate data. In *Marine Mammal Survey and Assessment Methods* (eds. G.W. Garner, S.C. Amstrup, J.L. Laake, B.F.J. Manly, L.L. McDonald and D.G. Robertson), Balkema, Rotterdam, pp. 87–97.

*Quang, P.X. and Lanctot, R.B. (1991) A line transect model for aerial surveys. *Biometrics*, **47**, 1089–102.

*Quinn, T.J., II (1977) *The effects of aggregation on line transect estimators of population abundance with application to marine mammal populations.* MS Thesis, University of Washington, WA, USA, 116pp.

*Quinn, T.J., II (1979) The effects of school structure on line transect estimators of abundance. In *Contemporary Quantitative Ecology and Related Ecometrics* (eds. G.P. Patil and M.L. Rosenzweig), International Co-operative Publishing House, Fairland, MD, USA, pp. 473–91.

*Quinn, T.J., II (1985) Line transect estimators for school populations. *Fisheries Research*, **3**, 183–99.

*Quinn, T.J., II and Gallucci, V.F. (1980) Parametric models for line transect estimators of abundance. *Ecology*, **61**, 293–302.

Raftery, A.E. and Schweder, T. (1993) Inference about the ratio of two parameters, with application to whale censusing. *American Statistician*, **47**, 259–64.

*Ralph, C.J. (1981) An investigation of the effect of seasonal activity levels on avian censusing. In *Estimating Numbers of Terrestrial Birds. Studies in Avian Biology No. 6* (eds. C.J. Ralph and J.M. Scott), Cooper Ornithological Society, pp. 265–70.

*Ralph, C.J. and Scott, J.M. (eds.) (1981) *Estimating Numbers of Terrestrial Birds. Studies in Avian Biology No. 6*, Cooper Ornithological Society.

*Ramsey, F.L. (1979) Parametric models for line transect surveys. *Biometrika*, **66**, 505–12.

*Ramsey, F.L. (1981) Introductory remarks: data analysis. In *Estimating Numbers of Terrestrial Birds. Studies in Avian Biology No. 6* (eds. C.J. Ralph and J.M. Scott), Cooper Ornithological Society, p. 454.

*Ramsey, F.L., Gates, C.E., Patil, G.P. and Taillie, C. (1988) On transect sampling to assess wildlife populations and marine resources. In *Handbook of Statistics 6: Sampling* (eds. P.R. Krishnaiah and C.R. Rao), Elsevier Publishers, Amsterdam, pp. 515–32.

*Ramsey, F.L. and Scott, J.M. (1978) Use of circular plot surveys in estimating the density of a population with Poisson scattering. Technical Report 60, Department of Statistics, Oregon State University, Corvallis, OR, USA.

*Ramsey, F.L. and Scott, J.M. (1979) Estimating population densities from variable circular plot surveys. In *Sampling Biological Populations* (eds. R.M. Cormack, G.P. Patil and D.S. Robson), International Co-operative Publishing House, Fairland, MD, USA, pp. 155–81.

*Ramsey, F.L. and Scott, J.M. (1981a) Tests of hearing ability. In *Estimating Numbers of Terrestrial Birds. Studies in Avian Biology No. 6* (eds. C.J. Ralph and J.M. Scott), Cooper Ornithological Society, pp. 341–5.

*Ramsey, F.L. and Scott, J.M. (1981b) Analysis of bird survey data using a modification of Emlen's methods. In *Estimating Numbers of Terrestrial Birds. Studies in Avian Biology No. 6* (eds. C.J. Ralph and J.M. Scott), Cooper Ornithological Society, pp. 483–7.

Ramsey, F.L., Scott, J.M. and Clark, R.T. (1979) Statistical problems arising from surveys of rare and endangered forest birds. *Proceedings of the 42nd Session of the International Statistical Institute*, pp. 471–83.

*Ramsey, F.L., Wildman, V. and Engbring, J. (1987) Covariate adjustments to effective area in variable-area wildlife surveys. *Biometrics*, **43**, 1–11.

*Rao, C.R. (1973) *Linear Statistical Inference and its Applications*, 2nd edn. Wiley, New York.

*Rao, P.V. (1984) Density estimation based on line transect samples. *Statistics and Probability Letters*, **2**, 1–57.

*Rao, P.V. and Portier, K.M. (1985) A model for line transect sampling clustered populations. *Statistics and Probability Letters*, **3**, 89–93.

Rao, P.V., Portier, K.M. and Ondrasik, J.A. (1981) Density estimation using line transect sampling. In *Estimating Numbers of Terrestrial Birds. Studies in Avian Biology No. 6* (eds. C.J. Ralph and J.M. Scott), Cooper Ornithological Society, pp. 441–4.

Raphael, M.G. (1987) Estimating relative abundance of forest birds: simple versus adjusted counts. *Wilson Bulletin*, **99**, 125–31.

Raphael, M.G., Rosenberg, K.V. and Marcot, B.G. (1988) Large-scale changes in bird populations of Douglas-fir forests, Northwestern California. In *Bird Conservation* (ed. J.A. Jackson), University of Wisconsin Press, Madison, WI, USA, pp. 63–83.

Ratti, J.T., Smith, L.M., Hupp, J.W. and Laake, J.L. (1983) Line transect estimates of density and the winter mortality of Gray Partridge. *Journal of Wildlife Management*, **47**, 1088–96.

Raum-Suryan, K.L. and Harvey, J.T. (1998) Distribution and abundance of and habitat use by harbor porpoise, *Phocoena phocoena*, off the northern San Juan Islands, Washington. *Fishery Bulletin*, **96**, 808–22.

*Redmond, R.L., Bicak, T.K. and Jenni, D.A. (1981) An evaluation of breeding season census techniques for long-billed curlews (*Numenius americanus*). In *Estimating the Numbers of Terrestrial Birds. Studies in Avian Biology No. 6* (eds. C.J. Ralph and J.M. Scott), Cooper Ornithological Society, pp. 197–201.

Remis, M.J. (2000) Preliminary assessment of the impacts of human activities on gorillas *Gorilla gorilla gorilla* and other wildlife at Dzanga-Sangha Reserve, Central African Republic. *Oryx*, **34**, 56–65.

*Reynolds, R.T., Scott, J.M. and Nussbaum, R.A. (1980) A variable circular-plot method for estimating bird numbers. *Condor*, **82**, 309–13.

*Richards, D.G. (1981) Environmental acoustics and censuses of singing birds. In *Estimating Numbers of Terrestrial Birds. Studies in Avian Biology No. 6* (eds. C.J. Ralph and J.M. Scott), Cooper Ornithological Society, pp. 297–300.

Ringvall, A., Patil, G.P. and Taillie, C. (2000) A field test of surveyors' influence on estimates in line transect sampling. *Forest Ecology and Management*, **137**, 103–11.

Rinne, J. (1985) Density estimates and their errors in line transect counts of breeding birds. *Ornis Fennica*, **62**, 1–8.

Rivera-Milan, F.F. (1999) Population dynamics of Zenaida doves in Cidra, Puerto Rico. *Journal of Wildlife Management*, **63**, 232–44.

*Robbins, C.S. (1981) Effect of time of day on bird activity. In *Estimating Numbers of Terrestrial Birds. Studies in Avian Biology No. 6* (eds. C.J. Ralph and J.M. Scott), Cooper Ornithological Society, pp. 275–86.

Robinet, O., Barre, N. and Salas, M. (1996) Population estimate for the Ouvea Parakeet *Eunymphicus cornutus uvaeensis*: its present range and implications for conservation. *Emu*, **96**, 151–7.

*Robinette, W.L., Jones, D.A., Gashwiler, J.S. and Aldous, C.M. (1956) Further analysis of methods for censusing winter-lost deer. *Journal of Wildlife Management*, **20**, 75–8.

*Robinette, W.L., Loveless, C.M. and Jones, D.A. (1974) Field tests of strip census methods. *Journal of Wildlife Management*, **38**, 81–96.

*Roeder, K., Dennis, B. and Garton, E.O. (1987) Estimating density from variable circular plot censuses. *Journal of Wildlife Management*, **51**, 224–30.

*Rosenstock, S.S. (1998) Influence of Gambel oak on breeding birds in ponderosa pine forests of northern Arizona. *Condor*, **100**, 485–92.

*Rosenstock, S.S., Anderson, D.R., Giesen, K.M., Leukering, T. and Carter, M.F. (in preparation) Estimating landbird abundance: current practices and an alternative.

Rotella, J.J. and Ratti, J.T. (1986) Test of a critical density index assumption: a case study with gray partridge. *Journal of Wildlife Management*, **50**, 532–9.

Routledge, R.D. and Fyfe, D.A. (1992) Confidence limits for line transect estimates based on shape restrictions. *Journal of Wildlife Management*, **56**, 402–7.

*Routledge, R.D. and Fyfe, D.A. (1992) TRANSAN – line transect estimates based on shape restrictions. *Wildlife Society Bulletin*, **20**, 455–6.

Royama, T. (1960) The theory and practice of line transects in animal ecology by means of visual and auditory recognition. *Yamashina Institute of Ornithology and Zoology*, **2**, 1–17.

*Sakamoto, Y., Ishiguro, M. and Kitawaga, G. (1986) *Akaike Information Criterion Statistics.* KTK Scientific Publishers, Tokyo, Japan.

*Satterthwaite, F.E. (1946) An approximate distribution of estimates of variance components. *Biometric Bulletin*, **2**, 110–4.

*Schwarz, C.J. and Seber, G.A.F. (1999) Estimating animal abundance: review III. *Statistical Science*, **14**, 427–56.

*Schweder, T. (1974) *Transformations of point processes: applications to animal sighting and catch problems, with special emphasis on whales.* PhD Thesis, University of California, Berkeley, CA, USA.

*Schweder, T. (1977) Point process models for line transect experiments. In *Recent Developments in Statistics* (eds. J.R. Barba, F. Brodeau, G. Romier and B. Van Cutsem), North-Holland Publishing Company, New York, USA, pp. 221–42.

Schweder, T. (1990) Independent observer experiments to estimate the detection function in line transect surveys of whales. *Report of the International Whaling Commission*, **40**, 349–55.

*Schweder, T. (1999) Line transecting with difficulties; lessons from surveying minke whales. In *Marine Mammal Survey and Assessment Methods* (eds. G.W. Garner, S.C. Amstrup, J.L. Laake, B.F.J. Manly, L.L. McDonald and D.G. Robertson), Balkema, Rotterdam, pp. 149–66.

Schweder, T., Hagen, G., Helgeland, J. and Koppervik, I. (1996) Abundance estimation of northeastern Atlantic minke whales. *Report of the International Whaling Commission*, **46**, 391–405.

*Schweder, T. and Høst, G. (1992) Integrating experimental data and survey data to estimate $g(0)$: a first approach. *Report of the International Whaling Commission*, **42**, 575–82.

*Schweder, T., Øien, N. and Høst, G. (1991) Estimates of the detection probability for shipboard surveys of northeastern Atlantic minke whales, based on a parallel ship experiment. *Report of the International Whaling Commission*, **41**, 417–32.

Schweder, T., Øien, N. and Høst, G. (1993) Estimates of abundance of northeastern Atlantic minke whales in 1989. *Report of the International Whaling Commission*, **43**, 323–31.

*Schweder, T., Skaug, H.J., Dimakos, X.K., Langaas, M. and Øien, N. (1997) Abundance of northeastern Atlantic minke whales, estimates for 1989 and 1995. *Report of the International Whaling Commission*, **47**, 453–83.

*Schweder, T., Skaug, H.J., Langaas, M. and Dimakos, X.K. (1999) Simulated likelihood methods for complex double-platform line transect surveys. *Biometrics*, **55**, 678–87.

Scott, J.M., Jacobi, J.D. and Ramsey, F.L. (1981a) Avian surveys of large geographical areas: a systematic approach. *Wildlife Society Bulletin*, **9**, 190–200.

*Scott, J.M. and Ramsey, F.L. (1981) Length of count period as a possible source of bias in estimating bird densities. In *Estimating Numbers of Terrestrial Birds. Studies in Avian Biology No. 6* (eds. C.J. Ralph and J.M. Scott), Cooper Ornithological Society, pp. 409–13.

*Scott, J.M., Ramsey, F.L. and Kepler, C.B. (1981b) Distance estimation as a variable in estimating bird numbers. In *Estimating Numbers of Terrestrial Birds. Studies in Avian Biology No. 6* (eds. C.J. Ralph and J.M. Scott), Cooper Ornithological Society, pp. 334–40.

*Seber, G.A.F. (1973) *The Estimation of Animal Abundance.* Hafner, New York.

*Seber, G.A.F. (1979) Transects of random length. In *Sampling Biological Populations* (eds. R.M. Cormack, G.P. Patil and D.S. Robson), International Co-operative Publishing House, Fairland, MD, USA, pp. 183–92.

*Seber, G.A.F. (1982) *The Estimation of Animal Abundance and Related Parameters.* Macmillan, New York.

*Seber, G.A.F. (1986) A review of estimating animal abundance. *Biometrics*, **42**, 267–92.

*Seber, G.A.F. (1992) A review of estimating animal abundance II. *International Statistical Review*, **60**, 129–66.

*Sedgwick, J.A. and Knopf, F.L. (1987) Breeding bird response to cattle grazing of a cottonwood bottomland. *Journal of Wildlife Management*, **51**, 230–7.

*Sen, A.R., Smith, G.E.J. and Butler, G. (1978) On a basic assumption in the line transect method. *Biometrical Journal*, **20**, 363–9.

Sen, A.R., Tourigny, J. and Smith, G.E.J. (1974) On the line transect sampling method. *Biometrics*, **30**, 329–41.

Sherman, D.E., Kaminski, R.M. and Leopold, B.D. (1995) Winter line-transect surveys of wood ducks and mallards in Mississippi Greentree Reservoirs. *Wildlife Society Bulletin*, **23**, 155–63.

Shirakihara, M., Shirakihara, K. and Takemura, A. (1994) Distribution and seasonal density of the finless porpoise *Neophocaena phocaenoides* in the coastal waters of western Kyushu, Japan. *Fisheries Science*, **60**, 41–6.

Short, J. and Turner, B. (1992) The distribution and abundance of the banded and rufous hare-wallabies, *Lagostrophus fasciatus* and *Lagorchestes hirsutus. Biological Conservation*, **60**, 157–66.

*Shupe, T.E., Guthery, F.S. and Beasom, S.L. (1987) Use of helicopters to survey northern bobwhite populations on rangeland. *Wildlife Society Bulletin*, **15**, 458–62.

*Silverman, B.W. (1982) Algorithm AS 176: kernel density estimation using the fast Fourier transform. *Applied Statistics*, **31**, 93–9.

*Silverman, B.W. (1986) *Density Estimation for Statistics and Data Analysis.* Chapman and Hall, London.

Skalski, J.R. (1994) Estimating wildlife populations based on incomplete area surveys. *Wildlife Society Bulletin*, **22**, 192–203.

Skaug, H.J. (1997) Perpendicular distance line transect methods based on integrated spatial hazard probability models. *Report of the International Whaling Commission*, **47**, 493–7.

*Skaug, H.J. and Schweder, T. (1999) Hazard models for line transect surveys with independent observers. *Biometrics*, **55**, 29–36.

Skellam, J.G. (1958) The mathematical foundations underlying the use of line transects in animal ecology. *Biometrics*, **14**, 385–400.

*Skirvin, A.A. (1981) Effect of time of day and time of season on the number of observations and density estimates of breeding birds. In *Estimating Numbers of Terrestrial Birds. Studies in Avian Biology No. 6* (eds. C.J. Ralph and J.M. Scott), Cooper Ornithological Society, pp. 271–4.

*Smith, G.E. (1979) Some aspects of line transect sampling when the target population moves. *Biometrics*, **35**, 323–9.

Smith, G.W. and Nydegger, N.C. (1985) A spot-light, line transect method for surveying jack rabbits. *Journal of Wildlife Management*, **49**, 699–702.

Smith, G.W., Nydegger, N.C. and Yensen, D.L. (1984) Passerine bird densities in shrubsteppe vegetation. *Journal of Field Ornithology*, **55**, 261–4.

Smith, T.D. (1981) Line-transect techniques for estimating density of porpoise schools. *Journal of Wildlife Management*, **45**, 650–7.

*Southwell, C. (1994) Evaluation of walked line transect counts for estimating macropod density. *Journal of Wildlife Management*, **58**, 348–56.

Southwell, C.J., Cairns, S.C., Palmer, R., Delaney, R. and Broers, R. (1997) Abundance of large macropods in the eastern highlands of Australia. *Wildlife Society Bulletin*, **25**, 125–32.

Southwell, C., Fletcher, M., McRae, P., Porter, B. and Broers, R. (1995) Abundance and harvest rate of the whiptail wallaby in southeastern Queensland, Australia. *Wildlife Society Bulletin*, **23**, 726–32.

Southwell, C. and Weaver, K. (1993) Evaluation of analytical procedures for density estimation from line transect data – data grouping, data truncation and the unit of analysis. *Wildlife Research*, **20**, 433–44.

Southwell, C.J., Weaver, K.E., Cairns, S.C., Pople, A.R., Gordon, A.N., Sheppard, N.W. and Broers, R. (1995) Abundance of macropods in north-eastern New South Wales, and the logistics of broad-scale ground surveys. *Wildlife Research*, **22**, 757–66.

*Spear, L., Nur, N. and Ainley, D.G. (1992) Estimating absolute densities of flying seabirds using analyses of relative movement. *Auk*, **109**, 385–9.

Ståhl, G., Ringvall, A. and Lämås, T. (2000) Guided transect sampling for assessing sparse populations. *Forest Science*, **46**, 108–15.

Stoate, C., Borralho, R. and Araujo, M. (2000) Factors affecting corn bunting *Miliaria calandra* abundance in a Portuguese agricultural landscape. *Agriculture Ecosystems and Environment*, **77**, 219–26.

*Stoyan, D. (1982) A remark on the line transect method. *Biometrical Journal*, **24**, 191–5.

*Stuart, A. and Ord, J.K. (1987) *Kendall's Advanced Theory of Statistics.* Volume 1, Griffin, London.

*Sugiura, N. (1978) Further analysis of the data by Akaike's information criterion and the finite correction. *Communications in Statistics, Theory and Methods A*, **7**, 13–26.

Summers, R.W., Mavor, R.A., Buckland, S.T. and MacLennan, A.M. (1999) Winter population size and habitat selection by crested tits *Parus cristatus* in Scotland. *Bird Study*, **46**, 230–42.

Szaro, R.C. and Jakle, M.D. (1982) Comparison of variable circular-plot and spot-map methods in desert riparian and scrub habitats. *Wilson Bulletin*, **94**, 546–50.

Tarvin, K.A., Garvin, M.C., Jawor, J.M. and Dayer, K.A. (1998) A field evaluation of techniques used to estimate density of Blue Jays. *Journal of Field Ornithology*, **69**, 209–22.

*Tasker, M.L., Hope Jones, P., Dixon, T. and Blake, B.F. (1984) Counting seabirds at sea from ships: a review of methods employed and a suggestion for a standardized approach. *Auk*, **101**, 567–77.

Thirgood, S.J., Leckie, F.M. and Redpath, S.M. (1995) Diurnal and seasonal variation in line transect counts of moorland passerines. *Bird Study*, **42**, 257–9.

*Thomas, L., Laake, J.L., Derry, J.F., Buckland, S.T., Borchers, D.L., Anderson, D.R., Burnham, K.P., Strindberg, S., Hedley, S.L., Marques, F.F.C., Pollard, J.H. and Fewster, R.M. (1998) Distance 3.5. Research Unit for Wildlife Population Assessment, University of St Andrews, St Andrews, UK.

Thomas, S.C. (1991) Population densities and patterns of habitat use among anthropoid primates of the Ituri Forest, Zaire. *Biotropica*, **23**, 68–83.

*Thompson, S.K. (1990) Adaptive cluster sampling. *Journal of the American Statistical Association*, **85**, 1050–9.

*Thompson, S.K. and Ramsey, F.L. (1987) Detectability functions in observing spatial point processes. *Biometrics*, **43**, 355–62.

Tilghman, N.G. and Rusch, D.H. (1981) Comparison of line-transect methods for estimating breeding bird densities in deciduous woodlots. In *Estimating Numbers of Terrestrial Birds. Studies in Avian Biology No. 6* (eds. C.J. Ralph and J.M. Scott), Cooper Ornithological Society, pp. 202–8.

Tomiatojc, L. and Verner, J. (1990) Do point counts and spot mapping produce equivalent estimates of bird densities? *Auk*, **107**, 447–50.

*Trenkel, V.M., Buckland, S.T., McLean, C. and Elston, D.A. (1997) Evaluation of aerial line transect methodology for estimating red deer (*Cervus elaphus*) abundance in Scotland. *Journal of Environmental Management*, **50**, 39–50.

*Turnock, B.J. and Quinn, T.J., II. (1991) The effect of responsive movement on abundance estimation using line transect sampling. *Biometrics*, **47**, 701–15.

van Hensbergen, H.J., Berry, M.P.S. and Juritz, J. (1996) Helicopter-based line transect estimates of some Southern African game populations. *South African Journal of Wildlife Research*, **26**, 81–7.

Varman, K.S. and Sukumar, R. (1995) The line transect method for estimating densities of large mammals in a tropical deciduous forest: an evaluation of models and field experiments. *Journal of Bioscience*, **20**, 273–87.

Velazquez, A. (1994) Distribution and population size of *Romerolagus diazi* on El Pelado volcano, Mexico. *Journal of Mammalogy*, **75**, 743–9.

*Verner, J. (1985) Assessment of counting techniques. In *Current Ornithology*, Volume 2 (ed. R.F. Johnston), Plenum Press, New York, USA, pp. 247–302.

Verner, J. (1988) Optimizing the duration of point counts for monitoring trends in bird populations. U.S. Forest Service Research Note PSW-395.

*Verner, J. and Ritter, L.V. (1985) A comparison of transects and point counts in oak–pine woodlands of California. *Condor*, **87**, 47–68.

Verner, J. and Ritter, L.V. (1986) Hourly variation in morning point counts of birds. *Auk*, **103**, 117–24.

Verner, J. and Ritter, L.V. (1988) A comparison of transect and spot mapping in oak–pine woodlands of California. *Condor*, **90**, 401–19.

Vidal, O., Barlow, J., Hurtado, L.A., Torre, J., Cendon, P. and Ojeda, Z. (1997) Distribution and abundance of the Amazon river dolphin (*Inia geoffrensis*) and the tucuxi (*Sotalia fluviatilis*) in the Upper Amazon river. *Marine Mammal Science*, **13**, 427–45.

Wade, P.R. and DeMaster, D.P. (1999) Determining the optimum interval for abundance surveys. In *Marine Mammal Survey and Assessment Methods* (eds. G.W. Garner, S.C. Amstrup, J.L. Laake, B.F.J. Manly, L.L. McDonald and D.G. Robertson), Balkema, Rotterdam, pp. 53–66.

Wade, P.R. and Gerrodette, T. (1993) Estimates of cetacean abundance and distribution in the eastern tropical Pacific. *Report of the International Whaling Commission*, **43**, 477–93.

Wallace, R.B., Painter, R.L.E. and Taber, A.B. (1998) Primate diversity, habitat preferences, and population density estimates in Noel Kempff Mercado National Park Santa Cruz Department, Bolivia. *American Journal of Primatology*, **46**, 197–211.

*Walsh, P.D. and White, L.J.T. (1999) What it will take to monitor forest elephant populations. *Conservation Biology*, **13**, 1194–202.

*Ward, A.J. and Hiby, A.R. (1987) Analysis of cue-counting and blow rate estimation experiments carried out during the 1985/86 IDCR minke whale assessment cruise. *Report of the International Whaling Commission*, **37**, 259–62.

Wardell-Johnson, G. and Williams, M. (2000) Edges and gaps in mature karri forest, south-western Australia: logging effects on bird species abundance and diversity. *Forest Ecology and Management*, **131**, 1–21.

*Warren, W.G. and Batcheler, C.L. (1979) The density of spatial patterns: robust estimation through distance methods. In *Spatial and Temporal Analysis in Ecology* (eds. R.M. Cormack and J.K. Ord), International Co-operative Publishing House, Fairland, pp. 240–70.

*Weir, B.S. (1990) *Genetic data analysis.* Sinauer Associates, Inc., Sunderland, MA, USA.

Welsh, C.J.E. and Capen, D.E. (1992) Availability of nesting sites as a limit to woodpecker populations. *Forest Ecology and Management*, **48**, 31–41.

Wetmore, S.P., Keller, R.A. and Smith, G.E.P. (1985) Effects of logging on bird populations in British Columbia as determined by a modified point-count method. *Canadian Field Naturalist*, **99**, 224–33.

White, D.H., Kepler, C.B., Hatfield, J.S., Sykes, P.W. and Seginak, J.T. (1996) Habitat associations of birds in the Georgia Piedmont during winter. *Journal of Field Ornithology*, **67**, 159–66.

*White, G.C., Anderson, D.R., Burnham, K.P. and Otis, D.L. (1982) Capture–recapture and removal methods for sampling closed populations. Los Alamos National Laboratory, Los Alamos, NM, USA. Rep. LA-8787-NERP.

*White, G.C., Bartmann, R.M., Carpenter, L.H. and Garrott, R.A. (1989) Evaluation of aerial line transects for estimating mule deer densities. *Journal of Wildlife Management*, **53**, 625–35.

White, L.J.T. (1994) Biomass of rain forest mammals in the Lopé Reserve, Gabon. *Journal of Animal Ecology*, **63**, 499–512.

Whitesides, G.H., Oates, J.F., Green, S.M. and Kluberdanz, R.P. (1988) Estimating primate densities from transects in a west African rain forest: a comparison of techniques. *Journal of Animal Ecology*, **57**, 345–67.

*Wiens, J.A. and Nussbaum, R.A. (1975) Model estimation of energy flow in northwestern coniferous forest bird communities. *Ecology*, **56**, 547–61.

Wigley, R.L., Theroux, R.B. and Murray, H.E. (1975) Deep-sea red crab, *Gervon quinquedens*, survey off northeastern United States. *Marine Fisheries Review*, **37**, 1–21.

Wiig, Ø. and Derocher, A.E. (1999) Application of aerial survey methods to polar bears in the Barents Sea. In *Marine Mammal Survey and Assessment Methods* (eds. G.W. Garner, S.C. Amstrup, J.L. Laake, B.F.J. Manly, L.L. McDonald and D.G. Robertson), Balkema, Rotterdam, pp. 27–36.

*Wildman, V.J. and Ramsey, F.L. (1985) Estimating effective area surveyed with the cumulative distribution function. Department of Statistics, Oregon State University, Corvallis, OR, USA., Technical Report No. 106, 37pp.

Williams, C.K., Applegate, R.D., Lutz, R.S. and Rusch, D.H. (2000) A comparison of raptor densities and habitat use in Kansas cropland and rangeland ecosystems. *Journal of Raptor Research*, **34**, 203–9.

*Wilson, K.R. and Anderson, D.R. (1985a) Evaluation of two density estimators of small mammal population size. *Journal of Mammalogy*, **66**, 13–21.

*Wilson, K.R. and Anderson, D.R. (1985b) Evaluation of a density estimator based on a trapping web and distance sampling theory. *Ecology*, **66**, 1185–94.

*Wood, S.N. (2000) Modelling and smoothing parameter estimation with multiple quadratic penalties. *Journal of the Royal Statistical Society B*, **62**, 413–28.

Woodroofe, M. and Zhang, R. (1999) Isotonic estimation for grouped data. *Statistics and Probability Letters*, **45**, 41–7.

Wywialowski, A.P. and Stoddart, L.C. (1988) Estimation of jack rabbit density: methodology makes a difference. *Journal of Wildlife Management*, **52**, 57–9.

*Yapp, W.B. (1956) The theory of line transects. *Bird Study*, **3**, 93–104.

Yoshida, H., Shirakihara, K., Kishino, H. and Shirakihara, M. (1997) A population size estimate of the finless porpoise, *Neophocaena phocaenoides*, from aerial sighting surveys in Ariake Sound and Tachibana Bay, Japan. *Researches on Population Ecology*, **39**, 239–47.

Yoshida, H., Shirakihara, K., Kishino, H., Shirakihara, M. and Takemura, A. (1998) Finless porpoise abundance in Omura Bay, Japan: estimation from aerial sighting surveys. *Journal of Wildlife Management*, **62**, 286–91.

*Zahl, S. (1989) Line transect sampling with unknown probability of detection along the transect. *Biometrics*, **45**, 453–70.

Zhang, S.P. and Karunamuni, R.J. (1998) On kernel density estimation near endpoints. *Journal of Statistical Planning and Inference*, **70**, 301–16.

Common and scientific names of plants and animals

Common name	Scientific name
American robin	*Turdus migratorius*
Antelope	Bovidae
Ape	Cercopithecidae
Badger	*Meles meles*
Baltic rush	*Juncus balticus*
Blackbird	*Turdus merula*
Blackcap	*Sylvia atricapilla*
Blue-winged teal	*Anas discors*
Bobolink	*Dolichonyx oryzivorus*
Box turtle	*Terrapene* spp.
Brown-headed cowbird	*Molothrus ater*
Bullrush	*Scirpus validus*
Cape buffalo	*Syncerus caffer*
Cattail	*Typha latifolia*
Chaffinch	*Fringilla coelebs*
Chiffchaff	*Phylloscopus collybita*
Cinnamon teal	*Anas cyanoptera*
Clark's nutcracker	*Nucifraga columbiana*
Coal tit	*Parus ater*
Dall's porpoise	*Phocoenoides dalli*
Darkling beetle	*Eleodes* spp.
Deer	Cervidae
Desert tortoise	*Gopherus agassizii*
Dolphin	Delphinidae
Dunnock	*Prunella modularis*
Eastern grey kangaroo	*Macropus giganteus*
Elephant	*Loxodonta* spp.
Field mouse	*Peromyscus* spp.
Fin whale	*Balaenoptera physalus*
Fruit bat	*Chiroptera* spp.
Gadwall	*Anas strepera*
Garden warbler	*Sylvia borin*
Goldcrest	*Regulus regulus*
Greasewood	*Sarcobatus vermiculatus*

Common name	Scientific name
Green-winged teal	*Anas carolinensis*
Grouse	Tetraoninae
Harbour porpoise	*Phocoena phocoena*
Hare	*Lepus* spp.
Hornbill	Bucerotidae
House wren	*Troglodytes aedon*
Jackrabbit	*Lepus* spp.
Kangaroo	Macropodidae
Kangaroo rat	*Dipodomys* spp.
Lake trout	*Salvelinus namaycush*
Lion	Felidae
Long-billed curlew	*Numenius americanus*
Long-tailed tit	*Aegithalos caudatus*
Mallard	*Anas platyrhynchos*
Minke whale	*Balaenoptera acutorostrata*
Mistle thrush	*Turdus viscivorus*
Mouse	Muridae
Northern bobwhite quail	*Colinus virginianus*
Northern pintail	*Anas acuta*
Northern right whale	*Eubalaena glacialis*
Northern shoveler	*Anas clypeata*
Parrot	Psittacidae
Pheasant	Phasianidae
Pig	*Sus scrofa*
Porpoise	Phocoenidae
Primate	Cercopithecidae
Pronghorn	*Antilocapra americana*
Quail	Odontophorinae
Rabbitbrush	*Chrysothamnus* spp.
Rabbits	Leporidae
Red crab	*Grapsus grapsus*
Red fox	*Vulpes vulpes*
Redhead	*Aythya americana*
Redpoll	*Carduelis flammea*
Red-winged blackbird	*Agelaius phoeniceus*
Robin	*Erithacus rubecula*
Rockfish	*Sebastes* spp.
Sagebrush	*Artemisia* spp.
Saltgrass	*Distichlis stricta*
Savannah sparrow	*Passerculus sandwichensis*
Scrub jay	*Aphelocoma coerulescens*
Seal	Otariidae/Phocidae
Sedge	*Carex* spp.
Sika deer	*Cervus nippon*
Siskin	*Carduelis spinus*

Common name	Scientific name
Song sparrow	*Melospiza melodia*
Song thrush	*Turdus philomelos*
Sperm whale	*Physeter macrocephalus*
Spikerush	*Eleocharis macrosachya*
Spotted dolphin	*Stenella attenuata*
Teal	*Anas* spp.
Tree pipit	*Anthus trivialis*
Vole	Microtidae
Whale	Balaenopteridae
White-crowned sparrow	*Zonotrichia leucophrys*
Whitethroat	*Sylvia communis*
Willow warbler	*Phylloscopus trochilus*
Wolf spider	*Atrax* spp.
Wren	*Troglodytes troglodytes*
Yellow warbler	*Dendroica petechia*

Glossary of notation and abbreviations

The following list is not exhaustive; notation is included here only if it is used through much of the text. Some of the notation listed below is occasionally used for another purpose; in such cases, the temporary definition is stated in the text. Standard mathematical and statistical symbols such as $\hat{}$, ∞ and $\sum$ are not listed.

μ	effective strip half-width $= 1/f(0) = \int_0^w g(x)\,dx$; the half-width of the strip extending either side of a transect centreline such that as many objects are detected outside the strip as remain undetected within it.
ν	effective area $= 2\pi/h(0) = 2\pi \int_0^w rg(r)\,dr$ (point transect sampling); the area such that as many objects are detected outside it as remain undetected inside it.
$\pi(s)$	probability distribution of cluster sizes in area A.
$\pi^*(s)$	probability distribution of sizes of detected clusters; this differs from $\pi(s)$ when sampling of clusters is size-biased.
ρ	effective radius $= \sqrt{\nu/\pi}$; the radius of the circle around each point such that as many objects are detected beyond ρ as remain undetected within ρ.
σ	a scale parameter, used primarily in the half-normal and hazard-rate detection functions.
θ	sighting angle (subscript i, if present, denotes the ith detection).
a	area within distance w of surveyed lines or points; the surveyed area.
A	size of study area, containing N objects; a sample of size a of this area is surveyed (subscript v, if present, denotes the vth stratum).
AIC	Akaike's Information Criterion, used for model selection.
b	dispersion parameter, also called variance inflation factor.
B	number of bootstrap resamples.
c	a multiplier, usually equal to one, but equal to 0.5 if just one side of the line is recorded (line transect sampling), or $\phi/2\pi$ if just an arc of ϕ radians is counted (point transect sampling and, especially, cue counting). It might also be the number of times each line or point is surveyed, or a parameter to be estimated, such as $g(0)$ for studies in which detection on the trackline is not certain.
c_i	cutpoint i, separating interval i from interval $i+1$, grouped distance data.
cov	sampling covariance
cv	coefficient of variation = (standard error)/(estimate). When expressed numerically, usually converted to a percentage by multiplying by 100.

D	density of objects in study area $= N/A$ (subscript v, if present, denotes the vth stratum).
$E(s)$	the mean size of the N_s clusters in the study area.
$f(y)$	the probability density function of perpendicular distances (line transects) or detection distances (point transects).
$f(y, s)$	the joint probability density function of distances y and cluster sizes s.
$f(y \mid s)$	the conditional probability density function of distances y given cluster size s.
$f(0)$	the value of the probability density function of perpendicular distances, evaluated at zero distance (line transect sampling).
$g(y)$	the detection function; the probability that an object at distance y from the line or point is detected.
$g(y, s)$	the bivariate detection function; the probability that a cluster of size s and at distance y from the line or point is detected.
$g(y \mid s)$	the conditional detection function; the probability that a cluster at distance y from the line or point is detected, given that it is of size s; functional expression is equivalent to $g(y, s)$.
$g(0)$	the probability that an object that is on the line or point ($y = 0$) is detected (usually $g(0) = 1$ by assumption).
$h(0)$	the slope of the probability density function of detection distances, evaluated at distance zero (point transect sampling) $= f'(0) = 2\pi/\nu = 1/\int_0^w r\, g(r)\, dr$.
k	number of replicate lines or points (subscript v, if present, denotes the vth stratum).
l_i	the length of line i in a line transect survey, $i = 1, \ldots, k$.
L	the total line length in a line transect survey $= \sum_{i=1}^{k} l_i$ (subscript v, if present, denotes the vth stratum).
$\mathcal{L}$	the likelihood function.
n	sample size; number of objects detected (subscript v, if present, denotes the vth stratum).
N	population size; total number of objects in the study area of size A (subscript v, if present, denotes the vth stratum).
N_s	when objects occur in clusters, the total number of clusters in the study area.
P_a	the probability that a randomly selected object in the surveyed area a is detected.
pdf	probability density function, for example $f(y)$.
r	the detection or radial distance; the distance of an object from the observer at the time the object is detected (subscript i, if present, denotes the ith detection).
$r_{1/2}$	the distance from a point at which probability of detection of an object is one half.
s	the size of a cluster of objects (subscript i, if present, denotes the ith detection).
sd	standard deviation.
se	standard error.

u number of groups for grouped distance data.

V number of strata.

var sampling variance.

w the truncation point; distances exceeding w either are not recorded or are truncated before analysis.

x the perpendicular distance; the distance of a detected object from the transect centreline (subscript i, if present, denotes the ith detection).

y the perpendicular distance x of a detected object from the centreline (line transect sampling) or the detection distance r of an object from the point (point transect sampling) (subscript i, if present, denotes the ith detection).

Index